Cambridge Studies in Biological and Evolutionary Anthropology 43

Bioarchaeology of Southeast Asia

Bioarchaeology of Southeast Asia is the first book to examine directly the biology and lives of the past people of this region. Bringing together the most active researchers in late Pleistocene/Holocene Southeast Asian human osteology, the book deals with two major approaches to studying human skeletal remains. Using analysis of the physical appearance of the region's past peoples, the first section explores such issues as the first peopling of the region, the evidence for subsequent migratory patterns (particularly between Southeast and Northeast Asia) and counter arguments centering on *in situ* microevolutionary change. The second section reconstructs the health of these same people in the context of major economic and demographic changes over time, including those caused by the adoption or intensification of agriculture. Written for archaeologists, bioarchaeologists and biological anthropologists, it is a fascinating insight into the bioarchaeology of this important region.

MARC OXENHAM is a lecturer in the School of Archaeology and Anthropology at the Australian National University in Canberra. For the last 10 years, he has been involved in bioarchaeological research in northern Vietnam, particularly in Vietnamese tropical and subtropical health during the Holocene, but he has recently extended his interests into the palaeohealth of sub-Arctic foragers in Northeast Asia.

NANCY TAYLES is a senior lecturer in the Department of Anatomy and Structural Biology at the University of Otago, New Zealand. She has been working as the bioarchaeologist in a multidisciplinary international team working on a series of prehistoric sites in southeast and northeast Thailand, but has also worked in Burma/Myanmar, Cambodia and Vietnam. Her research interests focus on issues of quality of life in prehistory, using indicators of health measured from human skeletal remains as evidence.

Cambridge Studies in Biological and Evolutionary Anthropology

Also available in the series

21 *Bioarchaeology* Clark S. Larsen 0 521 65834 9 (paperback)
22 *Comparative Primate Socioecology* P. C. Lee (ed.) 0 521 59336 0
23 *Patterns of Human Growth*, second edition Barry Bogin 0 521 56438 7 (paperback)
24 *Migration and Colonisation in Human Microevolution* Alan Fix 0 521 59206 2
25 *Human Growth in the Past* Robert D. Hoppa & Charles M. FitzGerald (eds.) 0 521 63153 X
26 *Human Paleobiology* Robert B. Eckhardt 0 521 45160 4
27 *Mountain Gorillas* Martha M. Robbins, Pascale Sicotte & Kelly J. Stewart (eds.) 0 521 76004 7
28 *Evolution and Genetics of Latin American Populations* Francisco M. Salzano & Maria C. Bortolini 0 521 65275 8
29 *Primates Face to Face* Agustín Fuentes & Linda D. Wolfe (eds.) 0 521 79109 X
30 *Human Biology of Pastoral Populations* William R. Leonard & Michael H. Crawford (eds.) 0 521 78016 0
31 *Paleodemography* Robert D. Hoppa & James W. Vaupel (eds.) 0 521 80063 3
32 *Primate Dentition* Daris R. Swindler 0 521 65289 8
33 *The Primate Fossil Record* Walter C. Hartwig (ed.) 0 521 66315 6
34 *Gorilla Biology* Andrea B. Taylor & Michele L. Goldsmith (eds.) 0 521 79281 9
35 *Human Biologists in the Archives* D. Ann Herring & Alan C. Swedlund (eds.) 0 521 80104 4
36 *Human Senescence – Evolutionary and Biocultural Perspectives* Douglas E. Crews 0 521 57173 1
37 *Patterns of Growth and Development in the Genus* Homo. Jennifer L. Thompson, Gail E. Krovitz & Andrew J. Nelson (eds.) 0 521 82272 6

38 *Neandertals and Modern Humans – An Ecological and Evolutionary Perspective* Clive Finlayson 0 521 82087 1

39 *Methods in Human Growth Research* Roland C. Hauspie, Noel Cameron & Luciano Molinari (eds.) 0 521 82050 2

40 *Shaping Primate Evolution* Fed Anapol, Rebella L. German & Nina G. Jablonski (eds.) 0 521 81107 4

41 *Macaque Societies – A Model for the Study of Social Organization* Bernard Thierry, Mewa Singh & Werner Kaumanns (eds.) 0 521 81847 8

Bioarchaeology of Southeast Asia

EDITED BY

MARC OXENHAM
Australian National University, Canberra, Australia

NANCY TAYLES
University of Otago, Dunedin, New Zealand

CAMBRIDGE UNIVERSITY PRESS
Cambridge, New York, Melbourne, Madrid, Cape Town, Singapore, São Paulo, Delhi

Cambridge University Press
The Edinburgh Building, Cambridge CB2 8RU, UK

Published in the United States of America by Cambridge University Press, New York

www.cambridge.org
Information on this title: www.cambridge.org/9780521120654

First published 2006
This digitally printed version 2009

A catalogue record for this publication is available from the British Library

ISBN 978-0-521-82580-1 hardback
ISBN 978-0-521-12065-4 paperback

Contents

Contents

Contributors

Hallie R. Buckley
Department of Anatomy and Structural Biology, University of Otago, PO Box 913, Dunedin, New Zealand

David Bulbeck
School of Archaeology and Anthropology, Australian National University, Canberra, ACT 0200, Australia

Fabrice Demeter
Laboratoire de Paléoanthropologie et Préhistoire, Collège de France, 11, Place Marcelin Berthelot, 75005 Paris, France

Kate Domett
School of Biomedical Sciences, James Cook University, Townsville, QLD 4811, Australia

Michele Toomay Douglas
Department of Anthropology, University of Hawaii at Manoa, Hawaii, USA

James F. Eder
Arizona State University, Tempe, AZ 85287-2402, USA

Tsunehiko Hanihara
Department of Anatomy, Saga Medical School, 5-1-1 Nabeshima, 849-8501, Japan

Etty Indriati
Laboratory of Bioanthropology and Paleoanthropology, Gadjah Mada University Faculty of Medicine, Yogyakarta 55281, Indonesia

Christopher A. King
Department of Anthropology, University of Hawaii at Manoa, Hawaii, USA

Adam Lauer
University of Hawaii at Manoa, USA

Hirofumi Matsumura
Department of Anatomy, Sapporo Medical University, Minami 1, Nishi 17, Chuo-ku, Sapporo 060-8556, Japan

Kim Thuy Nguyen
Institute of Archaeology, 61 phan Chu Trinh, Hanoi, Vietnam

Lan Cuong Nguyen
Institute of Archaeology, 61 phan Chu Trinh, Hanoi, Vietnam

Lynette Norr
University of Florida, Gainesville, FL 32611, USA

Marc Oxenham
School of Archaeology and Anthropology, Australian National University, Canberra, ACT 0200, Australia

Michael Pietrusewsky
Department of Anthropology, University of Hawaii, 2424 Maile Way, Saunders 346, Honolulu, Hawaii 96822, USA

Nancy Tayles
Department of Anatomy and Structural Biology, Otago School of Medical Sciences, University of Otago, Dunedin, New Zealand

Christy G. Turner II
Professor Emeritus, Arizona State University, Tempe, AZ 85287-2402, USA

Foreword *Emerging frontiers in the bioarchaeology of Southeast Asia*

CLARK SPENCER LARSEN

Much has changed in the two decades since Karl Hutterer (1982) lamented that tropical Southeast Asian archaeology had not advanced much beyond antiquated models based on limited empirical evidence. Indeed, in just the last few years, the scope of archaeology of this region has expanded in new and exciting ways (see Bellwood 1997, Higham and Thosarat 1998, Junker 1999, Higham 2003, Glover and Bellwood 2004). In reading the published archaeological literature on the region before the 1990s, one cannot help but ask why human remains – a highly visible part of this record, having been recovered from such well-known sites as Non Nok Tha and Ban Chiang in Thailand and Niah Cave in Borneo – have not been brought into the discussions about such issues as adaptation, landscape use, population history, settlement, subsistence practices and dietary shifts. Indeed, in this as in other areas of the world, the adoption of agriculture led to fundamental alterations of cultures and landscapes. Human remains provide an informative record of this important adaptive transition (e.g. Krigbaum 2003). Why not include them?

One reason for the lack of inclusion of human remains in developing an understanding of the prehistory of the region is that very little beyond descriptions of skeletons (e.g. Brooks *et al.* 1979) had been published prior to the mid 1990s. The other reason is that archaeologists have oftentimes viewed skeletons as not particularly informative about the past. Scientific reports on skeletons typically ended up as an appendix to an archaeological report, rarely read and not a part of the larger perspective about a region and its prehistoric occupants. This approach is changing, however. On the one hand, the results of new and comprehensive analyses of human remains from this region are appearing in the scholarly record (e.g. Tayles 1999, Oxenham 2000, Domett 2001, Pietrusewsky and Douglas 2002, Krigbaum 2001, 2003). These bioarchaeological studies provide a wealth of new information about population history, colonisation, lifestyle, foodways, nutrition, adaptive shifts, and specific and general aspects of health. On the other hand, we are beginning to see a greater appreciation for the role that analysis of skeletons can play in reconstructing and interpreting the past. Here and elsewhere, this transformation reflects the remarkable expansion of bioarchaeology since the early 1990s (see Larsen 1997, 2002), whereby human remains are increasingly seen as an important part of the design and completion of archaeological research.

Recognising the importance of the human skeletal record in Southeast Asia, Oxenham and Tayles have gathered leading experts to present their research on two areas of bioarchaeological enquiry: population history and health. Part I, the population history section, presents evidence from the analysis of cranial metric and non-metric data that speak to a complex population history involving both migration and *in situ* development. In the larger picture, Matsumura provides evidence to suggest that present-day Southeast Asian populations are genetically influenced by migrations from Northeast Asia (Ch. 2). In the last five millennia or so of prehistory, there is substantial evidence indicating local population continuity in inland Southeast Asia, perhaps more so than in coastal settings (Chs. 3 and 4). Within specific regions, we see biological change, such as involving a reduction in cranial length and facial robusticity, in peninsular Malaysia (Ch. 6), a pattern that has been well documented in other areas around the post-Pleistocene world. Arguably, these morphological changes reflect adaptation to new circumstances affecting the mastication and craniofacial development, such as the adoption of new foods and new ways of preparing them. Prior to the 1980s, such cultural and biological changes were largely seen as being externally driven (Hutterer 1982). The bioarchaeological record is showing that the origins of biological and cultural variability are complex and derive from both external and internal forces, ultimately resulting in the cultures and peoples that we see distributed about this vast region of the world today.

Part II deals with the palaeopathological record and adds much to the emerging picture of the history of human health in the Late Pleistocene and Holocene. Dental health – especially as it is represented by dental caries – does not appear to have declined substantially with the adoption of rice agriculture, either within particular settings (Ch. 9) or the region generally (Ch. 11). This seems to be the case because rice is not especially cariogenic, in contrast to plants domesticated in other areas of the world (e.g. maize in the Americas). Therefore, it should come as little surprise that health declines in the region are minimal (or non-existent) in comparison with other regions of the world where agriculture emerged (Larsen 1995), at least with respect to oral health. Although the evidence is more preliminary, other skeletal indicators (e.g. infection) also show lack of substantive health change (see Pietrusewsky and Douglas 2002).

There are some large skeletal samples in the region covered by this book (e.g. Ban Chiang), but overall, the samples are small, especially in comparison with settings from eastern North America and western Europe. Despite the limitations, the contributors to this volume present a fascinating picture of biological complexity, population history and health in prehistory. Understanding this bioarchaeological record of the past helps us to understand better the peoples and cultures that live in the region today.

References

Bellwood P. 1997. *Prehistory of the Indo-Malaysian Archipelago.* Honolulu: University of Hawaii Press.

Brooks S. T., Heglar R. and Brooks R. H. 1979. Radiocarbon dating and palaeoserology of a selected burial series from the Great Cave of Niah, Sarawak, Malaysia. *Asian Perspectives* **20**: 21–31.

Domett K. M. 2001. *British Archaeological Records International Series, No. 946: Health in Late Prehistoric Thailand.* Oxford: Archaeopress.

Glover I. and Bellwood P. 2004. *Southeast Asia: An Archaeological History.* London: Routledge.

Higham C. 2003. *Early Cultures of Mainland Southeast Asia.* Chicago, IL: Art Media Resources.

Higham C. and Thosarat R. 1998. *Prehistoric Thailand.* Bangkok: River Books.

Hutterer K. L. 1982. Early Southeast Asia: old wine in new skins? A review article. *Journal of Asian Studies* **41**: 559–570.

Junker L. 1999. *Raiding, Trading and Feasting: The Political Economy of Philippine Chiefdoms.* Honolulu: University of Hawaii Press.

Krigbaum J. S. 2001. Human paleodiet in tropical Southeast Asia: isotopic evidence from Niah Cave and Gua Cha. Ph.D. thesis, New York University.

2003. Neolithic subsistence patterns in northern Borneo reconstructed with stable carbon isotopes of enamel. *Journal of Anthropological Archaeology* **22**: 292–304.

Larsen C. S. 1995. Biological changes in human populations with agriculture. *Annual Review of Anthropology* **24**: 185–213.

1997. *Bioarchaeology: Interpreting Behavior from the Human Skeleton.* Cambridge, UK: Cambridge University Press.

2002. Bioarchaeology: the lives and lifestyles of past people. *Journal of Archaeological Research* **10**: 119–166.

Oxenham M. F. 2000. Health and behaviour during the mid Holocene and metal period of northern Viet Nam. Ph.D. thesis, Northern Territory University, Darwin, Australia.

Pietrusewsky M. and Douglas M. T. 2002 *University Museum Monograph 111: Ban Chiang, a Prehistoric Village Site in Northeast Thailand I: The Human Skeletal Remains*. Philadelphia, PA: University of Pennsylvania Press.

Tayles N. 1999. *Report of the Research Committee* LXI. *The Excavation of Khok Phanom Di: A Prehistoric Site in Central Thailand*, Vol. V: *The People*. London: Society of Antiquaries.

Preface

For every complex problem, there is a solution that is simple, neat, and wrong!

Attributed to H. L. Mencken

We hope that this book will spark wider interest in the bioarchaeology of Southeast Asia, including the neighbouring Pacific, and will spur the development of greater synthetic and collaborative research in the region. It is a delight to read about the culture, the society, the technology, the subsistence, the settlement patterns and the environment of prehistoric populations; however, central to all this are the people themselves. Nothing brings them to life as individuals, communities and populations like the physical remains themselves. They are, or should be, central to any archaeological endeavour. In particular, we hope that it will spark the interest of indigenous researchers and encourage them to consider the study of human skeletal remains as a central component to any archaeological project (even if it is only to show us that we have got it all wrong!).

The editors would like to thank all of the authors for their contributions to this volume. Further, we wish to thank all those anonymous reviewers who spent a considerable amount of their time and energy in providing a wealth of critical and constructive comment on each and all of these chapters.

1 *Introduction: Southeast Asian bioarchaeology past and present*

NANCY TAYLES
Otago School of Medical Sciences, University of Otago, Dunedin, New Zealand

MARC OXENHAM
Australian National University, Canberra, Australia

When originally discussing and formulating the idea that eventually led to this volume we asked ourselves 'is there a recognised need for a book on Southeast Asian bioarchaeology?' 'And does everyone know what bioarchaeology is anyway?' We address the second question, 'what is bioarchaeology?', first. Human remains provide the only direct record of the biology of the people and the populations who created the 'archaeological record' and these are, therefore, central to any archaeological research. This is not to deny the fascination of the material culture, the environmental context, the settlement patterns and the mortuary practices of past peoples, it is nevertheless (and despite the reluctance of many archaeologists to admit it) axiomatic that human remains *are* the people who created the pots, the tools, the houses, the middens and the modified landscapes. As such they must, or ought to be, recognised as central to any research of past society that uses archaeology as the means of data recovery. This recognition of the value of human remains as a window into past peoples has gained momentum over recent decades (e.g. Larsen 1997, Cox and Mays 2000). With the development of appropriate means of interpreting the skeletal evidence beyond the simple description that the term 'physical anthropology' implies, those of us who are captured by the challenge are now seeing ourselves as human biologists. Having moved beyond taxonomy into the wider and richer world of biology, an appropriate epithet was sought.

The field of human osteology is one that has existed for rather longer than what is referred to as bioarchaeology, but it carries with it the impression that the objective of practitioners is simply to study human bones; it is not. It is to study the people represented by the bones, which is

Bioarchaeology of Southeast Asia. Marc Oxenham and Nancy Tayles.
Published by Cambridge University Press.

an important shift in emphasis. Hence the adoption of the term bioarchaeology, which indicates the biological component of such research. As Clark Larsen (1997) noted, although the term was first applied to archaeozoology, it is now only used in reference to humans. It is not universally adopted (hence the title of the excellent volume by Margaret Cox and Simon Mays (2000): *Human Osteology in Archaeology and Forensic Science*). Nevertheless, and despite its meaning not being immediately obvious, it seems to us the most appropriate term to apply to the study of past people through the medium of their skeletal remains recovered from archaeological sites. It is conventionally used to refer to *Homo sapiens*; study of fossils of earlier hominin species is palaeoanthropology.

Now to the issue of Southeast Asia and why it merits a book like this. There are multiple reasons why we deemed the time to be right for this exercise. Despite the fame, or even notoriety, of some of the earliest hominin specimens in Asia, Southeast Asia as a region has been relatively invisible in the broader sweep of world prehistory and in the centres of bioarchaeological research in Europe and North America. Beyond the Indonesian fossils, Southeast Asia and the southwest Pacific have unique characteristics which will mean that research on the area will not only clarify issues about regional prehistoric peoples but also contribute to an understanding of prehistoric human biology worldwide. These include a very long human settlement; geographic variation over time and space, including marked changes in sea level; a climate that varies from the hot, humid tropical equatorial region to a cooler subtropical climate away from the equator; and a current rice-based subsistence system with a deep antiquity in many areas. This suite of characteristics does not occur anywhere else in the world and justifies this effort to take the first step in what should be a long process synthesising research on human biology to address the issues of human evolution, variation and biocultural development in this unique environment. Its singularity is further enhanced by it being a crossroads between the major, influential and very different cultural and biological regions of China, India and Melanesia.

Beyond its geographical boundaries, the influence of Southeast Asia spreads far to the east as the biocultural origin of the eastern Pacific populations. Eastern Melanesia and Polynesia, designated Remote Oceania, in Roger Green's (1991) very appropriate definition, have had a very short human settlement but, for this very reason, have attracted large sums of money, and a considerable amount of polemic, in the search for the origins of their peoples. This origin is still unclear but inevitably includes a reference to 'Southeast Asia' as if this were a well-defined, homogeneous region. The reality, of course, is that it is anything but well defined and certainly it is gloriously varied in its human biology (and in its cultures, but that seems to be an issue that people happily recognise).

Southeast Asia has been an area that has stimulated much discussion and some grand and interesting theories about human origins and migrations (e.g. Oppenheimer 1998). Nevertheless, despite the ongoing significance of, controversy about and research into the Indonesian *Homo erectus* remains, which have attracted international attention from the time of Dubois in the nineteenth century, it is almost as if the later human biological prehistory of the region is of no concern on the world stage. As an example, it does not figure in the comprehensive review by Mark Cohen and George Armelagos (1984) of the human biological response to the development of agriculture, despite at the time having been acknowledged to have adopted agriculture as early as the fifth or even sixth millennium BP (Higham and Bannanurag 1990). Nor does it appear in standard texts of biological or physical anthropology other than in the context of human evolution and the Javan *H. erectus* specimens.

There can be numerous explanations for this, not the least of which has been the political instability of a number of states over the last half century. The insecurity of the region and individual countries has worked against the development of a research ethos among the local scholarly communities that is secure enough to develop objectives in the field of prehistoric human biology beyond the ratification of the unity and duration of the peoples living in each country. Western scholars have only relatively recently begun working here and local archaeologists have joined them in research that transcends modern political boundaries. There is now a cohort of bioarchaeologists that has developed expertise in the area and has produced a corpus of literature which has prompted this first attempt to draw together a benchmark publication. This book is not a comprehensive treatise but a starting point showing the breadth of research in the area, which we hope will serve as a stimulus for further consolidation of the topic.

Southeast Asia and the southwest Pacific is for all these worthy reasons a place to be reckoned with in human bioarchaeology. It is overdue for attention on a broader scale. It is well past time to bring together a group of authors in an effort to 'mark a line in the sand' in bioarchaeological research and to encourage people away from their local specialities into a regional synthesis. In an ideal world, this book would include everyone who works in this topic as contributors and provide an integrated overview of the whole region. This, of course, was our initial aim. The reality is something different and we have collected here a selection of papers that cover the two themes of skeletal studies: one on the evolution and variation in morphology and relationships among groups (biodistance) and the second with contributions on the quality of life of the prehistoric inhabitants as represented by their health.

Beyond the consideration of the people themselves, it is also time to integrate the findings of bioarchaeology into the general archaeological literature on the region. Publications on prehistory in general are primarily focused on archaeology and linguistics, although some refer to human biology, for example Bellwood (1997) in his comprehensive review of Indo-Malaysia and Higham in his introductory chapter to Jin, Seielstad and Xiao's edited volume (2001). Others tend to skirt around the issue, although clearly desirous of including human variation in a triumvirate with linguistics and archaeological evidence in the search for clarification of the prehistory of the region. This lack of integration we see as probably a consequence of the lack of a summarising publication on human biology. We hope this volume will begin the process.

Where is Southeast Asia? Figure 1.1 shows the core, which is mainland Southeast Asia: Burma (Myanmar), Thailand, Laos, Cambodia, Vietnam and, ideally, southwest China (but the reality is that there is little or nothing available to us from this area) and island Southeast Asia: the Indo-Malaysian archipelago (Malaysia and Indonesia), the island of Borneo (East Malaysia, Sarawak and Sabeh, Kalimantan and Brunei) and the Philippines. The authors of the first section of the book have also spread their nets to include samples from the wider east Asian and Pacific region (Fig. 1.2). Both maps identify locations referred to in the chapters. Buckley's samples (Ch. 13) are from the southwest Pacific and are located in Fig. 13.1 (p. 310).

Development and current state of bioarchaeology in Southeast Asia

The first bioarchaeological study of significance in Southeast Asia was carried out by the Belgian anatomist Eugene Dubois following his discovery of Pleistocene hominin specimens at Trinil, in the Solo Valley, Java in 1891. These finds, comprehensively described in his monograph (Dubois 1894), were subsequently paraded throughout Europe and promoted as a form of human forebear or missing link. The interest in what became known as *Pithecanthropus erectus* was such that at least 80 papers and books were published on these specimens by the turn of the twentieth century (Trinkaus and Shipman 1993; Shipman 2001).

From the beginnings of palaeoanthropological research, explanations for observed modern human diversity have been developed and debated.

Figure 1.1. Core Southeast Asian study area encompassed by this volume showing the location of the major sites referred to in the text.

Figure 1.2. Expanded study area including the location of those specimens examined in the population history studies. Refer to Fig. 13.1 for southwest Pacific sites.

Some early theories that were quite bizarre included the polygenetic model proposed by the German anatomist Hermann Klaatsch (1923) in which an early ancestral species was considered to give rise to a human-like form, which subsequently diverged into separate gorilla, Neandertal and Black African branches, and another form that diverged into separate Orang-utang and Asiatic/Aurignacian branches. Klaatsch's research took him to Australia, where he studied aboriginal peoples for several years, and for a brief visit to Trinil, where he contracted the malarial parasites that were to contribute to his premature death in 1916 (Heilborn 1923).

It was only the rapid accumulation of fossil specimens, particularly from China and Indonesia, in the 1930s and 1940s that led to the development of more sophisticated theories of human origins and diversity. Two molars discovered at the close of 1926 by the Austrian palaeontologist Otto Zdansky (1927) at Dragon Bone Hill, or Zhoukoudian, northeastern China saw the beginnings of the most spectacular period of hominin fossil discoveries ever to be seen in the region. Davidson Black, director of the Cenozoic Research Laboratory in China, christened one of the finds, a single lower permanent molar, *Sinanthropus*. This rather daring move paid off when in late 1929 a skull of *Sinanthropus pekinensis* (Peking man) was uncovered at Zhoukoudian by the Chinese palaeontologist Wenzhong Pei (1929). The following decade revealed the wealth of material at Zhoukoudian. Much is owed to Franz Weidenreich (e.g. 1936, 1943) for his thorough description of this material.

At the same time that the significance of the Chinese fossils was being debated, a Dutch geologist, Ter Haar, discovered terraces containing a group of individuals, now popularly known as Ngandong or Solo man, in Java. A Dutch engineer, W. F. F. Oppenoorth, described these fossils and named them *Homo 'javanthropus' soloensis*. A palaeontologist for the Geological Survey (Holland) in Java, G. H. R. von Koenigswald (1937, 1956), interpreted them as a Javanese Neandertaler. Others (e.g. Oppenoorth 1937) rejected this appellation and allied them with *Pithecanthropus*, a position generally accepted today (Santa Luca 1980). More than 40 years after the original Trinil discovery Dubois continued to engage in the debates surrounding the new Javan finds and was particularly opposed to suggestions that there was an ancestral relationship between *Pithecanthropus* and the Ngandong specimens (Dubois 1937) or even that newly discovered pithecanthropine specimens could be assigned to the genus (Dubois 1938a,b, 1940a,b). He even went to the extent of designating as incorrect the reconstructions of pithecanthropine material by von Koenigswald (Dubois 1938b) and Weidenreich (Dubois 1940a). Dubois delivered his last paper defending his views against Weidenreich and Koenigswald on 30 November 1940 (Dubois 1940b) and died two weeks later at the age of 82.

The combined Chinese and Javan finds led the most prolific writer on Asian hominin material of the time to formulate a model of human evolution that still reverberates today in one form or another. Weidenreich (1936) argued that mosaic evolution was the mechanism responsible for the levels of variation he perceived within each human type or stage (living and fossil). He noted that many scholars placed every new hominin into a dead-end side branch and justified this by citing particular specialisations that precluded these forms from direct human ancestry (Weidenreich 1936). He asked where one draws the line between irreversible specialisations and variability within a polymorphic form. In developing the idea of mosaic evolution, Weidenreich (1940) suggested that the high level of variation seen in modern populations was the same in fossil populations. This allowed him to argue that evolution was marked by general stages in development that were differentially represented both spatially and temporally (Weidenreich 1947). He thus rejected the single origin and dispersal model for human evolution and migration as articulated by Howells (1944, 1948) and argued for local continuity scenarios. In this way, the scene was set for modern palaeoanthropological debates centred around Multiple Origins/Regional Continuity (e.g. Thorne and Wolpoff 1992, Wolpoff *et al.* 2000) and Single Origin/Out-of-Africa (e.g. Wilson and Cann 1992; Stringer 2003) models of human origins and diversity. Over recent years, the Asian hominin sample has seen further increases in sample size in both Indonesia (see Wolpoff (1999, pp. 453–465), for a useful overview) and China (see Wu and Poirier (1995) for an excellent review and detailed descriptions of the specimens and sites). These specimens continue to tax researchers in human evolution to this day with a diverse literature devoted to the subject (see Lewin 1999, Wolpoff 1999).

Leaving the deeper reaches of the Pleistocene and focusing on Southeast Asian bioarchaeological research on anatomically modern humans, by the early twentieth century, colonialism had inspired prehistoric research in the region, with bioarchaeology almost as a 'by-catch'. The Ecole Française d'Extrême Orient was established in Southeast Asia and this stimulated research and excavation in what was then French Indochina, now the modern political entities of Laos, Cambodia and Vietnam. Of the prehistorians who were working in the region and recovered human skeletal remains in the course of excavation, Henri Mansuy and Madelaine Colani are the best known in archaeological circles. The sites from which they reported skeletal remains included Pho Binh Gia, Dong Thûoc, Keophay, Khackiem, Hamrong and Lang Cuom in northern Vietnam (Mansuy 1924, 1925a,b, Mansuy and Colani 1925; see also Verneau 1909; Saurin 1939). Further, Patte (1932, 1965) reported on Da But period remains from northern Vietnam and on others from Minh-Cam in central Vietnam (Patte 1925). Two cave sites in northern Laos, Tam Hang and Tam Pong, had skeletal remains (Fromaget and Saurin 1936, Fromaget 1940,

Saurin 1966) and this material, or at least those with whole crania, has been included in numerous skeletal analyses since then (e.g. Pietrusewsky 1988; see also Chs. 2, 3 and 5).

In Malaya, now the states of Malaysia and Singapore, first the Dutch and later the English were also developing an interest in prehistory. Human remains were found at Gua Cha by de Sieveking (1954) and documented by Trevor and Brothwell (1962). The site was later revisited by Adi and Bellwood (Bellwood and Adi 1981, Adi 1985) while Bulbeck (1981) reported on aspects of the skeletons. Fragmentary remains were found at sites in Perak: at Gua Kerbau by van Stein Callenfels and Evans (van Stein Callenfels 1936a) and reported by Duckworth (1934), and at Gua Kajang by Evans (1918). Van Stein Callenfels (1936b) and Mijsberg (1940) also found fragmentary secondary burials in middens at Guar Kepah, Pulau Pinang, some of which were published by Mijsberg (1940). On the island of Borneo, in Sarawak, Niah Cave has been excavated over several decades beginning in the 1950s. This early excavation yielded the so-called 'deep skull' (Kennedy 1977) likely dating from *c.* 43,000–44,000 years BP (Barker *et al.* 2002a). From Palawan Island, Philippines, there are also remains from caves: cranial fragments from Tabon Cave (Fox 1970) and specimens from other caves with a report on dental morphology by Winters (1974). More recently, excavations led by Zuraina Majid of Universiti Sains Malaysia led to the discovery of Perak man (Zuraina 1994). Barker has recently undertaken extensive re-excavations of Niah Cave (Barker *et al.* 2000, 2001, 2002a,b, 2004) and reported on further significant human skeletal remains. Krigbaum (2001, 2003) has produced detailed analyses of isotopic evidence regarding diet at Niah Cave while Manser (2003) has begun a preliminary re-examination of skeletal remains from the West Mouth burials.

In Thailand, the one modern state with no history of colonisation, there was not such an early start, but Quaritch-Wales (1937, 1964) excavated a small skeletal sample in the 1920s. By the 1960s, the attention of non-colonialist western prehistorians turned to Southeast Asia. The Thai–Danish excavations in Kanchanaburi province led by van Heekeren and by Sørenson (Sørensen and Hatting 1967, van Heekeren and Knuth 1967) resulted in the first comprehensive documentation of a substantial sample of human skeletal remains (Sangvichien *et al.* 1969). Khok Charoen was excavated by Watson and Loofs-Wissowa (Higham 2002) but no report on the poorly preserved skeletal remains has been published. William Solheim II and Donn Bayard excavated at Non Nok Tha in the late 1960s, with publications of the skeletal material by Pietrusewsky (1974a,b) and a comprehensive analysis by Michele

Toomay Douglas (1996; see also Ch. 8). Chester Gorman, with Pisit Charoenwongsa, excavated at the now famous Ban Chiang site in the mid 1970s, with skeletal evidence incorporated into demographic and morphometric studies by Pietrusewsky (e.g. 1978, 1981, 1982, 1984) in the succeeding decade and again in a comprehensive analysis by Douglas (1996). This site has now been fully documented in an exemplary publication by Pietrusewsky and Douglas (2002). King and Norr report on an isotopic study of diet at Ban Chiang in Ch. 10 of this volume.

Further excavations in northeast Thailand continued in the early 1980s with Ban Na Di by Charles Higham and Amphan Kijgnam and skeletal reports by Warrachai Wiriyaromp (1984a,b) and Houghton and Wiriyaromp (1984). This collection was also reviewed by Domett (2001). In 1985, Higham turned his attention to the coast and excavated Khok Phanom Di with Rachanie Bannanurag (Thosarat) with skeletal remains analysed by Choosiri (1988, 1991), Tayles (1999) and Domett (2001). More recently, Bentley (2004) has analysed stable isotopes in search of evidence for migration at the site. In the 1990s, excavations in Thailand expanded exponentially. Higham and Thosarat excavated Nong Nor, near Khok Phanom Di, with the large but poorly preserved skeletal collection analysed by Tayles *et al.* (1998) and Domett (2001).

In central Thailand, the Thailand Archaeometallurgy Project led by Vince Pigott and Surapon Natapintu, with Roberto Ciarla and Fiorella Rispoli, recovered human remains from several sites in the Khao Wong Prachan Valley including Non Pa Wai, Non Mak La, Nil Kham Haeng and Ban Phu Noi. Natapintu, Ciarlo and Rispoli also excavated human remains from Ban Phu Noi, north of Khao Wong Prachan; Ciarlo and Rispoli found further burials at Ban Tha Kae (Higham 2002). Agnosti Agelarakis (1996, 1997) researched the human skeletal remains from Non Pa Wai.

In the mid 1990s, Higham and Thosarat again turned their attention to the northeast and the Origins of Angkor Project was born, resulting in excavations at a number of sites: Ban Lum Khao (skeletal report by Domett (2001)), Noen U-Loke and Ban Non Wat. The Noen U-Loke skeletal analysis is in preparation by Tayles but an overview is provided in a comparison with Ban Lum Khao in Ch. 9. At the time of writing, Ban Non Wat has three excavations completed in 2002–2004 and three more planned. Also in the 1990s, Jean-Pierre Pautreau and his team excavated the site of Ban Wang Hai in northern Thailand (Pautreau *et al.* 2004). In southern Thailand, the cave site of Moh Kiew (Pookajorn 1992, 1994) yielded a late Pleistocene skeleton documented by Choosiri (in Pookajorn 1994). Matsumura reports on the morphometry of this skeleton in Ch. 2. In addition, the Thai archaeological community has advanced greatly in their expertise and there have been numerous excavations by Thai researchers although, understandably, their work is published in Thai.

In the 1990s, the political situation in Cambodia became relatively stable and Dougald O'Reilly excavated at a heavily looted iron age site, Phum Snay (O'Reilly and Sytha 2001), with skeletal reports currently in preparation by Kate Domett. Miriam Stark is now leading a project around the area of Angkor Borei (Stark *et al.* 1999, 2003, Stark 2001, Bishop *et al.* 2003), with Michael Pietrusewsky coordinating the skeletal analysis; Christophe Pottier has been excavating at Prei Khmeng, near Angkor, with skeletal analysis by a group led by Rethy Chhem.

In Vietnam, Nguyen (1994) noted that 33 of an estimated 100 Hoabinhian period sites have provided human skeletal remains. During the French occupation, Colani (1927) published an account of 20 crania and some associated postcranial material from Lang Gao Cave, Luong Son District, Hoa Binh Province in northern Vietnam. While the whereabouts of this material is no longer known, Nguyen (1976) has published metrical observations made by Colani on one cranium. The only other reference to human material discovered during the French occupation is to some jaw and teeth fragments from Lang Bon, Thanh Hoa Province (see Nguyen 1994). More recently, a complete Hoabinhian period cranium was excavated at Mai Da Nuoc (Nguyen 1985, 1986) and the remains of at least eight individuals were discovered at Mai Da Dieu (Nguyen 1986) only a kilometre distant from Mai Da Nuoc. Another relatively complete Hoabinhian cranium has been reported from Dong Can, Ky Son District, Hoa Binh Province (Nguyen 1991). Finally, the skull cap and some associated maxillary remains of an individual from Hang Muoi cave, Tan Lac District, Hoa Binh Province was described by Nguyen (1967). Re-dating by accelerator mass spectroscopy analysis of this material, now to 12,020 $\pm$ 40 years BP, is provided in Bulbeck *et al.* (2003).

Post-Hoabinian human remains make up the vast majority of the prehistoric skeletal sample in Vietnam and some mention of these has already been made in reference to French Indochina. The most important and largest pre-metal age assemblage, with a minimum number of 94 individuals, was excavated at a Da But period cemetery at Con Co Ngua, Ha Trung District, Thanh Hoa Province in 1979–1980 by the Institute of Archaeology, Hanoi (Bui Vinh 1980; Nguyen *et al.* 1980). Early bronze period remains are rare, although the recent recovery of well-preserved human remains from Man Bac, Ninh Binh Province, promises to fill this under-represented period in Vietnamese prehistory (Nguyen 2001). Later bronze period remains are comprehensively surveyed by Nguyen (1996), with the more complete specimens being described in some detail, while

Oxenham (2000) examined the majority of these from northern Vietnam in a bioarchaeological study. Although many of the remains from northern Vietnam are isolated or form only small subsamples, Nui Nap (dated to between 2,400 and 1,700 years BP), Dong Hieu District, Thanh Hoa Province, supplied a minimum of 32 individuals and Quy Chu (dated to between 3,300 and 2,400 years BP), Hoang Hoa District, Thanh Hoa Province had a minimum of 20 individuals. While very little in the way of human material has been recovered from central coastal Vietnam, large excavations in the south at Giong Phet and Giong Ca Vo, Can Gio District, Ho Chi Minh City, dated to between 2,450 and 2,100 years BP, uncovered over 400 jar burials of which 241 contained skeletal material (Nguyen 1995; Nguyen *et al.* 1995).

No recent reports of human skeletal remains are available from Laos. Myanmar (Burma), despite being a British colony for a significant period and having an archaeology community, failed to attract the attention of international researchers, with no bioarchaeological evidence forthcoming. However, a recent surge of interest has resulted in a series of excavations where human remains were uncovered but the local archaeological community have not allowed them to be removed from the ground (Tayles *et al.* 2001, Pautreau *et al.* 2001a,b, 2003).

In Indonesia, apart from the palaeoanthropological finds, which we will not detail here as they are numerous (see Wolpoff 1999, pp. 453–465), there is also a sample of temporally more recent skeletal remains (review by Jacob 1967). Bulbeck (1981) has reported on remains from elsewhere in Indonesia in his M.A. thesis and material from northern Sumatra was documented by Budhisampurno (1985 cited in Bellwood 1997). Bulbeck has continued to work in the region and Bellwood (1997, p. 86) cited unpublished examinations by him of skeletal remains dated 2,000 years BP from Morotai Island, north of Halmahera.

The consequence of this pattern of excavation and recovery of human skeletal remains is that there is patchy coverage of the Southeast Asian region, with some geographic areas yielding comprehensive analyses and others very poorly understood, if at all. The temporal coverage also varies dramatically, with, as would be expected given both the sea-level fluctuations and the increasing population density, an exponential increase in representative samples over time. Overlying this geographic and temporal heterogeneity is a further variable: that of the development of research aims over time from an almost purely morphological focus, addressing issues of typology and origins of populations, to a wider appreciation of both the basis for morphological variation (Bulbeck 1981, 1982) and the richness of evidence for health and quality of life. The last became evident from the late 1960s with Sangvichien *et al.* (1969) and has since the 1990s become the focus of a wide range of research.

Until the move from purely typological studies, with the single exception of Michael Pietrusewsky's prolific works, Southeast Asia was given little focused attention. Data from the area were included in broad regional studies by, for example, Christy Turner (e.g. 1987, 1989, 1990) and Loring Brace (e.g. 1976, 1980, Brace and Hinton 1981). Interpretations were based on assumptions of variation reflecting a series of migrations, culminating in a southern Mongoloid expansion in the last millennium, bringing Thai speakers into mainland Southeast Asia. The interpretation of the morphological evidence was largely based on extrapolation from linguistics (Blust 1996, Bellwood 1997, Spriggs 1998, Diamond and Bellwood 2003). This has now been superseded in the minds of many archaeologists by genetic evidence extrapolated from modern populations: hence the exclusion of consideration of the morphological evidence in favour of genetics in Diamond and Bellwood (2003) and Jin *et al.* (2001). Nevertheless, morphology is still the focus of many researchers, as evidenced by the first section of this book. These studies are aimed at the big, regional questions, not the local or national issues that tend to be the focus of the site-based studies.

Current perspectives on the region

What do we know about prehistoric Southeast Asians from their skeletal remains? How does what we know fit with the theories of archaeologists and linguists?

Peter Bellwood (1997) concluded his comprehensive review of Indo-Malaysian prehistory with a series of outstanding issues, many of which relate to human biology and, in particular, population history. The first of these concerns Javan *H. erectus* and for our purposes the question of its relationship to modern populations is the most important. He asks about the origin of earliest *H. sapiens* immigrants and their possible relationship with resident *H. erectus*. Clearly there were migrations, with sparse bio-archaeological evidence in Southeast Asia itself from about 40,000 years BP, as Matsumura details in Ch. 2. The presence of human populations in Sahul (New Guinea and Australia) by this time is irrefutable evidence of long-distance movements of humans – and the assumption must be that they arrived in these areas from, or via, island Southeast Asia.

In the century since Dubois, a significant number of other hominin remains have been found in Java, with '... 23 skulls and the teeth and bones of more than 100 individuals' (Gibbons 2003, p. 1293) by early 2003.[1] These are mostly from central and east Java, from the middle Pleistocene, although there are holders of the opinion that some may date from the early Pleistocene, even to 1.8 million years BP (Swisher *et al.* 1994). Then there are the Ngandong, Sambungmacan and the remains from eastern Java. The dating of these is again controversial, with dates from 900,000 years BP for Sambungmacan and older than 80,000–60,000 years BP for Ngandong, although there are suggestions that Ngandong could have been as late as 53,000–27,000 years BP (Swisher *et al.* 1996), which raises issues of possible contemporaneity with *H. sapiens*.

Clearly there are issues here of major importance to the human biological story for both Southeast Asia and the western Pacific. Did *H. erectus* genes contribute to the morphology of the so-called 'indigenous' populations of the region, the Australo-Melanesians? Or were these people immigrants from somewhere unknown who overran *H. erectus*, resulting in its extinction? There is a yawning chronological gulf (but see Swisher *et al.* (1996)) between the Indonesian *H. erectus* remains and the next earliest evidence of modern human activity in Southeast Asia. The earliest known *H. sapiens* remains in Java itself, from Wajak (also spelt Wadjak), are dated to 6,560 years BP (Storm 2001). Clearly, we can make no assumptions based on this material. There are earlier human remains from other areas in Southeast Asia (e.g. Niah Cave, Borneo) but again the issue of *H. erectus* genetic contributions to modern populations cannot be answered by these. The issue is an important one. First, because it stimulates controversy about the origins of the first settlers of Sahul and in particular the Australian Aborigines (Storm 2001). Second, on a much grander scale, it is a significant component (along with the similar issue of the relationship between the Zhoukoudian *H. erectus* specimens and modern Chinese) of the major contentious issue of the origin of modern human diversity. The two camps in this debate, as mentioned earlier in this chapter, are the Out-of-Africa or

[1] As this book was going to press, an article in the journal *Nature* (Brown *et al.* 2004) reported the finding of hominin skeletal remains from the site of Liang Bua, on the island of Flores, Indonesia, dating to approximately 18,000 years BP. The hominin remains, consisting of an incomplete skeleton, partly articulated (and provocatively named in the media as a 'hobbit'), together with fragments from other individuals, were identified on the basis of cranial morphology as *Homo* sp. The partial skeleton was from a very short individual with a cranial capacity below that of smaller Australopithecines. Brown *et al.* argued that this is a new species, *Homo floriensis*, and suggested that the find is highly significant for human evolution in general and particularly that in Asia. If the attribution of these remains to a new species of *Homo* at such a late date with such a small cranial capacity is confirmed, it will indeed be a very influential find. Further examination and analysis of the material is at a very early stage and we look forward to robust analysis and discussion of its significance.

Single Origin model and the Regional Continuity or Multiple Origins model. As has already been noted, it is not appropriate for this debate to be detailed here and it has received a vast amount of attention in the literature since the early 1990s (e.g. Thorne and Wolpoff 1992, Wilson and Cann 1992, Wolpoff *et al.* 2000, Stringer 2003).

This volume does not enter into the specifics of the skeletal evidence for the possible contribution of *H. erectus* to modern Southeast Asians. The identification on the basis of phenotype of two separate populations in the region, the Australo-Melanesians (Australoids) and the Mongoloids, and the identification of these populations, respectively, as descendents of the Javan and Chinese *H. erectus* has had support in the past from Weidenreich (1946) and Coon (1962). This issue was thoroughly reviewed by Bellwood (1997) and is not in need of further discussion here. Suffice to say that we do not know the answer (yet) to this question.

However, allied to the global question of the evolutionary origins of modern human diversity is the origins and relationship of diversity as revealed in anatomically modern human skeletal remains from Southeast Asia. This is addressed in Part I, with a range of approaches by various authors. Matsumura (Ch. 2), Pietrusewsky (Ch. 3), Hanihara (Ch. 4) and Demeter (Ch. 5) have all contributed research evidence relevant to this question on a broad geographic scale by examining the dental and skeletal morphological evidence among contemporary and past populations in the region, and between these populations and a wider sweep of human populations, particularly those from Northeast Asia. Bulbeck and Lauer (Ch. 6) and Turner and Eder (Ch. 7) have approached the issue by looking at the Negrito populations in Malaysia and the Philippines. The Negrito are particularly interesting as they are often cited as 'relict' populations of the Australo-Melanesians (Bellwood 1997).

The identification of skeletal remains as either those of Australo-Melanesians or Mongoloids, of course, begs the question of the reality of these two populations ever having existed as discrete entities. Has it ever been possible to identify them from dental and skeletal morphology? Research on modern population history uses multivariate statistics to compare morphologies and discriminate between groups – but the statistics themselves are designed to differentiate and will continue until they do. Does this differentiation create discrete groups that have any biological reality? Or any genetic reality?

Collections of skulls such as that from Lang Cuom, North Vietnam, (Mansuy and Colani 1925) were variously classified and reclassified by Mansuy and Colani as Melanesian/Australoid/Indonesian, by Saurin (1939) as Melanesian/Negrito/Indonesian, and by Coon (1962) as 'modified Australoid', 'more or less Mongoloid' and 'mixed Negrito'. Classifications such as these, and the interpretations of timing and routes of migrations and 'racial hybridisation' based upon them, formed the basis of early theories of the origins of Southeast Asian peoples (e.g. Barth 1952, Coedes 1968, Coon 1962, Jacob 1967). The answer to this could, in theory, be determined by extraction of 'ancient' DNA, either mitochondrial or nuclear, from the skeletons in question, but the reality is that the preservation in the tropics is generally so poor that the results of such endeavours have, at least until now, produced more questions than answers. An exception is the study by Oota *et al.* (2001) of mitochondrial DNA extracted from six prehistoric skeletons from the cave sites of Moh Kiew and Sakai on the Malay Peninsula. These individuals show differing phylogenetic relationships when compared with the Semang and Senoi, modern Orang Asli hunter–gatherers and slash-and-burn agriculturalists, respectively. The earlier (late palaeolithic) of two Moh Kiew specimens most closely clusters with the Semang and the later preneolithic Moh Kiew and all four neolithic Sakai specimens cluster with the Senoi (see also Ch. 2 for a morphometric analysis of the Moh Kiew specimens).

There is, of course, a burgeoning interest in DNA research on modern populations and the extrapolation back into prehistory of the results of such analyses. Yet, again, the reality is that there are so many factors other than migration that affect the individual genome and the gene pool of even the most isolated population that such analyses almost inevitably raise more questions than answers – and certainly stimulate a lot of debate among competing interpretations (see Su and Jin (2001) for a useful review).

Given the reality that each and every human has a unique phenotype, and that that phenotype has contributions from ancestral genes (which are themselves products of the raft of evolutionary processes including selection, drift, flow and mutation) acted on during the individual's lifetime by a range of environmental effects (using environment in its broadest sense, both natural and cultural), can we interpret an individual's morphology as evidence of genetic relationships with another individual from a different geographic area, widely separated in time? No wonder we find it very difficult to interpret the evidence, and answers differ whether the basis of comparison is dental or skeletal, metric or 'non-metric'. This does not negate the method, clearly, as the contributors to this volume have shown that it has wide support.

The issue is the continued use of the descriptors Australo-Melanesian and Mongoloid as if they describe clearly distinctive populations. Even Bellwood (1997, p. 75, italics original), who acknowledged that '... the terms *Australo-Melanesian* and *Mongoloid* themselves are idealised models', nevertheless finds the need to continue to use these terms as descriptors. Categorisation is a natural human approach to making sense of the variation in our world (hence the power of taxonomy), but in many situations it creates expectations of discrete variation where only continuous variation exists. It also raises expectations of identifiable bases for such variation where the reality is, as always, much too complex to be simply explained. There is, of course, in Southeast Asia and western Melanesia an extraordinary variety of phenotypes and it is this variation, as much as anything, that stimulates the search for understanding of relationships among the populations.

To return to the focus of many of the chapters in Part I, the expectation of relationships between Northeast and Southeast Asians is an outcome of the long-standing Southern Mongoloid Migration model, which has at its base, in fact, the Regional Continuity model as it has been previously described in this part of the world. There is seen to be an 'indigenous' population of 'Australo-Melanesians' exemplified by the dark-skinned Negritos in Southeast Asia and the dark-skinned Melanesians to the east. This is alongside the lighter-skinned 'Mongoloid' populations such as the Thai, Mon Khmer, Malay and western Indonesians. In order for the latter groups to have been found in this part of Asia, they must have migrated southwards as descendents of peoples influenced by the genes of the northern Chinese *H. erectus*. This model of southward migration is supported by linguistics, as they are predominantly speakers of Austric (Austroasiatic and Austronesian) languages, with their linguistic origins in southern China (Higham 2001). It is also supported by the evidence of the chronology of the development of agriculture, which was much earlier in China than in Southeast Asia (Higham 2003). The movements of these languages and technologies were much later than the appearance of the first populations in the region identified as Mongoloid on the basis of skeletal morphology. The varied and troublesome morphology of the remains from Tam Hang Cave in Laos, dated to the early Holocene, includes a variety of morphologies variously defined as either Northeast Asian or Australo-Melanesian (see Ch. 2). The migration is, therefore, seen as a constant process, not starting with the development of agriculture but strongly stimulated by it in the neolithic.

It is acknowledged that languages and ideas can migrate without being accompanied by a wholesale movement of peoples – in other words independently of morphology or genes. The role of agriculture as a driver of migration means that many of the publications on this issue are looking for explanations of movements of technology rather than focusing on migration of people (Diamond and Bellwood 2003). This, in part, reflects the relative ease of identifying agriculture from the archaeological record and the difficulty of identifying the biological relationships of people, as already discussed, although it is also reflective of the importance of the control of food supply in the development of human societies and, therefore, of the attention accorded it by prehistorians. Languages are often considered in association with technology, so linguistic and archaeological evidence drive models, with human biology, including genetics, inferred accompaniments.

To summarise the role of the first section of this book; the overriding question is who was living in Southeast Asia in prehistory? Who were they descended from? How were populations related to one another? What drove the development of the diversity of human biology evident in the region today?

Part II addresses a second important question. What was the quality of life of these people? This is determined by considering the health, both of individuals and of populations. Quality of life is, of course, not independent of the population history, as population growth and the vigour of the people, and their exposure to infectious diseases, have a significant part to play in the behaviour of people, again both as individuals and communities. This, together with the extent to which they are able to exploit their environment, influences population history.

The determination of individual health from the skeleton is based on evidence for achievement of growth potential, from evidence of growth disruption, from evidence for dental and skeletal pathology and from age at death, as a measure of longevity. The data from individuals can then be amalgamated into measures of population health through comparisons with other samples. Measures of mean body size, quality of bone structure, prevalence of dental and skeletal pathologies and demographic parameters of mortality and fertility are all considered, while bone chemistry provides insight into diet and hence nutrition.

The issues considered by the contributors to Part II cover a range of specific topics within this broad spectrum of measures of health. Douglas (Ch. 8) examines the dental health of the people of Non Nok Tha, a site in northeast Thailand covering the period from before and during the Bronze Age. Domett and Tayles (Ch. 9) compare change over time in health in general in the Mun River Valley, also in northeast Thailand, using samples from the bronze age site of Ban Lum Khao and the iron age site of Noen U-Loke. The concentration on the geographic area of northeast Thailand, which is relatively well covered compared with other regions of mainland Southeast Asia, continues in Ch. 10 where King and Norr describe the use of bone chemistry to examine diet, using stable isotopes measured in the samples from the site of Ban Chiang. Oxenham, Nguyen Lan Cuong and Nguyen Kim Thuy (Ch. 11) explore the issue of dental health on a broader scale, through the consideration of evidence from a range of samples from a wide geographic area of mainland Southeast Asia. In Ch. 12, Indriati presents evidence for health from a different perspective, a different geographic region and a very much earlier time period by reviewing the cranial lesions of one of the *H. erectus* specimens from Ngandong in Java. The nature of the *H. erectus* samples precludes a population study approach, but the reporting of individual pathology as a case study provides an insight into the lives of people from these ancient communities. The final chapter (Ch. 13) is again from a different region and has a different focus; Buckley examines a sample from Melanesia and considers the impact of an influential disease, malaria, which has profound implications not only for Melanesia but for the wider geographic area of Southeast Asia.

The range of approaches to the issue of human health illustrates the diversity of evidence for this topic. It has received far less attention from the archaeologists and linguists, whose focus has been on population history, so there has not been the concentration on the development of the explanatory models that underpin research into population history. The most recent and most comprehensive reviews of regional health and disease are those of Tayles (1999), Oxenham (2000), Domett (2001) and Pietrusewsky and Douglas (2002), all concentrating on mainland Southeast Asia. It is superfluous for us to duplicate these here, but, to summarise them, the over-riding conclusions are, unsurprisingly, that the evidence is complex. Between-site comparisons are difficult because these require confidence that interobserver error is minimised. The field of palaeopathology is relatively new and, unlike biodistance studies where methods have been developed and refined over a long period, there are no standardised, universally accepted data collection methods for the variety of data. Further, unlike the biodistance data, which are recorded in the two categories of metric and non-metric, there is also the issue of the variety of the data. Nevertheless, pictures are beginning to emerge of regional evidence for health over time and between geographic locations.

To summarise these, what is shown is that the passage of time appears not to have resulted in the deterioration in health identified elsewhere in the world during late prehistory (Cohen and Armelagos 1984, Steckel and Rose 2002). This is generally ascribed to the adoption and intensification of agriculture, with its associated implications of reduction in nutritional variety, increase in population density and migrations of peoples who may have transported infectious diseases along with their skills and languages. The picture that dimly emerges from the regional evidence available so far from mainland Southeast Asia is that there was indeed an increase in infection over time (Oxenham *et al.* 2005, Tayles and Buckley 2004) but health in general did not deteriorate. The picture is quite different for the Pacific however, where Buckley (Ch. 13) has demonstrated a decrease in infectious disease over time with the colonisation of Polynesia. This is an important exception to the general rule, and the prehistory of Polynesian health is intimately related to the ecology of a single disease. An important issue in prehistoric health in the regions covered by this volume is the contribution of malaria to human behaviour. Samerchai Poolsuwan (1995) has reviewed the evidence for the role of malaria in prehistory in mainland Southeast Asia. He has demonstrated the diversity of malaria ecology among populations and environments in prehistory. The interaction of the malarial parasites, the mosquito vectors, the human and animal hosts and the environment produces a complex pattern. Human modifications to the environment further enhanced the complexity, particularly during the development of rice agriculture with its accompanying alterations to forest cover and drainage patterns and the creation of standing bodies of water in wet rice production. Overlying this is the diversity of human red cell genetic variants, which have evolved in response to malarial endemicity. These have again been reviewed by Poolsuwan (2003). He has shown the complexity of the Southeast Asian picture, this time using data from modern populations but relating this to prehistory. While acknowledging the influence of migrations that have occurred in historic times, he has been able to present a plausible argument for the modern diversity having been a long-standing development, occurring since prehistory. This genetic diversity, of course, is indicative of the difficulty of defining genetic relationships among prehistoric populations.

There is no direct skeletal evidence for malaria, other than that of skeletal responses to the severe anaemia that can develop in homozygotes for some of the genetic variants (e.g. β-thalassaemia) (see Tayles 1999, pp. 178–183). A population exposed to malaria, either endemic or epidemic, will not necessarily carry any of the haemoglobin variants that confer protection against the disease in heterozygotes. There is, therefore, not necessarily any indication that the disease has created a health problem or stimulated behavioural changes. The complexity of the situation in mainland Southeast Asia means that malarial influences on prehistoric populations is almost impossible to demonstrate without these clues. However, in the South Pacific, the situation is different, with a relatively short prehistory in a series of archipelagos with well-described and understood vector ecology and parasite distribution. Buckley (Ch. 13) demonstrates how it is possible to consider the effects of malaria on human settlement and survival, even in the presence of other infectious diseases. This chapter contains significant issues for those of us working in Southeast Asia to consider in our determination of prehistoric health.

Returning to the observations for mainland Southeast Asia, the explanation for the evidence for health and infection over time may lie in the environment. While it is undeniable that the starchy staple in Southeast Asia, rice, was widely adopted and plays to this day an unqualified central role in the diet, it was nevertheless complemented by a varied and nutritious diet. This diet was available for two reasons. First, the tropical environment is very rich in food resources; plant foods grow rapidly and easily and protein is easily available, not the least from fish, which thrive in the many ponds and waterways in even the seasonally driest parts of the region such as the Khorat Plateau in northeast Thailand. Second, these easily acquired resources meant that, although the people of mainland Southeast Asia might have experienced a transition to, and intensification of, agriculture, they nevertheless continued to hunt, gather and fish for a varied diet. This can be observed today, where villagers living in the rice-growing areas and subsisting on rice continue to supplement their diet with wild vegetable foods, fish from local waterways and trapping small mammals and insects. The environment is the key; they did not find themselves largely reliant on the starchy staple once they had cleared land for agriculture. Their diet remained varied and nutritious, keeping them in relatively good health despite the proliferation of infections.

Clearly, there is much to consider in defining the human biology of prehistoric Southeast Asians. Migrations, changes in subsistence mode, cultural, political, social behaviour all compete against a background of major environmental variation over time. To expect that a single volume will furnish the answers to all of these questions is obviously unrealistic. Our hope is that as you read through the following chapters you will gain new insights into the complexities of the issues and be stimulated into adding your own contribution to the picture.

References

Adi H. T. 1985. The re-excavation of Gua Cha, Ulu Kelantan, West Malaysia. *Federation Museums Journal Monograph* **30**.

Agelarakis A. 1996. The archaeology of human bones: prehistoric copper producing peoples in the Khao Wong Prachan Valley, central Thailand. *Bulletin of the Indo-Pacific Prehistory Association* **14**: 133–139.

1997. Some reconstructions of human bio-cultural conditions during the 3rd and 2nd millennia BC. In Ciarla R. and Rispoli F. eds., *South-east Asian Archeology 1992*. Rome: Istituto Italiano per L'Africa e L'Oriente, pp. 99–117.

Barker G., Barton H., Beavitt P. *et al.* 2000. The Niah Caves project: preliminary report on the first (2000) season. *Sarawak Museum Journal* **55**(n.s. 76): 111–149.

Barker G., Badang D., Barton H. *et al.* 2001. The Niah Cave project: the second (2001) season of fieldwork. *Sarawak Museum Journal* **56**(n.s. 77): 37–119.

Barker G., Barton H., Beavitt P. *et al.* 2002a. Prehistoric foragers and farmers in southeast Asia: renewed investigations at Niah Cave, Sarawak. *Proceedings of the Prehistoric Society* **68**: 147–164.

Barker G., Barton H., Bird M. *et al.* 2002b. The Niah Cave project: the third (2002) season of fieldwork. *Sarawak Museum Journal* **57**(n.s. 78): 87–177.

2004. The Niah Cave project: the fourth (2003) season of fieldwork. *Sarawak Museum Journal*, **58**, in press.

Barth E. 1952. The southern mongoloid migration. *Man* **52**: 5–8.

Bellwood P. 1997. *Prehistory of the Indo-Malaysian Archipelago*, revised edn. Honolulu: University of Hawaii Press.

Bellwood P. and Adi bin H. T. 1981. A home for ten thousand years. *Hemisphere* **25**: 310–313.

Bentley R. A. 2004. Characterising human mobility by strontium analysis of the skeletons. Appendix. In Higham C. F. W. and Thosarat R., eds. *Report of the Research Committee* LXXII: *The Excavation of Khok Phanom Di, A Prehistoric Site in Central Thailand*, Vol. VII: *Summary and Conclusions.* London: Society of Antiquaries, pp. 159–166.

Bishop P., Penny D., Stark M. and Scott M. 2003. A 3.5 Ka record of paleoenvironments and human occupation at Angkor Borei, Mekong Delta, southern Cambodia. *Geoarchaeology* **18**: 359–393.

Blust R. 1996. Beyond the Austronesian homeland: the austric hypothesis and its implications for archaeology. In Goodenough W. H., ed., *Prehistoric Settlement of the Pacific. Transactions of the American Philosophical Society* **86**: 117–140.

Brace C. L. 1976. Tooth reduction in the Orient. *Asian Perspectives* **19**: 203–219.

1980. Tooth-size and Austronesian origins. Naylor P., ed., *Austronesian Studies: Papers from the second Eastern Conference on Austronesian Languages. Michigan Papers on South and Southeast Asia* **15**: 167–180.

Brace C. L. and Hinton R. J. 1981. Oceanic tooth size variation as a reflection of biological and cultural mixing. *Current Anthropology* **22**: 549–569.

Brown P., Sutikna T., Morwood M. J. *et al.* 2004. A new small-bodied hominin from the late Pleistocene of Flores, Indonesia. *Nature* **431**: 1055–1061.

Bui Vinh 1980. *Bao cao khai quat, di tich van hoa Da But, dia diem khao co hoc: Con Co Ngua (Thanh-Hoa)*. Unpublished Excavation Report, Institute of Archaeology, Hanoi, Vietnam.

Bulbeck D. 1981. Continuities in Southeast Asian evolution since the late Pleistocene. M.A. thesis, Australian National University, Canberra, Australia.

1982. A re-evaluation of possible evolutionary processes in Southeast Asia since the late Pleistocene. *Bulletin of the Indo-Pacific Prehistory Association* **3**: 1–21.

Bulbeck D., Oxenham M., Nguyen L. C. and Nguyen K. T. 2003. Implications of the terminal Pleistocene skull from Hang Muoi, North Vietnam. In *Proceedings of the 2003 Australian Archaeological Association Conference*, Jindabyne, New South Wales, December 2003.

Choosiri P. 1988. An analysis of human remains from Khok Phanom Di. M.A. thesis, University of Otago, New Zealand.

1991. *The Human Remains from the Excavation of Khok Phanom Di.* [in Thai] Bangkok, Fine Arts Department.

Coedes G. 1968. *The Indianized States of Southeast Asia.* Honolulu: East-West Centre Press.

Colani M. 1927. La grotte sepulcrale de Lang Gao. *L'Anthropologie* **37**: 227–229.

Cohen M. N. and Armelagos G. J. (eds.). 1984. *Paleopathology at the Origins of Agriculture*. Orlando, FL: Academic Press.

Coon C. 1962. *The Origin of Races.* London: Cape.

Cox M. and Mays S. (eds.) 2000. *Human Osteology in Archaeology and Forensic Science*. London: Greenwich Medical Media.

de Sieveking G. 1954. Excavations at Gua Cha, Kelantan 1954. Part 1. *Federation Museums Journal* **1–2**: 75–143.

Diamond J. and Bellwood P. 2003. Farmers and their languages: the first expansions. *Science* **300**: 597–603.

Domett K. M. 2001. *British Archaeological Reports International Series*, No. 946: *Health in Late Prehistoric Thailand.* Oxford: Archaeopress.

Douglas M. T. 1996. Paleopathology in human skeletal remains from the pre-metal, bronze, and iron ages, Northeastern Thailand. Ph.D. thesis, University of Hawaii-Manoa. Ann Arbor, MI: University Microfilms.

Dubois E. 1894. Pithecanthropus erectus. *Eine menschenaehliche Ubergangsform aus Java.* Batavia: Landsdrukkerij.

1937. On the fossil human skulls recently discovered in Java and *Pithecanthropus erectus. Man* **xxxvii**: 1–7.

1938a. The mandible recently described and attributed to the *Pithecanthropus* by G. H. R. von Koenigswald, compared with the mandible of *Pithecanthropus erectus* described in 1924 by Eug. Dubois. *Proceedings of the Section of Sciences of the Koninklijke Akadamie van Wetenschappen* **41**: 139–147.

1938b. On the fossil human skull recently described and attributed to *Pithecanthropus erectus* by G. H. R. von Koenigswald. *Proceedings of the Section of Sciences of the Koninklijke Akadamie van Wetenschappen* **41**: 380–386.

1940a. The fossil human remains discovered in Java by Dr. G. H. R. von Koenigswald and attributed by him to *Pithecanthropus erectus*, in reality remains of *Homo wadjakensis* (syn. *Homo soloensis*). *Proceedings of the Section of Sciences of the Koninklijke Akadamie van Wetenschappen* **43**: 494–496.

1940b. The fossil human remains discovered in Java by Dr. G. H. R. von Koenigswald and attributed by him to *Pithecanthropus erectus*, in reality remains of *Homo wadjakensis* (conclusion). *Proceedings of the Section of Sciences of the Koninklijke Akadamie van Wetenschappen* **43**: 842–851.

Duckworth W. L. H. 1934. Human remains from rock-shelters and caves in Perak, Pahang and Perlis and from Selinsing. *Journal of Malayan Branch Royal Asiatic Society* **XII**: 149–167.

Evans I. H. N. 1918. Preliminary report on cave exploration near Lenggong, Upper Perak. *Journal of the Federated Malay States Museum* **7**: 227–234.

Fromaget J. 1940. Les récentes découvertes anthropologiques dans les formations préhistoriques de la chaîne annamitique. *Proceedings of the Third Far Eastern Prehistory Congress*, Singapore, pp. 60–70.

Fromaget J. and Saurin E. 1936. Note préliminaire sur les formations cénozoïques et plus récentes de la Chaîne annamitique septentrionale et du Haut-Laos. *Bulletin du Service Géologique de l'Indochine* **22**: 36–48.

Fox R. B. 1970. *National Museum Monograph 1: The Tabon Caves*. Manila: National Museum.

Gibbons A. 2003. Paleoanthropology: Java skull offers new view of *Homo erectus*. *Science* **299**: 1293.

Green R. C. 1991. Near and Remote Oceania: disestablishing 'Melanesia' in culture history. In Pawley A., ed., *Man and a Half: Essays in Pacific Anthropology and Ethnobiology in Honour of Ralph Bulmer*. Auckland: The Polynesian Society, pp. 491–502.

Heilborn A. 1923. Introduction. In Klaatsch H., ed., *The Evolution and Progress of Mankind*. London: T. Fisher Unwin, pp. 15–29.

Higham C. F. W. 2001. Prehistory, language and human biology. In Jin L., Seielstad M. and Xiao C., eds., *Genetic, Linguistic and Archaeological Perspectives on Human Diversity in Southeast Asia*. Singapore: World Scientific, pp. 3–16.

2002. *Early cultures of mainland Southeast Asia*. Bangkok: River Books.

2003. Languages and farming dispersals: Austroasiatic languages and rice cultivation. In Bellwood P. and Renfrew C., eds., *Examining the Language/Farming Dispersal Hypothesis*. Cambridge: McDonald Institute for Archaeological Research, pp. 223–232.

Higham C. F. W. and Bannanurag R. (eds.). 1990. *Report of the Research Committee* XLVII. *The Excavation of Khok Phanom Di, A Prehistoric Site in Central Thailand*, Vol. 1: *The Excavation, Chronology and Human Burials*. London: Society of Antiquaries.

Houghton P. and Wiriyaromp W. 1984. The people of Ban Na Di. In Higham C. F. W., and Kijngam A., eds., *British Archaeological Reports International Series*, No. 231: *Prehistoric Investigations in Northeast Thailand*, Vol. 2. London: Archaeopress, pp. 391–412.

Howells W. W. 1944. Fossil man and the origin of races. *American Anthropologist* **42**: 182–193.

1948. *Mankind So Far*. London: Sigma Books.

Jacob T. 1967. *Some Problems Pertaining to the Racial History of the Indonesian Region*. Utrecht: Netherlands Bureau for Technical Assistance.

Jin L., Seielstad M. and Xiao C. (eds.) 2001. *Genetic, Linguistic and Archaeological Perspectives on Human Diversity in Southeast Asia*. Singapore: World Scientific.

Kennedy K. A. R. 1977. The deep skull of Niah. *Asian Perspectives* **20**: 32–50.
Klaatsch H. 1923. *The Evolution and Progress of Mankind*. London: T. Fisher Unwin.
Krigbaum J. S. 2001, Human paleodiet in tropical Southeast Asia: isotopic evidence from Niah Cave and Gua Cha. Ph.D. thesis, Department of Anthropology, New York University.
2003. Neolithic subsistence patterns in northern Borneo reconstructed with stable carbon isotopes of enamel. *Journal of Anthropological Archaeology* **22**: 292–304.
Larsen C. S. 1997. *Bioarcheology: Interpreting Behavior from the Human Skeleton*. Cambridge, UK: Cambridge University Press.
Lewin R. 1999. *Human Evolution: An Illustrated Introduction*. Malden, MA: Blackwell Science.
Manser J. 2003. Geometric morphometric analysis of the human burial series from Niah Cave, Borneo. *American Journal of Physical Anthropology* **120**: 145–146.
Mansuy H. 1924. Contribution à l'étude de la préhistoire de l'Indochine IV: stations préhistoriques dans les cavernes du massif calcaire de Bac-son (Tonkin). *Mémoires du Service Géologique de l'Indochine* **11**: 15–20.
1925a. Contribution à l'étude de la préhistoire de l'Indochine caverne sépulcrale de Ham rong. Description d'un crâne Indonésien de Cho ganh (Tonkin). Complément a l'étude des cranes recueillis dans la caverne sepulcrale de Lang cuom, Massif de Bac-son. *Bulletin du Service Géologique de l'Indochine* **14**: 5–28.
1925b. Contribution à l'étude de la préhistoire de l'Indochine VI: stations préhistorique de Kêo-Phay (suite), de Khac-Kiêm (suite), de Lai-Ta et de Bang-Mac, dans le massif calcaire de Bac-son (Tonkin). *Mémoires du Service Géologique de l'Indochine* **12**: 6–17.
Mansuy H. and Colani M. 1925. Contribution à l'étude de la préhistoire de l'Indochine VII: néolithique inférieur (Bac-Sonien) et néolithique supérieur dans le haut-Tonkin (dernières recherches), avec la description des crânes du gisement de Lang-Cuom. *Mémoires du Service Géologique de l'Indochine* **12**: 1–45.
Mijsberg W. A. 1940. On a Neolithic Paleo-Melanesian lower jaw fragment found in a kitchen midden at Guak Kepah, Province Wellesley, Straits Settlements. In Chasen F. N. and Tweedie M. W. F., eds., *Proceedings of the Third Congress of Prehistorians of the Far East*. Singapore: The Government of the Straits Settlements, pp. 100–118.
Nguyen D. 1967. Nguoi co o Hang Muoi. *Tin Tuc Hoat Dong Khoa Hoc* **6**: 7–12.
Nguyen K. D., Trinh C., Dang V. T., Vu Q. H. and Nguyen T. H. 1995. Do trang suc trong cac mo chum o Can Gio (Thanh pho Ho Chi Minh). *Khao Co Hoc* **2**: 27–46.
Nguyen L. 1976. So M16 o hang Lang Gao (Ha Son Binh). *Nhung Phat Hien Moi Ve Khao Co Hoc* 112–116.
Nguyen L. C. 1985. Two precious ancient crania discovered in the west of Thanh Hoa Province. *Vietnam Social Sciences* **2**: 125–129.
1986. Two early Hoabinhian crania from Thanh Hoa province, Vietnam. *Zeitschrift für Morphologie und Anthropologie* **77**: 11–17.

1991. Tim thay di cot nguoi co o Dong Can (Ha Son Binh). *Hinh Thai Hoc* **1**: 35–36.

1994. On human remains from the Hoabinh culture in Vietnam. *Vietnam Social Sciences* **5**: 64–69.

1995. Nghien cuu nhung di cot nguoi co tim thay o hai dia diem Giong Phet, Giong Ca Vo, Huyen Can Gio (Thanh pho Ho Chi Minh). *Khao Co Hoc* **2**: 20–26.

1996. *Anthropological Characteristics of Dong Son Population in Vietnam*. Hanoi: Social Science Publishing House.

2001. Ve nhung di cot nguoi co o dia diem Man Bac (Ninh Binh). *Khao Co Hoc* **1**: 47–67.

Nguyen L. C., Nguyen K. T., and Vo H. 1980. Di cot nguoi co o Con Co Ngua (Thanh Hoa). *Nhung Phat Hien Moi Ve Khao Co Hoc* 56–59.

Oota H., Kurosaki K., Pookajorn S., Ishida T., and Ueda S. 2001. Genetic study of the palaeolithic and neolithic Southeast Asians. *Human Biology* **73**: 225–231.

Oppenheimer S. 1998. *Eden in the East*. London: Phoenix.

Oppenoorth W. F. F. 1937. The place of *Homo soloensis* among fossil men. In MacCurdy G. G., ed., *Early Man: As Depicted by Leading Authorities at the International Symposium of the Academy of Natural Sciences, Philadelphia, March 1937*. New York: Lippincott, pp. 349–360.

O'Reilly D. J. W. and Sytha P. 2001. Recent excavations in northwest Cambodia. *Antiquity* **75**: 265–266.

Oxenham M. F. 2000. Health and behaviour during the mid Holocene and metal period of northern Viet Nam. Ph.D. thesis, Northern Territory University, Darwin.

Oxenham M. F., Nguyen K. T. and Nguyen L. C. 2005. Skeletal evidence for the emergence of infectious disease in bronze and iron age northern Viet Nam. *American Journal of Physical Anthropology*, **126**: 359–376.

Patte E. 1925. Etude anthropologique du crâne néolithique de Minh Cam (Annam). *Bulletin du Service Géologique de l'Indochine* **13**: 3–26.

1932. Le kjökkenmödding de Dabút et ses sépultures (Province de Thanh Hoa, Indochine). *Bulletin du Service Géologique de l'Indochine*, Vol. 19, part 3.

1965. Les ossements du kjökkenmodding de Da But. *Bulletin et Mémoires de la Société d'Anthropologie* **40**: 1–87.

Pautreau J.-P., Pauk Pauk U. and Domett K. 2001a. Le cimetière de Hnaw Khan. *Aséanie* **8**: 73–102.

2001b. Paving the way towards a prehistoric chronology: the Hnaw Khan burial site. *Archèologia* **383**: 58–65.

Pautreau J.-P., Mornais P., Coupey A.-S., Pellé F. and Aung Aung Kyaw. 2003. Une protohistoire méconnue, le Cimetière de Ywa Htin. *Archéologia* **404**: 48–56.

Pautreau J.-P., Mornais P. and Doy-Asa T. 2004. *Ban Wang Hai: Excavations of an Iron-age Cemetery in Northern Thailand.* Bangkok: Silkworm Books.

Pei W. C. 1929. An account of the discovery of an adult *Sinanthropus* skull in the Chou Kou Tien deposit. *Bulletin of the Geological Society of China* **8**: 203–205.

Pietrusewsky M. 1974a. *University of Otago Studies in Prehistoric Anthropology, Vol. 6. Non Nok Tha: The Human Skeletal Remains from the 1966 Excavations at Non Nok Tha, N.E. Thailand.* Dunedin: University of Otago Press.

1974b. Neolithic populations of Southeast Asia studied by multivariate craniometric analysis. *Homo* **25**: 207–230.

1978. A study of early metal age crania from Ban Chiang, Northeast Thailand. *Journal of Human Evolution* **7**: 383–392.

1981. Cranial variation in early metal age Thailand and Southeast Asia studied by multivariate procedures. *Homo* **32**: 1–26.

1982. The ancient inhabitants of Ban Chiang: the evidence from human skeletal and dental remains. *Expedition* **24**: 42–50.

1984. Pioneers on the Khorat plateau: the prehistoric inhabitants of Ban Chiang. *Journal of the Hong Kong Archaeological Society* **X**: 90–106.

1988. Multivariate comparisons of recently excavated neolithic human crania from the Socialist Republic of Vietnam. *International Journal of Anthropology* **3**: 267–283.

Pietrusewsky M. and Douglas M. T. 2002. *University Museum Monograph* 111: *Ban Chiang, A Prehistoric Village Site in Northeast Thailand.* Philadelphia, PA: University of Pennsylvania Press.

Pookajorn S. 1992. Recent evidence of a late Pleistocene to a middle Holocene archaeological site at Moh Khiew Cave, Krabi province, Thailand. *Silpakorn Journal* **35**: 93–119.

1994. *Final Report of Excavations at Moh Khiew Cave, Krabi Province; Sakai Cave Trang Province and Ethnoarcheological Research of Hunter–Gatherer Group, So-called Sakai or Semang at Trang Province.* Bangkok: Department of Archaeology, Silpakorn University.

Poolswan, S. 1995. Malaria in prehistoric southeastern Asia. *Southeast Asian Journal of Tropical Medicine and Public Health* **26**: 3–22.

2003. Testing the 'malaria hypothesis' for the case of Thailand: a genetic appraisal. *Human Biology* **75**: 585–605.

Quaritch-Wales H. G. 1937. Some ancient human skeletons excavated in Siam. *Man* **37**: 113–114.

1964. Some ancient human skeletons excavated in Siam: a correction. *Man* **64**: 121.

Sangvichien S., Sirigaroon P. and Jorgensen J. B. 1969. *Archaeological Excavations in Thailand*, Vol. 3: *Ban Kao.* Part 2: *The prehistoric Thai Skeletons.* Copenhagen: Munksgaard.

Santa Luca A. 1980. *The Ngandong Fossil Hominids: A Comparative Study of a Far Eastern* Homo erectus *Group.* New Haven, CT: Yale University Press.

Saurin E. 1939. Crânes préhistoriques inédits de Lang cuom. In *Proceedings of the 10th Congress of the Far Eastern Association of Tropical Medicine*, Hanoi, Vol. 1, pp. 815–831.

1966. Le mobilier préhistorique de l'Abri-sous-roche de Tam Pong (Haut-Laos). *Bulletin de la Société d'Etudes Indochinoises* **41**: 107–118.

Shipman P. 2001. *The Man Who Found the Missing Link: Eugene Dubois and His Lifelong Quest to Prove Darwin Right.* New York: Simon & Schuster.

Sørensen P. and Hatting T. 1967. *Archaeological Excavations in Thailand,* Vol. II: *Ban Kao*, Part I: *The Archaeological Materials from the Burials*. Copenhagen: Munksgaard.

Spriggs M. 1998. From Taiwan to the tuamotus: absolute dating of Austronesian language spread and major sub groups. In Blench R. and Spriggs M., eds., *Archaeology and Language II: Correlating Archaeological and Linguistic Hypotheses*. London: Routledge, pp. 115–127.

Stark M. T. 2001. Some preliminary results of the 1999–2000 archaeological field investigations at Angkor Borei, Takeo Province. *Udaya: Journal of Khmer Studies* **1**: 19–36.

Stark M. T., Griffin P. B., Phoeurn C. *et al.* 1999. Results of the 1995–1996 archaeological field investigations at Angkor Borei, Cambodia. *Asian Perspectives* **38**: 7–36.

Stark M. T., Sanderson D. C. W., Bishop P. and Spencer J. Q. 2003. Luminescence dating of anthropogenically reset canal sediments from Angkor Borei, Mekong Delta, Cambodia. *Quaternary Science Reviews* **22**: 1111–1121.

Steckel R. H. and Rose J. C. 2002. *The Backbone of History. Health and Nutrition in the Western Hemisphere*. Cambridge, UK: Cambridge University Press.

Storm P. 2001. The evolution of humans in Australasia from an environmental perspective. *Palaeogeography, Palaeoclimatology and Palaeoecology* **171**: 363–383.

Stringer C. 2003. Human evolution: out of Ethiopia. *Nature* **423**: 692–695.

Swisher C. G. III, Curtis G. H., Jacob T. *et al.* 1994. Age of the earliest known hominids in Java, Indonesia. *Science* **263**: 1118–1121.

Swisher C. G. III, Rink W. J., Anton S. C. *et al.* 1996. Latest *Homo erectus* of Java: potential contemporaneity with *Homo sapiens* in southeast Asia. *Science* **274**: 1870–1874.

Su B. and Jin L. 2001. Origins and prehistoric migrations of modern humans in East Asia. In Jin L., Seielstad M. and Xiao C., eds., *Genetic, Linguistic and Archaeological Perspectives on Human Diversity in Southeast Asia*. Singapore: World Scientific, pp. 107–132.

Tayles N. 1999. *Report of the Research Committee* LXI. *The Excavation of Khok Phanom Di, a Prehistoric Site in Central Thailand,* Vol. V: *The People*. London: Society of Antiquaries.

Tayles N. and Buckley H. R. 2004. Leprosy and tuberculosis in iron age Southeast Asia. *American Journal of Physical Anthropology*, **125**: 239–256.

Tayles N., Domett K. and Hunt V. 1998. The people of Nong Nor. In Higham C. F. W., and Thosarat R., eds. *University of Otago Studies in Prehistoric Anthropology*: No. 18: *The Excavation of Nong Nor, a Prehistoric Site in Central Thailand.* Dunedin: University of Otago Press, pp. 321–368.

Tayles N., Domett K. and Pauk Pauk 2001. Bronze age Myanmar (Burma): a report on the people from the cemetery of Nyaunggan, Upper Myanmar. *Antiquity* **75**: 273–278.

Thorne A. G. and Wolpoff M. H. 1992. The multiregional evolution of humans. *Scientific American* **266**: 76–79,82–83.

Trevor J. C. and Brothwell D. R. 1962. The human remains of Mesolithic and Neolithic date from Gua Cha, Kelantan. *Federation Museums Journal* **7**: 6–22.

Trinkaus E. and Shipman P. 1993. *The Neandertals: Changing the Image of Mankind.* New York: Knopf.

Turner C. G. II. 1987. Late Pleistocene and Holocene population history of East Asia based on dental variation. *American Journal of Physical Anthropology* **73**: 305–332.

1989. Teeth and prehistory in Asia. *Scientific American* **260**: 88–96.

1990. Major features of sindodonty and sundadonty. *American Journal of Physical Anthropology* **82**: 295–317.

van Heekeren H. R. and Knuth E. 1967. *Archaeological excavations in Thailand,* Vol. 1: *Sai Yok.* Copenhagen: Munksgaard.

van Stein Callenfels S. 1936a. The Melanesoid civilizations of eastern Asia. *Bulletin of the Raffles Museum, Series B* **1**: 41–51.

1936b. An excavation of three kitchen middens at Guak Kepah, Province Wellesley, Straits Settlements. *Bulletin of the Raffles Museum, Series B* **1**: 27–37.

Verneau R. 1909. Les crânes humains du gisement préhistorique de Pho-Binh-Gia (Tonkin). *L'Anthropologie* **20**: 545–559.

von Koenigswald G. H. R. 1937. A review of the stratigraphy of Java and its relations to early man. In MacCurdy G. G., ed., *Early Man: as Depicted by Leading Authorities at the International Symposium of the Academy of Natural Sciences, Philadelphia, March 1937.* New York: Lippincott, pp. 23–33.

1956. *Meeting Prehistoric Man.* London: The Scientific Book Club.

Weidenreich F. 1936. The mandibles of *Sinanthropus pekinensis*: a comparative study. *Palaeontologica Sinica* Series D, Vol. VII, Fascile **3**: 1–132.

1940. Some problems dealing with ancient man. *American Anthropologist* **42**: 375–383.

1943. The skull of *Sinanthropus pekinensis*: a comparative study on a primitive hominid skull. *Palaeontologica Sinica New Series D* **10** (whole series **127**): 1–184.

1946. *Apes giants and man.* Chicago, IL: University of Chicago Press.

1947. Facts and speculations concerning the origin of *Homo sapiens. American Anthropologist* **49**: 187–203.

Wilson A. C. and Cann R. L. 1992. The recent African genesis of humans. *Scientific American* **266**: 68–73.

Winters N. J. 1974. An application of dental anthropological analysis to the human dentition of two early metal age sites, Palawan, Philippines. *Asian Perspectives* **17**: 28–35.

Wiriyaromp W. 1984a. The human skeletal remains from Ban Na Di. M.A. thesis, University of Otago, New Zealand.

1984b. A prehistoric population from North East Thailand. In Bayard D., ed. *University of Otago Studies in Prehistoric Anthropology*, Vol. 16: *Southeast Asian Archaeology at the XV Pacific Science Congress.* Dunedin: University of Otago Press, pp. 42–49.

Wolpoff M. H. 1999. *Paleoanthropology.* Boston: McGraw-Hill.

Wolpoff M. H., Hawks J. and Caspari R. 2000. Multiregional, not multiple origins. *American Journal of Physical Anthropology* **112**: 129–136.

Wu X. Z. and Poirier F. E. 1995. *Human Evolution in China: A Metric Description of the Fossils and a Review of the Sites.* New York: Oxford University Press.

Zdansky O. 1927. Preliminary notice on two teeth of a hominid from a cave in Chihli (China). *Bulletin of the Geological Society of China* **5**: 281–284.

Zuraina M. (ed.). 1994. *The Excavation of Gua Gunung Runtuh and the Discovery of the Perak Man in Malaysia.* Kuala Lumpur: Department of Museums and Antiquities.

Part I *Morphological diversity, evolution and population relationships*

2 *The population history of Southeast Asia viewed from morphometric analyses of human skeletal and dental remains*

HIROFUMI MATSUMURA
Sapporo Medical University, Sapporo, Japan

Introduction

The population history of Southeast Asia is complex because of the various migration processes and the intermixing of populations since prehistoric times. The limited number of prehistoric human remains and uncertainty over their dating are additional problems in studies of this region. In general terms, Southeast Asia is thought to have been originally occupied by indigenous people (sometimes referred to as Australo-Melanesians), who exchanged genes with immigrants from north and/or east Asia leading to the formation of present day Southeast Asians (van Stein Callenfels 1936, Mijsberg 1940, von Koenigswald 1952, Coon 1962, Jacob 1967, Bellwood 1987, 1989, 1996, 1997, Brace *et al.* 1991). It should be noted that the term Australo-Melanesian is commonly used to refer to either the recent indigenous people of Australia, New Guinea and island Melanesia or the people of that regional phenotype. This population history scenario for Southeast Asia is known as the Immigration or Two-layer model. The Two-layer hypothesis is supported by a wide range of genetic, linguistic and archaeological evidence. Classic genetic markers and recent mitochondrial DNA analyses (Ballinger *et al.* 1992, Cavalli-Sforza *et al.* 1994, Omoto and Saitou 1997, Tan 2001) have found many biological similarities between Chinese and Southeast Asian samples. Linguistic and archaeological studies have linked the premodern expansion of the Austronesian and Austroasiatic language families with the dispersal of rice-cultivating populations during the neolithic period (Renfrew 1987, 1989, 1992, Bellwood 1991, 1993, 1996, 1997, Bellwood *et al.* 1992, Hudson 1994, 1999, 2003, Blust 1996a,b, Glover

Bioarchaeology of Southeast Asia. Marc Oxenham and Nancy Tayles.
Published by Cambridge University Press.

and Higham 1996, Higham 1998, 2001, Bellwood and Renfrew 2003, Diamond and Bellwood 2003). Both linguistic and archaeological considerations suggest that southern China and Taiwan were the ultimate sources of these language and population dispersals.

There are, however, different interpretations regarding these peoples based on recent studies of dental and cranial morphology. Studies by Turner (1989, 1990, 1992) based on non-metric dental traits demonstrated that both early and modern Southeast Asians display the so-called 'sundadont' dental complex. Turner concluded that early sundadont populations migrated into Northeast Asia and evolved into 'sinodont' populations. Hanihara (1993, 1994), using craniometric techniques, argued that the Proto-Malaysians, who are morphologically similar to the present-day Dayak, were the original source population for modern Southeast Asians. Both Turner and Hanihara proposed that the evolution of many present-day Southeast Asians was by local adaptation and not by significant admixture with Northeast Asians. In addition, Hanihara's assessment did not support the hypothesis of the presence of the Australo-Melanesian lineage in early Southeast Asia. Pietrusewsky (1992, 1994, 1999; Pietrusewsky and Douglas 2002; Ch. 3) has undertaken a number of large-scale craniometric analyses of Southeast Asian material and argued for regional continuity in Southeast Asia and relatively close affinities between modern East and Southeast Asians, coupled with a distinct dissimilarity to Australo-Melanesians. However, these studies did not include extensive mesolithic or Hoabinhian samples from Southeast Asia.

The hypotheses proposed by Turner and Hanihara can be called Regional Continuity or Local Evolution models. Such models take the opposite approach to that seen in the Immigration or Two-layer model mentioned above. Consequently, debates over the population history of Southeast Asia revolve around two main questions. The first is whether the preneolithic occupants of Southeast Asia have an Australo-Melanesian affinity. The second is the scale of the dispersal of Northeast Asians into this region from China and whether they have mixed with or replaced the preexisting peoples since the neolithic period. In order to consider the first issue, I will review previous discoveries and studies of preneolithic human remains from Southeast Asia and also present results of evaluations of Australo-Melanesian affinity for more recently excavated late Pleistocene and early Holocene fossils from Thailand, Malaysia and Vietnam. Finally, in order to address both the claim of an Australo-Melanesian affinity of the preneolithic Southeast Asians and the degree of genetic influence of Northeast Asians in this region, I will present my own results on population affinities between prehistoric and modern Southeast Asians based on dental morphological data.

The comparative prehistoric samples examined in this chapter are listed in Table 2.1, and the sample locations are shown in Figs. 1.1 (p. 5) and 1.2 (p. 6). Craniometric data collection methods follow Martin and Saller (1957) while maximum tooth crown diameters follow the recording protocols of Fujita (1949).

Preneolithic human skeletons in Southeast Asia

By the end of the nineteenth century, a considerable number of prehistoric human remains had been recovered in various regions of Southeast Asia. As far as the pre-ceramic cultural sites are concerned, most of the skeletal remains were found in Malaysia, Indonesia, the Philippines and Laos. In Malaysia, several sites in the north of Peninsular Malaysia have produced mesolithic (Tampanian) human skeletons. At Gua Kajang, Lenggong district, Evans (1918) found a fragment of human jaw with some teeth. Furthermore, Gordon, Evans and van Stein Callenfels collected fragmentary human bones of several individuals from Guar Kerbau in Gunung Pondok (Evans 1928). Duckworth (1934) interpreted these remains as having Australo-Melanesian features. Further away, in Guar Kepah, Mijsberg (1940) found human jaws from a shell midden site and identified them as 'Palaeo-Melanesian', while a later excavation by van Steins Callenfels (1936) discovered more remains from the subsequent neolithic period. The Gua Cha remains excavated by de Sieveking (1954) were from a rock shelter located in the State of Kelantan dating to the Hoabinhian and neolithic periods (10,000 to 2,000 years BP). More recently, some additional human remains were discovered by Adi and Bellwood (Adi 1985). Trevor and Brothwell (1962) concluded that the Gua Cha skeletons have a close Melanesian affinity. Furthermore, Bulbeck (2000), in his study of non-metric dental traits, found common features between the Gua Cha, Melanesian and neolithic Jomon people of Japan.

In East Malaysia, Niah Cave in Sarawak is the well-known site of the earliest appearance of humans in Southeast Asia. The so-called 'deep skull' from Niah Cave has an associated radiocarbon date of *c.* 40,000 years BP (Kennedy 1977). Brothwell (1960) examined this skull and concluded that it bore the closest similarity to Tasmanians.

In Indonesia, two skulls were found in 1888 and 1890 at Wajak in Gunung Lawa, central Java. Dubois (1922), who first examined these

Table 2.1. *List of prehistoric Southeast/East Asian samples*

Series	N	Period (years BP)	Provenance	Institution
Author's data				
Late Pleistocene Moh Khiew Cave	1	*c.* 25,000	Krabi Province, Southern Thailand	Silpakorn University
Early Hoabinhian Mai Da Nuoc	1	*c.* 10,000	Thanh Hoa Province, Northern Vietnam	Institute of Archaeology, Hanoi
Early Hoabinhian Mai Da Dieu	1	*c.* 10,000	Thanh Hoa Province, Northern Vietnam	Institute of Archaeology, Hanoi
Early Holocene Wajak (teeth)	2	*c.* 6,500?	Java, Indonesia	Natural History Museum, Leiden
Early Holocene Gua Gunung Runtuh	1	*c.* 10,000	State of Perak, Peninsula Malay	University Sains Malaysia
Early Holocene Laos	18	–	Tam Hang, Tam Pong sites, Northern Laos	Musée de l'Homme
Mesoneolithic Gua Cha	20	*c.* 10,000–2,000	Kelantan, Malay peninsula	University of Cambridge
Mesolithic Guar Kepah	27	–	Mainland Penang, Peninsula Malay	British Museum
Mesolithic Liang Momer	1	*c.* 7,000–4,000	Flores Island, Indonesia	Natural History Museum, Leiden
Mesolithic Liang Toge	1	*c.* 7,000–4,000	Flores Island, Indonesia	Natural History Museum, Leiden
Other mesolithic Flores	7	*c.* 7,000–4,000	Flores Island, Indonesia	Natural History Museum, Leiden
Neolithic Bac Son Vietnamese	17	*c.* 10,000–6,000	Northern Vietnam	Musée de l'Homme
Neolithic Da But Vietnamese	46	*c.* 5,000	Northern Vietnam	Institute of Archaeology, Hanoi
Neolithic Ban Kao Thai	35	*c.* 4,100–3,600	Kanchanaburi Province, Thailand	Mahidol University, Bangkok
Neolithic Southern Chinese	108	*c.* 5,000	Weidun, Songze sites, Jiangnan region, southern China	Shanghai Museum of Natural History; Nanjing Museum

Neolithic Jomon	711	*c.* 5,000–2,300	Japan	See Matsumura (1995)
Metal age Dong Son Vietnamese	44	*c.* 3,000–1,700	Northern Vietnam	Institute of Archaeology, Hanoi
Metal age Leang Tjadang	100	–	Sulawesi Island, Indonesia	Natural History Museum, Leiden
Metal age Yayoi Japanese	212	*c.* 2,800–1,700	Kyushu and Yamaguchi districts, western Japan	See Matsumura (1994)
Bronze age Anyang Chinese	21	3,300–3,100	Henan Province, China	Academia Sinica, Taipei
Bronze age Chifeng Chinese	38	127–221(BC)	Chifeng, Inner Mongolia	See Matsumura (1995)
Published data				
Late Pleistocene Coobool Creek	31	*c.* 14,300	Murray Valley, southeastern Australia	Brown (1989)
Late Pleistocene Keilor	1	*c.* 12,000	Victoria, southeastern Australia	Brown (1989)
Late Pleistocene Upper Cave	2	–	Zhoukoutien, northern China	Wu (1961)
Late Pleistocene Liujiang	1	–	Guanxi Province, southern China	Woo (1959)
Late Pleistocene Minatogawa	1	*c.* 18,000	Okinawa Island, Japan	Suzuki (1982)
Early Holocene? Wajak (skulls)	2	*c.* 6,500?	Java, Indonesia	Storm (1995)
Bronze–iron age Ban Chiang	55	4,100–1,800	Northeast Thailand	Pietrusewsky and Douglas (2002)

specimens, labelled them Proto-Australian. Later, Weidenreich (1945) regarded the Wajak skulls as akin to the terminal Pleistocene Keilor specimen from Australia. Jacob (1967) and Wolpoff *et al.* (1984), in their analyses of the same material, supported these conclusions. In contrast, Storm (1995) examined more details of the Wajak skeletons and found a similarity to modern Indonesians rather than to Australians. Therefore, the assessment of morphological affinity of the Wajak remains differs between researchers. In addition to this problem, the chronological framework of these fossils has been unclear until recently. Dubois (1922) first estimated the date of the Wajak remains to be in the late Pleistocene. Later, Jacob (1967) also suggested a late Pleistocene age. More recent dating of the skeleton using accelerator mass spectroscopy (AMS) indicates the middle Holocene, *c.* 6,500 years BP (Storm 1995).

From the western part of Flores Island, some human remains associated with mesolithic artefacts have been recovered from various sites. The Liang Momer and Liang Toge sites produced five individuals, including nearly complete crania. Many further fragmentary remains were unearthed from Liang X, Gua Alo, Gua Nempong and other sites. Jacob (1967) identified these Flores populations as Australo-Melanesian.

Tabon Cave on Palawan Island is a well-known site that has produced the oldest human skeletal remains in the Philippines, consisting of a frontal bone and two mandibular fragments. The frontal bone has been AMS dated to *c.*16,500 years BP. Macintosh (1978) considered this specimen to be linked to the Australo-Melanesians.

In conclusion, early researchers of the human remains discovered from preneolithic cultural levels in Southeast Asia often described the morphological features of these specimens as being typical of Australo-Melanesians, citing their dolichocranic skulls, protruding glabellas, massive jaws with relatively large teeth, alveolar prognathism and long, slender limbs.

New evidence for the existence of Australo-Melanesians

Since the early 1990s, more human skeletal remains from the late Pleistocene and early Holocene have been discovered from various regions of Southeast Asia. At least four cave sites have produced relatively well-preserved preneolithic human remains, which may prove to be extremely important for our understanding of the initial peopling of Southeast Asia. The oldest skeleton from these four sites is from the Moh Khiew Cave in southern Thailand (Pookajorn 1991, 1994). The next oldest is the Gua Gunung skeleton, the so-called Perak man (Zuraina 1994), from mainland

Malaysia, while the other two skulls come from the Mai Da Nuoc and Mai Da Dieu sites in northern Vietnam (Cuong 1986).

Moh Khiew Cave

A human skeleton was discovered during excavations undertaken in 1990 and 1991 by Pookajorn of Silpakorn University at Moh Khiew Cave, Krabi Province, southern Thailand (Pookajorn 1991, 1994). An adult female (Fig. 2.1) buried in the preneolithic cultural level has been AMS dated to 25,800 $\pm$ 600 years BP (TK-933Pr) on a charcoal sample. A description of the preservation and morphology was first provided by Choosiri (Pookajorn 1994). Oota *et al.* (2001) analysed mitochondrial DNA from the skeleton and revealed close affinities with the present Semang aborigines in the Malay Peninsula. However, Lauer (2002) calculated Penrose's distances on the basis of tooth measurements and found a close affinity with the later Hoabinhian people in neighbouring regions.

Following these studies, Pookajorn and I carried out statistical comparisons of the Moh Khiew skeletal measurements with early and modern females from Southeast Asia and Australia (Matsumura and Pookajorn 2005). Twelve cranial and mandibular measurements (Martin Nos. 46, 48, 51, 54, 55, 60, 61, 65, 66, 69, 70 and 71) were used for calculating biological distances owing to the absence of many portions of the facial skeleton. Figure 2.2a shows the results of a cluster analysis applied to the Mahalanobis' generalised distances on the basis of these cranial measurements. This dendrogram demonstrates that the Moh Khiew Cave specimen is most similar to the Australian samples, including the late Pleistocene Coobool Creek samples and present-day Aborigines.

In addition, biological distances were computed using dental measurements. Since heavy attrition obviously reduced the mesiodistal dimensions of the tooth crowns, only 14 buccolingual crown diameters were used for the distance computations. The dendrogram of Q-mode correlation coefficients calculated using these dental measurements indicated a close affinity to the mesolithic and neolithic samples from Flores and Indochina together with the Coobool Creek specimens (Fig. 2.2b). Although both the cranial and dental affinities are based on a limited portion of the skeleton, these findings suggest that the Moh Khiew specimen, as well as other late Pleistocene fossils from Southeast Asia such as the Niah and Tabon specimens, is representative of an early group of people in Southeast Asia who may have originated from the common ancestor of the Australian Aborigines.

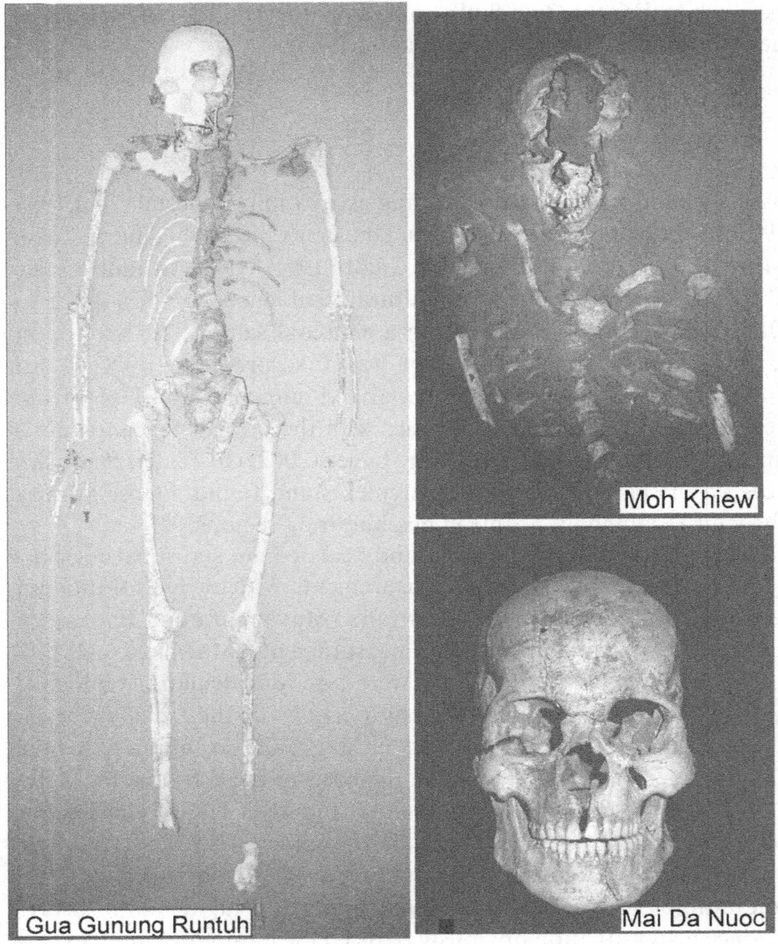

Figure 2.1. Recently excavated human remains from late Pleistocene and early Holocene sites in Southeast Asia (Gua Gunung Runtuh, Malaysia; Moh Khiew Cave, Thailand; Mai Da Nuoc, Vietnam).

Gua Gunung Runtuh

The discovery in 1987 of Kota Tampan, an undisturbed palaeolithic tool workshop site, led to a large-scale survey of the Lenggong valley, resulting in the identification of many potential sites. Among these sites was a cave, Gua Gunung, in the Kepala Gajah limestone massif. The site was excavated in 1990 and 1991 by a team led by Zuraina of Universiti Sains

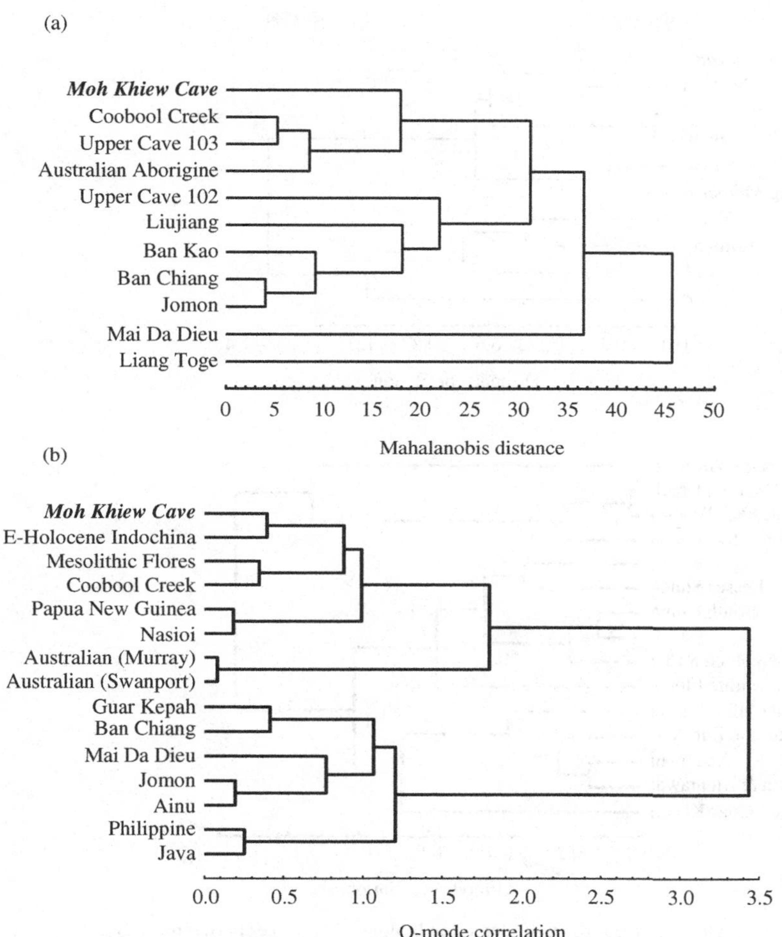

Figure 2.2. Dendrograms depicting biological distances between the Moh Khiew specimens and comparative female samples. (a) Mahalanobis distances based on cranial metrics. (b) Q-mode correlations $(1 - r)$ based on dental metrics. E-Holocene, early Holocene.

Malaysia. The excavation of Gua Gunung in 1990 revealed a 10,000–11,000 years BP primary burial of an adult male (Fig. 2.1), associated with stone tools and food remains (Zuraina 1994). A study of the Gua Gunung specimen by Jacob and Soepriyo (1994) demonstrated Australo-Melanesian characteristics in the cranial and limb bones, strengthening the possibility of an Australo-Melanesian occupation in this region prior to the early Holocene.

(a)

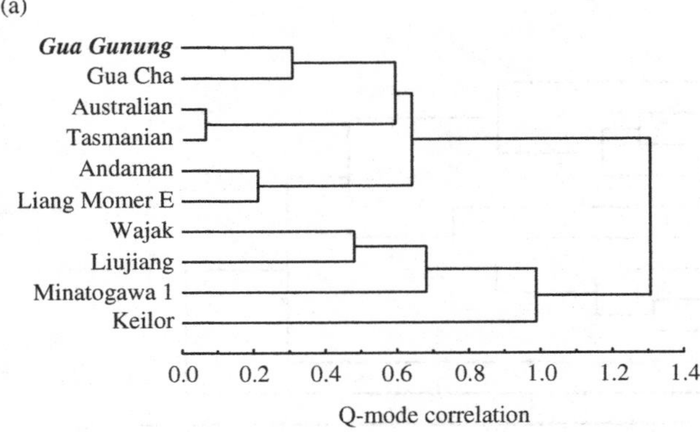

(b)

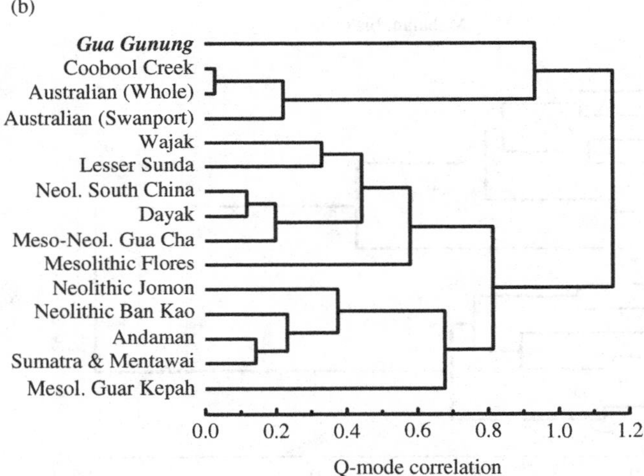

Figure 2.3. Dendrograms depicting biological distances between the Gua Gunung specimens and comparative male samples based on cranial (a) and dental (b) metrics. Meso-Neol., mesoneolithic; Mesol., mesolithic; Neol., neolithic.

Matsumura and Zuraina (1995, 1999) carried out a statistical analysis of the skeletal measurements in order to evaluate the Gua Gunung specimen's Australo-Melanesian affinity. Cranial measurements were restricted to the parietal, temporal and occipital bones and some parts of the facial skeleton. Figure 2.3a depicts a dendrogram of Q-mode correlation coefficients based on seven cranial measurements (Martin Nos. 8, 11, 12, 31, 54, 61 and 48d). The close similarity of the Gua Gunung specimen to the Gua Cha (H12) sample is demonstrated in this figure. These two samples are secondly clustered with the Australian Aborigines and Tasmanians. The other late Pleistocene and early Holocene samples, Liujiang from southern China, Keilor from Australia, Wajak from Indonesia and Minatogawa from Okinawa, are distinctly separated from the Gua Gunung specimen, forming another major cluster.

To evaluate the dental affinity, only buccolingual crown diameters were used because of heavy attrition. Figure 2.3b shows a dendrogram of a Q-mode correlation matrix based on these crown diameters. The Gua Gunung specimen is joined with the Australian Aborigines, including the late Pleistocene Coobool Creek sample. Limb bone measurements of the Gua Gunung specimen were analysed and results indicated that the Gua Gunung limbs are also proportionally much closer to the Australian Aborigines (see Matsumura and Zuraina 1999).

Mai Da Nuoc and Mai Da Dieu

Large numbers of prehistoric and historic human remains have been accumulated from recent excavations conducted by Vietnamese archaeologists. Among them, nearly complete skulls from the Hoabinhian period (*c.* 8,000–10,000 years BP) were excavated from the Mai Da Nuoc (Fig. 2.1) and Mai Da Dieu sites in Thanh Hoa province, northern Vietnam. Cuong (1986) studied these two skulls and found characteristics similar to the Australians, despite also partially revealing some Asiatic features.

The author in collaboration with Cuong carried out statistical comparisons with some representative early Southeast Asians using nine cranial measurements (Martin Nos. 1, 8, 17, 45, 48, 51, 52, 54 and 55) taken from the Mai Da Nuoc male skull. Figure 2.4 shows a dendrogram on the basis of Q-mode correlation coefficients of these cranial measurements. Two major clusters are formed, with Mai Da Nuoc specimens being connected with Gua Cha (No. H12), the Liang Momer specimens and neolithic Vietnamese from the Con Co Ngua site of the Da But culture. Two Australian Aborigine samples and Loyalty Islanders are also clustered with these early Malay, Flores and Vietnamese samples. By comparison, two late Pleistocene samples from Liujiang and Minatogawa form another major cluster together with the pre-neolithic Tam Hang series from Laos, neolithic Ban Kao from Thailand and the Jomon people from Japan. Thus, the cranial affinities observed in the former cluster clearly indicate that the early inhabitants of Southeast Asia, including the Mai Da Nuoc specimen, possessed characteristics similar to those of Australo-Melanesian skulls.

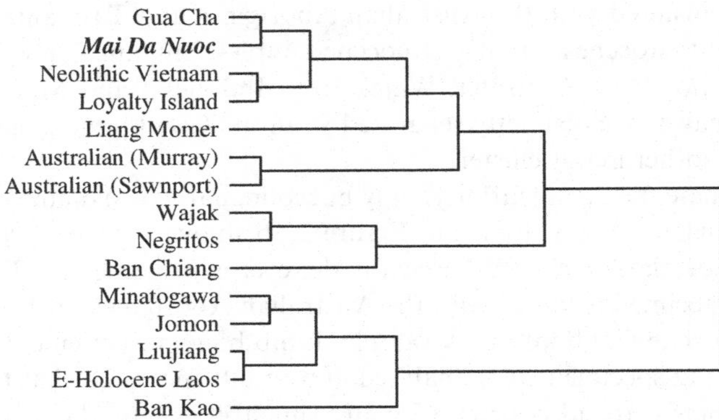

Figure 2.4. Dendrogram depicting biological distances using Q-mode correlation (1 – *r*) between the Mai Da Nuoc specimens and comparative male samples based on cranial metrics.

Population affinities of Southeast Asians based on dental morphology

In this final section, I present population relationships based on non-metric and metric tooth morphology (data source: Matsumura 1994, 1995, 2002, Matsumura *et al.* 2001, Matsumura and Hudson 2005) in order to discuss the population history of Southeast Asia from the perspective of dental characteristics. This analysis leads to a somewhat different interpretation from Turner's Local Evolution model based on his non-metric dental classification of sundadont and sinodont.

Non-metric dental traits

I investigated 21 non-metric dental traits that are regarded as relevant in studying population affinity (Table 2.2. and Fig. 2.5; Matsumura 1995). Smith's mean measures of divergence (Berry and Berry 1967) were calculated using these traits between the 25 prehistoric samples, which consist of the original data listed in Table 2.1 in addition to several other modern and historic samples (see Matsumura and Hudson 2005). In order to facilitate statistically meaningful comparisons, some smaller samples were either combined or excluded from the non-metric comparisons. The mesolithic Flores and Malay group included the Gua Cha and Guar Kepah series from Malaysia and the Flores series from Indonesia. The neolithic Vietnamese included the Bac Son and Da But series.

Table 2.2. *Criteria for presence in the 21 non-metric dental characteristics*

Trait	Tooth	Description	Presence
Shovelling	UI1	Hanihara *et al.* (1970)	DLF ⩾ 0.5 mm
	UI2	Hanihara *et al.* (1970)	DLF ⩾ 0.5 mm
Double shovelling	UI1	Suzuki and Sakai (1973)	+++, ++
	UI2	Suzuki and Sakai (1973)	+++, ++
Dental tubercle	UI1	Turner *et al.* (1991)	3–6
	UI2	Turner *et al.* (1991)	3–6
Spine	UI1	0: none; 1: present	1
Interruption groove	UI2	Turner *et al.* (1991)	1–3
Winging (bilateral)	UI1	Enoki and Dahlberg (1958)	Bilateral
De Terra's tubercle	UP1	Saheki (1958)	+, ++, +++
Double roots	UP1	Turner *et al.* (1991)	2–3
	UP2	Turner *et al.* (1991)	2–3
Carabelli's trait	UM1	Dahlberg's P-plaque	d–g
Hypocone reduction	UM2	Dahlberg (1949)	3+
Sixth cusp	LM1	Turner *et al.* (1991)	1–5
Seventh cusp	LM1	Turner *et al.* (1991)	2–4
Protostylid	LM1	Dahlberg's P-plaque	3–5
Deflecting wrinkle	LM1	Turner *et al.* (1991)	2–3
Groove pattern Y	LM1	Jørgensen (1955)	Y
Groove pattern X	LM2	Jørgensen (1955)	X
Hypoconulid reduction	LM2	Turner *et al.* (1991)	Cusp number, 4

DLF, depth of lingual fossa. See Table 2.3 for tooth abbreviations.

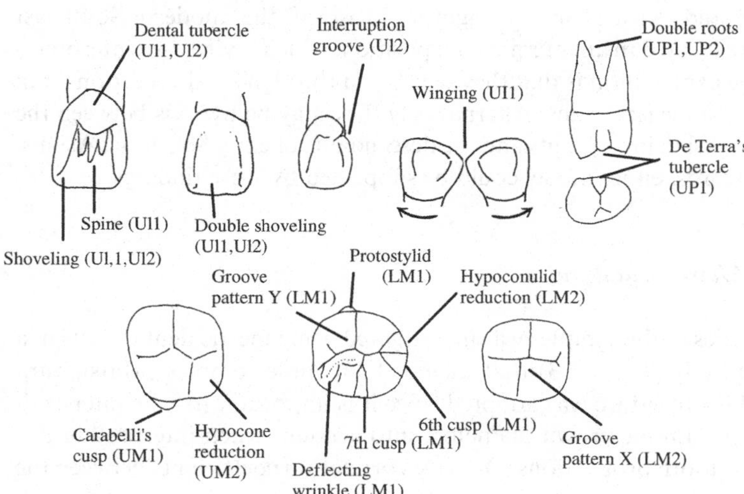

Figure 2.5. Twenty-one non-metric dental traits examined. U, upper; L, lower; I, incisor; P, premolar; M, molar.

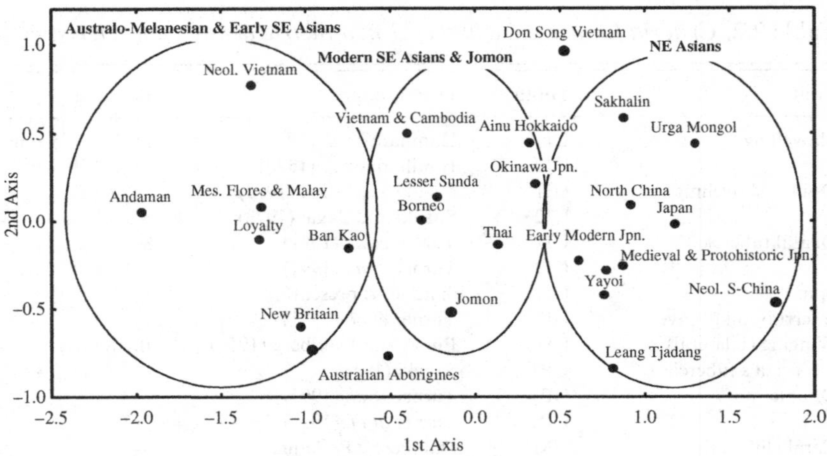

Figure 2.6. Two-dimensional expression of Smith's distances based on 21 non-metric dental traits (sexes combined). Neol., neolithic; Mes., mesolithic.

Figure 2.6 is a two-dimensional expression of the multidimensional scaling (MDS) method (Torgerson 1958) applied to the computed distance matrix. At highest values on the 1st axis, Northeast Asians such as modern Mongolians, Chinese and Japanese are clustered together showing their close affinities. At lowest values on the 1st axis, preneolithic and neolithic Southeast Asians are grouped together with the Andaman and Loyalty Islanders and Australian Aborigines. Most of the modern Southeast Asians are located in an intermediate position, which invites two interpretations. One explanation is that they developed through local evolution from the early Southeast Asians. Alternatively they may be hybrids between the Northeast Asian immigrants and the descendants of early Southeast Asians. In other words, either model could be supported by these findings.

Metric dental traits

The affinities of the samples was investigated using metric dental traits in a factor analysis (Table 2.3) that included Japanese samples (Matsumura 1994). This procedure was adopted since it is important to take inter-trait correlations among crown diameters into account when interpreting differences in tooth proportions. Q-mode correlation coefficients between the 36 samples were computed using the factor scores based on the tooth crown diameters to measure the similarity in the tooth size proportions.

Table 2.3. *Factor loadings after varimax rotation for the six factors based on the tooth crown diameters of the Japanese sample*

Factor 1		Factor 2		Factor 3		Factor 4		Factor 5		Factor 6	
MD UI2	0.620	MD UP1	0.689	MD UC	0.704	MD LM1	0.789	MD UI1	0.724	MD UM2	0.778
BL UI1	0.685	MD UP2	0.811	MD LC	0.698	BL UM1	0.650	MD LI1	0.806	MD LM2	0.731
BL UI2	0.817	MD LP1	0.617	BL UC	0.663	BL LM1	0.771	MD LI2	0.747		
BL LI1	0.678	MD LP2	0.720	BL LC	0.754						

UI, upper incisor; LI, lower incisor; UP, upper premolar; LP, lower premolar; UC, upper canine; LC, lower canine; UM, upper molar; LM, lower molar; MD, mesiodistal diameter; BL, buccolingual diameter.
From Matsumura 1994.

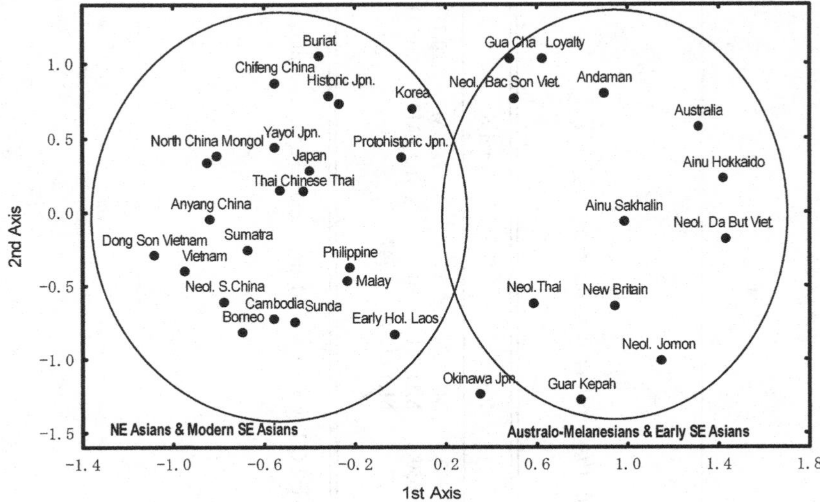

Figure 2.7. Two-dimensional expression of the Q-mode correlation coefficients based on dental measurements (males). Neol., neolithic; Hol., Holocene.

Figure 2.7 displays a two-dimensional expression of MDS applied to the correlation matrix. So-called Mongoloid peoples, including the modern Chinese, Mongolian, Buriats and Japanese, are situated at lowest values on the 1st axis. In addition to these typical Northeast Asians, most of the modern Southeast Asian samples such as the Borneans, Filipinos, Malays, Thais, Indonesians, Cambodians, Vietnamese and Sunda Islanders are also grouped together. At higher values, the Australian Aborigines, Loyalty Islanders, Jomon, Ainu, Hoabinhian and mesolithic Malay and Flores samples, neolithic Thai and Vietnamese and Andaman Islanders are grouped together, indicating their close affinities. The population affinities based on metric tooth traits demonstrate that many present-day Southeast Asians share common features with Northeast Asians, suggesting that the modern people of Southeast Asia were genetically influenced by Northeast Asians.

Discussion and conclusions

This chapter initially reviewed early studies of preneolithic human remains from Southeast Asia. Many of these early studies described the morphological features of these specimens as typical of Australo-Melanesians. Next, a summary was presented of the results of my morphometric analyses of more recently discovered skeletons, which may provide further evidence that a people akin to the Australo-Melanesians existed in early Southeast Asia. The biological distances based on the skeletal and dental measurements demonstrated that the Moh Khiew Cave remains from Thailand, the Gua Gunung remains from Malaysia and the Mai Da Nuoc skull from Vietnam were morphologically linked with the Australo-Melanesian lineage. In these analyses, the early Malay and Flores specimens, such as Gua Cha and Liang Momer, also displayed close affinities with the Australo-Melanesians. These findings suggest that these specimens, as well as other fossils from Tabon and Niah, can be regarded as an early group of Southeast Asians who originated in late Pleistocene Sundaland and were the ancestors of the modern Melanesian and Australian aboriginal peoples. Until the end of the last glacial stage, lowered sea levels reduced the water barrier between Australia and Southeast Asia. Against this geographic background, during the late Pleistocene, genetic separation between the inhabitants of Australia and Southeast Asia is considered to have been still incomplete. Some traits exhibited in this skeletal and dental morphology may have been retained in Southeast Asians until the early Holocene.

Finally, I demonstrated population affinities based on dental morphology, addressing the issue of the genetic influence of Northeast Asians. This work, based on the non-metric dental traits of the various populations of East Asia, indicated a different scenario from Turner's theory (Turner 1989, 1990, 1992). From my dental analyses, it seems reasonable to reaffirm that the early inhabitants of Southeast Asia were of the Australo-Melanesian lineage, at least before the neolithic period.

With regard to the affinities of present-day Southeast Asians, however, MDS analysis using my non-metric dental data demonstrated that many present-day Southeast Asians are intermediately located between the early Southeast Asians and the recent Northeast Asians. Cluster analysis confirmed that the population samples examined in another study (Matsumura and Hudson 2005) can be divided into two major clusters that correspond to Turner's sinodont and sundadont classifications. A dendrogram of the cluster analysis is, in general, quite useful in emphasising the closeness of certain samples or in highlighting distantly linked samples, as shown in this chapter. However, this method has a drawback in that we cannot recognise samples intermediately positioned between the major clusters because such samples occasionally join with one or another of the major clusters. When the population affinities are examined cautiously using the MDS, it can be seen that the non-metric dental traits observed in early Southeast Asians and Australo-Melanesians are regarded as the

(a)

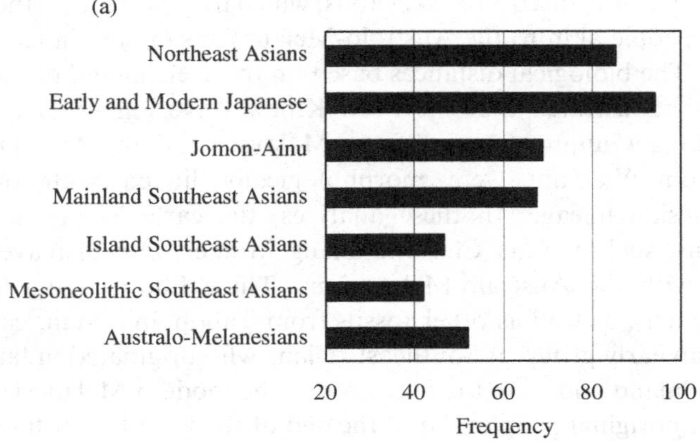

(b)

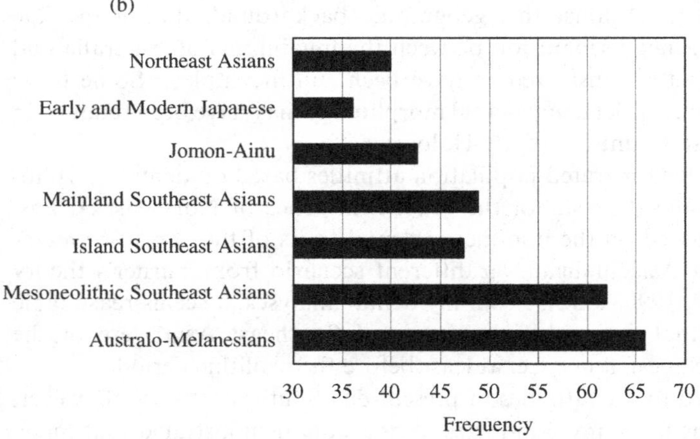

Figure 2.8. Frequencies of the shovel-shaped maxillary central incisor (a) and the double-rooted maxillary first premolar (b) in Northeast and Southeast Asians.

original sundadont dental complex, while the dental traits of modern Southeast Asians have been formed by hybridisation between the sundadont and sinodont complexes. Although I cannot describe the tendencies of all the investigated non-metric traits in detail here, I will show a few representative dental characters that may most accurately reflect the biological relationships of modern Southeast Asians with other populations. The shovel-shaped incisor and the double-rooted maxillary first premolar are key representative traits that classify the dental series into the sundadont and sinodont types, as already noted by Turner. As shown in Fig. 2.8 and Table 2.4, the occurrence of shovel-shaped incisors, on the one hand, is

Table 2.4. *Frequencies of 21 non-metric dental traits for seven major regional and chronological groups in Southeast and Northeast Asia*

Trait	Tooth	Frequency of trait (sample size)						
		Prehistoric Southeast Asians	Island Southeast Asians	Mainland Southeast Asians	Australo-Melanesians	Jomon-Ainu	Early and Modern Japanese	Northeast Asians
Shoveling	UI1	42.1 (76)	46.7 (15)	67.7 (62)	52.1 (94)	69.0 (97)	94.3 (363)	85.5 (55)
	UI2	21.9 (64)	40.9 (22)	52.8 (70)	38.8 (129)	42.7 (236)	75.6 (434)	81.8 (66)
Double shoveling	UI1	0.0 (89)	0.0 (16)	7.6 (66)	2.6 (116)	0.4 (240)	5.0 (475)	7.3 (69)
	UI2	0.0 (75)	0.0 (24)	2.6 (77)	0.6 (161)	0.0 (232)	0.2 (467)	0.0 (79)
Dental tubercle	UI1	11.6 (86)	0.0 (24)	3.0 (68)	25.0 (124)	6.8 (254)	3.9 (489)	1.3 (76)
	UI2	12.5 (72)	6.9 (29)	3.9 (76)	9.6 (167)	20.5 (254)	7.1 (491)	1.2 (88)
Spine	UI1	32.9 (79)	20.0 (20)	36.4 (66)	44.5 (108)	24.4 (201)	19.9 (497)	53.0 (68)
Interruption groove	UI2	12.9 (70)	19.2 (26)	35.5 (76)	20.3 (148)	58.3 (218)	42.1 (505)	32.9 (79)
Winging (bilateral)	UI1	13.8 (65)	18.5 (27)	23.1 (65)	17.1 (123)	8.3 (227)	14.1 (320)	11.5 (78)
De Terra's tubercle	UP1	7.3 (82)	11.9 (67)	22.7 (119)	10.4 (183)	13.0 (233)	27.1 (559)	52.0 (123)
Double roots	UP1	61.9 (63)	47.4 (95)	48.9 (141)	65.8 (152)	42.6 (216)	36.1 (556)	40.0 (115)
	UP2	9.3 (54)	12.1 (91)	8.1 (136)	20.6 (136)	1.6 (182)	7.0 (300)	7.1 (98)
Carabelli's trait	UM1	15.5 (116)	33.6 (113)	25.4 (185)	20.8 (279)	5.4 (392)	17.5 (729)	18.0 (195)
Hypocone reduction	UM2	24.8 (125)	12.1 (107)	15.3 (183)	4.5 (286)	8.3 (242)	12.3 (648)	18.8 (186)
Sixth cusp	LM1	20.9 (91)	25.4 (63)	25.6 (125)	39.9 (198)	24.5 (408)	26.9 (643)	17.5 (103)
Seventh cusp	LM1	5.0 (101)	6.2 (65)	7.7 (142)	11.6 (224)	9.2 (414)	4.6 (695)	13.7 (110)
Protostylid	LM1	4.7 (106)	4.6 (65)	6.6 (136)	8.0 (224)	5.0 (475)	5.3 (686)	8.4 (108)
Deflecting wrinkle	LM1	33.3 (45)	16.0 (50)	26.1 (96)	37.7 (154)	14.1 (130)	40.7 (449)	32.0 (75)
Groove pattern Y	LM1	81.3 (75)	67.8 (59)	75.6 (123)	74.5 (212)	70.3 (454)	75.9 (673)	71.7 (106)
Groove pattern X	LM2	30.7 (101)	34.8 (69)	33.3 (153)	22.5 (254)	14.9 (449)	44.6 (561)	45.9 (111)
Hypoconulid reduction	LM2	47.2 (108)	40.3 (67)	44.4 (142)	49.3 (231)	44.6 (242)	36.9 (574)	41.2 (97)

See Table 2.3 for tooth abbreviations.

lower in many Southeast Asians, than in Northeast Asians. At the same time, the occurrence is higher in mainland compared with island and prehistoric Southeast Asians. On the other hand, the frequency of double-rooted premolar shows a reversal of this tendency.

There remains the possibility that the non-metric dental traits observed in present-day Southeast Asians arose by local evolution. However, the results of my metric dental comparisons clearly show that the Northeast Asian migration genetically influenced many postneolithic Southeast Asians, whose MDS values exhibit quite different population relationships from those displayed by Turner's (1992) non-metric dental study. Taking into consideration both the dental metric and non-metric traits, together with the osteometrics, it is not inconsistent to conclude that the people associated with the Hoabinhian culture were akin to Australo-Melanesians, while many modern Southeast Asians can be interpreted as hybrids of Northeast Asians and such preexisting people. Cases of drastic population change or large-scale admixture with Northeast Asians are especially evident in northern Vietnam and Japan during the early metal age (Matsumura 1994, 1995, 1998, 2001, Matsumura *et al.* 2001). Consequently, it can be concluded that the skeletal and dental remains support the hypothesis stated by Bellwood (1987, 1996, 1997) that there has been a diffusion of migrants from the Asian continent, probably through southern China, into Southeast Asia since the neolithic period. These people intermixed with the indigenous Australo-Melanesian stock as they diffused through the region. Although Bellwood stated that the dispersion of the Chinese began in the late neolithic period associated with rice cultivation, the genetic influence of the northern source seems to be stronger in the postneolithic rather than the neolithic period. My study does not take into account sufficient data from the late neolithic period and further research on skeletal remains from that period is required to determine the exact timing of these prehistoric population dispersals into Southeast Asia.

It is noteworthy that in Japan, the Jomon and their descendants the Ainu possess non-metric features intermediate between the Northeast Asian sinodont and the early Southeast Asian sundadont patterns, although their metric traits are shared with the latter. Furthermore, craniometric analyses demonstrated close affinities between the Jomon, Minatogawa, Liujiang and early Holocene Laotians (Fig. 2.4). Contrary to expectations, the skeletal remains from the Tam Hang site in Laos, recorded by Huard and Saurin (1938), displayed fewer Australo-Melanesian affinities. Indeed, from my own observations, their cranial morphology shows considerable variation. A portion of this cranial series shows Australo-Melanesian features, while others have Northeast Asian characteristics, suggesting that the Tam Hang specimens do not consist of a single population lineage. If this perspective is appropriate, admixture with Northeast Asians might have occurred since the late Pleistocene in some northern regions of mainland Southeast Asia. People in other marginal regions of East Asia, such as Minatogawa, Jomon, Liujiang and Ban Kao, may also be considered to have been formed through an earlier diffusion of the northern people, because they have close affinities to the Tam Hang series. These findings suggest that this early group, which inhabited the marginal regions of the Asian continent, arose through the diffusion of Northeast Asians during the late Pleistocene prior to their dramatic expansion to southern regions in the post-neolithic period.

Acknowledgements

I am grateful to the late Professor Surin Pookajorn, Silpakorn University, Dato, Professor Zuraina Majid, Universiti Sains Malaysia, and Dr Nguyen Lan Cuong, Institute of Archaeology, Hanoi for permission to investigate their excavated skeletal collections. Thanks are also due to Dr Marc F. Oxenham, Australian National University, Dr Nancy G. Tayles, University of Otago, and Dr Mark Hudson, University of Tsukuba, for their editing and manuscript correction.

This study was supported by a Grant-in-Aid for Scientific Research in 2002 (No. 14540064) and 2003 (15405018) from the Japan Society for the Promotion of Science.

References

Adi H. T. 1985. The re-excavation of the rock shelter of Gua Cha, Ulu Kelantan, West Malaysia. *Journal of the Federated Malay States Museums* **56**: 47–63.

Ballinger S. W., Schurr T. G., Torroni A. *et al.* 1992. Southeast Asian mitochondrial DNA analysis reveals genetic continuity of ancient Mongoloid migrations. *Genetics* **130**: 139–152.

Bellwood P. 1987. The prehistory of island Southeast Asia: a multidisciplinary review of recent research. *Journal of World Prehistory* **1**: 171–224.

1989. The colonization of the Pacific: some current hypotheses. In Hill A. V. S. and Serjeantson S. W., eds., *The Colonization of the Pacific: A Genetic Trail.* Oxford: Clarendon Press, pp. 1–59.

1991. The Austronesian dispersal and the origin of languages. *Scientific American* **265**: 88–93.

1993. An archaeologist's view of language macrofamily relationships. *Bulletin of the Indo-Pacific Prehistory Association* **13**: 46–60.

1996. Early agriculture and the dispersal of the southern Mongoloids. In Akazawa T. and Szathmáry E. J. E., eds., *Prehistoric Mongoloid Dispersals*. Oxford: Oxford University Press, pp. 287–302.

1997. *Prehistory of the Indo-Malaysian Archipelago*, revised edn. Honolulu: University of Hawaii Press.

Bellwood P. and Renfrew C. (eds.). 2003. *Examining the Farming/Language Dispersal Hypothesis*. Cambridge: McDonald Institute for Archaeological Research.

Bellwood P., Gillespie R., Thompson G. B. *et al.* 1992. New dates for prehistoric Asian rice. *Asian Perspectives* **31**: 161–170.

Berry A. C. and Berry R. J. 1967. Epigenetic variation in the human cranium. *Journal of Anatomy* **101**: 361–379.

Blust R. A. 1996a. Austronesian culture history: the window of language. In Goodenough W. H., ed., *Prehistoric Settlement of the Pacific*. Philadelphia: American Philosophical Society, pp. 28–35.

1996b. Beyond the Austronesian homeland: the Austric hypothesis and its implications for archaeology. [In Goodenough W. H., ed., *Prehistoric Settlement of the Pacific*.] *Transactions of the American Philosophical Society*, **86**: 117–140.

Brace C. L., Tracer D. P. and Hunt K. D. 1991. Human craniofacial form and the evidence for the peopling of the Pacific. *Bulletin of the Indo-Pacific Prehistory Association* **12**: 247–269.

Brothwell D. R. 1960. Upper Pleistocene human skull from Niah Caves. *Sarawak Museum Journal* **9**: 323–349.

Brown P. 1989. *Terra Australia 13. Coobool Creek. A Morphological and Metrical Analysis of the Crania, Mandibles and Dentitions of a Prehistoric Australian Human Population*. Canberra: Department of Prehistory, Research School of Pacific Studies, Australian National University.

Bulbeck D. 2000. Dental morphology at Gua Cha, West Malaysia, and the implications for 'Sundadonty'. *Bulletin of the Indo-Pacific Prehistory Association* **19**: 17–41.

Cavalli-Sforza L. L., Menozoi P. and Piazza A. 1994. *The History and Geography of Human Genes*. Princeton, CT: Princeton University Press.

Coon G. S. 1962. *The Origin of Races*. New York: Alfred A. Knopf.

Cuong N. L. 1986. Two early Hoabinhian crania from Thanh Hoa province, Vietnam. *Zeitschrift für Morphologie und Anthropologie* **77**: 11–17.

Dahlberg A. A. 1949. The dentition of the American Indian. In Laughlin W. S., ed., *The Physical Anthropology of the American Indian*. New York: Viking Foundation, pp. 138–176.

de Sieveking G. 1954. Excavations at Gua Cha, Kelantan 1954. Part 1. *Federation Museums Journal* **1–2**: 75–143.

Diamond J. and Bellwood P. 2003. Farmers and their languages: the first expansions. *Science* **300**: 597–603.

Dubois E. 1922. The proto-Australian fossil man of Wadjak, Java. *Koninklijke Akademie van Wetenschappen te Amsterdam* **B23**: 1013–1051.

Duckworth W. L. H. 1934. Human remains from rock-shelters and caves in Perak, Pahang and Perlis and from Selinsing. *Journal of Malayan Branch Royal Asiatic Society* **XII**: 149–167.

Enoki K. and Dahlberg A. A. 1958. Rotated maxillary central incisors. *Orthodontic Journal in Japan* **17**: 157.

Evans I. H. N. 1918. Preliminary report on cave exploration near Lenggong, upper Perak. *Journal of the Federated Malay States Museums* **7**: 227–234.

1928. Further excavations at Gunung Pondok. *Journal of the Federated Malay States Museums* **12**: 136–142.

Fujita T. 1949. On the standards for measurements of teeth. [In Japanese] *Journal of the Anthropological Society of Nippon* **61**: 27–32.

Glover I. C. and Higham C. F. W. 1996. New evidence for early rice cultivation in South, Southeast and East Asia. In Harris D. R., ed., *The Origins and Spread of Agriculture and Pastoralism in Eurasia*. London: UCL Press, pp. 413–441.

Hanihara K., Tanaka T. and Tamada M. 1970. Quantitative analysis of the shovel-shaped character in the incisors. *Journal of the Anthropological Society of Nippon* **78**: 90–93.

Hanihara T. 1993. Craniofacial features of Southeast Asians and Jomonese: a reconsideration of their microevolution since the late Pleistocene. *Anthropological Science* **101**: 25–46.

1994. Craniofacial continuity and discontinuity of Far Easterners in the late Pleistocene and Holocene. *Journal of Human Evolution* **27**: 417–441.

Higham C. F. W. 1998. Archaeology, linguistics and the expansion of the East and Southeast Asian Neolithic. In Blench R. and Spriggs M., eds., *Archaeology and Language II: Archaeological Data and Linguistic Hypotheses*. London: Routledge, pp. 103–114.

2001. Prehistory, language and human biology: is there a consensus in East and Southeast Asia? In Jin L., Seielstad M. and Xiao C. J., eds., *Genetic, Linguistic and Archaeological Perspectives on Human Diversity in Southeast Asia*. Singapore: World Scientific, pp. 3–16.

Huard P. and Saurin E. 1938. État actuel de la craniologie Indochinoise. *Bulletin du Service Géologique de l'Indochine* **25**: 1–103.

Hudson M. J. 1994. The linguistic prehistory of Japan: some archaeological speculations. *Anthropological Science* **102**: 231–255.

1999. *Ruins of Identity: Ethnogenesis in the Japanese Islands*. Honolulu: University of Hawaii Press.

2003. Agriculture and language change in the Japanese islands. In Bellwood P. and Renfrew C., eds., *Examining the Farming/Language Dispersal Hypothesis*. Cambridge, UK: McDonald Institute for Archaeological Research, pp. 311–318.

Jacob T. 1967. Some problems pertaining to the racial history of the Indonesian region. Ph.D. thesis, University of Utrecht, Utrecht.

Jacob T. and Soepriyo A. 1994. A preliminary palaeoanthropological study of the Gua Gunung Runtuh human skeleton. In Zuraina M., ed., *The Excavation of Gua Gunung Runtuh and the Discovery of the Perak Man in Malaysia*. Kuala Lumpur: Department of Museums and Antiquity of Malaysia, pp. 48–69.

Jørgensen K. D. 1955. The *Dryopithecus* pattern in recent Danes and Dutchmen. *Journal of Dental Research* **34**: 195–208.

Kennedy K. A. R. 1977. The deep skull of Niah: an assessment of twenty years of speculation concerning its evolutionary significance. *Asian Perspectives* **20**: 32–50.

Lauer A. J. 2002. Craniometric measurements and tooth morphology of archaeologically derived skeletal remains from the Malay Peninsula. M.A. thesis, Australian National University, Canberra.

Macintosh N. W. G. 1978. The Tabon Cave mandible. *Archaeology and Physical Anthropology in Oceania* **13**: 143–159.

Martin R. and Saller K. 1957. *Lehrbuch der Anthropologie*, Vol 1. Stuttgart: G. Fischer.

Matsumura H. 1994. A microevolutional history of the Japanese people from a dental characteristics perspective. *Anthropological Science* **102**: 93–118.

1995. Dental characteristics affinities of the prehistoric to the modern Japanese with the East Asians, American natives and Australo-Melanesians. *Anthropological Science* **103**: 235–261.

1998. Native or migrant lineage? The Aeneolithic Yayoi people in western and eastern Japan. *Anthropological Science* **106**(Supplement): 17–25.

2001. Differentials of Yayoi immigration to Japan as derived from dental metrics. *Homo* **52**: 135–156.

2002. The possible origin of the Yayoi migrants based on the analysis of the dental characteristics. In Nakahashi T. and Li M., eds., *Ancient People in the Jiangnan Region, China*. Fukuoka: Kyushu University Press, pp. 61–72.

Matsumura H. and Zuraina M. 1995. Metrical analysis of the dentition of Perak man from Gua Gunung Runtuh in Malaysia. *Bulletin of the National Science Museum, Tokyo Series D* **21**: 1–10.

1999. Metric analyses of the early Holocene human skeleton from Gua Gunung Runtuh in Malaysia. *American Journal of Physical Anthropology* **109**: 327–340.

Matsumura H. and Hudson M. 2005. Dental perspectives on the population history of Southeast Asia. *American Journal of Physical Anthropology* **127**: 182–209.

Matsumura H. and Pookajorn S. 2005. A morphometric analysis of the late Pleistocene human skeleton from the Moh Khiew Cave in Thailand. *Homo*, **56**: 93–118.

Matsumura H., Cuong N. L., Thuy N. K. and Anezaki T. 2001. Dental morphology of the early Hoabinhian, the Neolithic Da But and the Metal Age Dong Son cultural people in Vietnam. *Zeitschrift für Morphologie und Anthropologie* **83**: 59–73.

Mijsberg W. A. 1940. On a Neolithic Paleo-Melanesian lower jaw found in kitchen midden at Guar Kepah, province Wellesley, Straits Settlements. *Proceedings of the Third Congress of Prehistorians of the Far East, Singapore*, pp. 100–118.

Omoto K. and Saitou N. 1997. Genetic origins of the Japanese: a partial support for the 'dual structure hypothesis'. *American Journal of Physical Anthropology* **102**: 437–446.

Oota H., Kurosaki K., Pookajorn S., Ishida T. and Ueda S. 2001. Genetic study of the Paleolithic and Neolithic Southeast Asians. *Human Biology* **73**: 225–231.

Pietrusewsky M. 1992. Japan, Asia and the Pacific: a multivariate craniometric investigation. In Hanihara K., ed., *Japanese as a Member of the Asian and Pacific Populations*. Kyoto: International Research Center for Japanese Studies, pp. 9–52.

1994. Pacific–Asian relationships: a physical anthropological perspective. *Oceanic Linguistics* **33**: 407–429.

1999. A multivariate craniometric study of the inhabitants of the Ryukyu Islands and comparison with cranial series from Japan, Asia and the Pacific. *Anthropological Science* **107**: 255–281.

Pietrusewsky M. and Douglas M. T. 2002. *Ban Chiang, A Prehistoric Village Site in Northeast Thailand I: The Human Skeletal Remains*. Philadelphia: University of Pennsylvania Museum of Archaeology and Anthropology.

Pookajorn S. 1991. *Preliminary Report of Excavations at Moh Khiew Cave, Krabi Province, Sakai Cave, Trang Province and Ethnoarchaeological Research of Hunter-Gatherer Group, So-call Sakai or Semang at Trang Province*. Bangkok: Department of Archaeology, Silpakorn University.

1994. *Final Report of Excavations at Moh Khiew Cave, Krabi Province; Sakai Cave Trang Province and Ethnoarcheological Research of Hunter–Gatherer group, So-called Sakai or Semang at Trang Province*. Bangkok: Department of Archaeology, Silpakorn University.

Renfrew C. 1987. *Archaeology and Language: the Puzzle of Indo-European Origins*. London: Jonathan Cape.

1989. Models of change in language and archaeology. *Transactions of the Philological Society* **87**: 103–155.

1992. World languages and human dispersals: a minimalist view. In Hall J. A. and Jarvie I. C., eds., *Transition to Modernity: Essays on Power, Wealth and Belief*. Cambridge, UK: Cambridge University Press, pp. 11–68.

Saheki M. 1958. On the heredity of the tooth crown configuration studied in twins. [In Japanese with English summary.] *Acta Anatomica Nipponica* **33**: 456–470.

Storm P. 1995. The evolutionary significance of the Wajak skulls. *Scripta Geologica* **110**: 1–247.

Suzuki H. 1982. Skulls of the Minatogawa man. *Bulletin of the University Museum of the University of Tokyo* **19**: 7–49.

1973. *The Japanese Dentition*. Matsumoto: Shinshu University.

Tan S. G. 2001. Genetic relationships among sixteen ethnic groups from Malaysia and Southeast Asia. In Jin L., Seielstad M. and Xiao C., eds., *Genetic, Linguistic and Archeological Perspectives on Human Diversity in Southeast Asia*. Singapore: World Scientific, pp. 83–91.

Torgerson W. S. 1958. *Theory and Methods of Scaling*. New York: Wiley.

Trevor J. C. and Brothwell D. R. 1962. The human remains of Mesolithic and Neolithic date from Gua Cha, Kelantan. *Federation Museums Journal* **7**: 6–22.

Turner C. G. II. 1989. Teeth and prehistory in Asia. *Scientific American* **260**: 70–77.

1990. Major features of sundadonty and sinodonty, including suggestions about East Asian microevolution, population history and late Pleistocene relationships with Australian aborigines. *American Journal of Physical Anthropology* **82**: 295–317.

1992. Microevolution of East Asian and European populations: a dental perspective. In Akazawa T., Aoki K. and Kimura T., eds., *The Evolution and Dispersal of Modern Humans in Asia*. Tokyo: Hokusensha, pp. 415–438.

Turner C. G. II, Nichol C. R. and Scott G. R. 1991. Scoring procedures for key morphological traits of the permanent dentition: the Arizona State University dental anthropology system. In Kelly M. A. and Larsen C. S., eds., *Advances in Dental Anthropology*. New York: Wiley-Liss, pp. 13–31.

van Stein Callenfels S. 1936. The Melanesoid civilizations of Eastern Asia. *Bulletin of Raffles Museum Series B* **1**: 41–51.

von Koenigswald G. H. R. 1952. Evidence of a prehistoric Austral–Melanesoid population in Malaya and Indonesia. *Southwestern Journal of Anthropology* **8**: 92–96.

Weidenreich F. 1945. The Keilor skull: a Wadjak type from South-East Australia. *American Journal of Physical Anthropology* **3**: 225–236.

Wolpoff M. H., Wu X. and Thorne A. G. 1984. Modern *Homo sapiens* origins: a general theory of hominid evolution involving the fossil evidence from East Asia. In Smith F. H. and Spencer F., eds., *The Origins of Modern Humans*. New York: Liss, pp. 411–484.

Woo J. 1959. Human fossils found in Liukiang, Kwangsi, China. *Vertebrata Palasiatica* **3**: 108–118.

Wu X. 1961. On the racial types of the Upper Cave Man of Choukoutien. *Scientia Sinica* **10**: 998–1005.

Zuraina, M. (ed.). 1994. *The Excavation of Gua Gunung Runtuh and the Discovery of the Perak Man in Malaysia*. Kuala Lumpur: Department of Museums and Antiquities.

3 *A multivariate craniometric study of the prehistoric and modern inhabitants of Southeast Asia, East Asia and surrounding regions: a human kaleidoscope?*

MICHAEL PIETRUSEWSKY
University of Hawaii, Hawaii, USA

Introduction

The peoples and cultures of Southeast Asia, a region with boundaries that are more influenced by its past inhabitants than by today's geopolitical borders, have been described as representing a human kaleidoscope (Bowles 1977). Broadly defined, this region now includes present-day Myanmar (Burma), Thailand, former French Indochina (Cambodia, Laos and Vietnam), Malaysia and the islands of Indonesia and the Philippines. The people and the prehistory of this region are often portrayed as a southern division of eastern Asia (Bellwood 1997). While making sense of the biology of the modern-day inhabitants of this region has proved daunting, including its earlier inhabitants adds yet another dimension, one that lends itself to addressing issues including the origins of the people in the region, ancient and modern.

In recent years, new archaeological and linguistic perspectives on the evolution and prehistory of Southeast Asia and East Asia have emerged, positions that frequently centre on rice domestication, the development of agriculture and the dispersal of languages hypothesised as having most likely emanated from southern China. Archaeologists (e.g. Bellwood 1996, 2000, Glover and Higham 1996, Higham 1996, 2001) as well as historical linguists (e.g. Bayard 1996–97, Blust 1996) now argue against both *in situ* agricultural development and diffusion of agricultural technology to the indigenous hunter–gathering populations in late Holocene Southeast Asia, in favour of an agricultural colonisation model. Bellwood (1997) has

Bioarchaeology of Southeast Asia. Marc Oxenham and Nancy Tayles.
Published by Cambridge University Press.

argued most strenuously for a population displacement to account for the people who now inhabit Southeast Asia. In this view, the indigenous inhabitants of Southeast Asia were replaced by an immigrant group of people of a more northern origin, or, using his terminology, Australoids were displaced by Mongoloids (Bellwood 1997, pp. 83–87). Theoretically, such a scenario should result in the presence of a somewhat hybridised population living in this region. A similar view of an agriculturally driven demic expansion of more northern populations into Southeast Asia, replacing the earlier groups residing there, has been espoused by Higham (2001).

In contrast to this model of demic expansion of agriculturalists into the region, some physical anthropologists (e.g. Turner 1987, 1990, Hanihara 1993; see Ch. 4) have proposed the Population Continuity Model, which argues that the present-day inhabitants of Southeast Asia evolved from earlier groups living within this region from the late Holocene onward. Turner's work focused on dental non-metric traits and the recognition of two polar dental complexes, sundadonty for Southeast Asia and Polynesia and sinodonty for the inhabitants of East Asia (Turner 1987, 1990). Using features of cranial and dental morphology in Southeast Asian skeletal series, Bulbeck (1982; but see Ch. 6) provided further support for the continuity model.

Studies in physical anthropology, especially those that utilise human skeletal and dental remains, should help in evaluating these models of displacement and continuity, or suggest alternative explanations for the biological variability of the region's inhabitants, past and present.

Crania and biodistance studies

As a result of both flawed theory and unrefined methodology, earlier studies in physical anthropology often fell short of their goal of understanding biological relatedness and human evolution. In recent years, however, the field has seen a resurgence of interests under the guise of biodistance studies. While morphological variation, including craniometric variation, is subject to non-genetic, environmental influences, this category of variation is generally viewed as reflecting a genetic similarity that provides the basis for biodistance studies (Buikstra *et al.* 1990). Improvements in statistical methods, especially the development of multivariate statistical procedures, and breakthroughs in evolutionary and population biology theories have made possible an unprecedented objectivity for comparing human groups and measuring biological relatedness.

As a result, metric data continue to be an important and valuable source of information for examining relatedness between and within populations (e.g. van Vark and Howells 1984, Pietrusewsky 2000). The precision and repeatability of measurement techniques, the generally conservative nature of this category of variation, its direct link with the past, and the demonstration that craniometric traits have a genetic component (e.g. Sjøvold 1984) have all contributed to this renewed interest.

There have been numerous attempts, including work by Pietrusewsky (1974, 1981, 1988, 1990, 1994, 1997, 1999, 2000, 2005, Pietrusewsky and Ikehara-Quebral 2001) and others (e.g. Brace and Hunt 1990, Hanihara 1993, 1994, 1996; Ch. 4), to apply multivariate statistical procedures to craniometric data from Southeast Asia and neighbouring regions. Summaries of these previous analyses, which include some of the same archaeological cranial series from Southeast Asia used in the present study, are presented in Pietrusewsky and Douglas (2002, pp. 230–233). Only limited references are made to this earlier multivariate work in discussing the present results. Although sample sizes are often quite small, and different statistical methods and comparative series are used, this body of multivariate craniometric investigations has provided some of the first glimpses of the internal and external relationships of the region's earlier and modern inhabitants.

In this chapter, measurements recorded in early as well as modern crania from Southeast Asia and beyond are used to assess the patterning of biological relationships among the region's inhabitants. This new craniometric analysis expands on an earlier comprehensive multivariate craniometric comparison (Pietrusewsky 1997) in a number of ways. First, this study includes prehistoric crania from archaeological sites in northeast Thailand, such as the specimens from Non Nok Tha. Second, a pooled sample from former French Indochina, used in the previous analysis, is separated into an early Holocene and mid Holocene series in the present analysis. Third, cranial series not used in previous investigations (e.g. Burma, Gambier Islands, Loyalty Islands, New Caledonia, Santa Cruz Island, Solomon Islands, Dawson Strait Islands, etc.) are included in the present study. Finally, the results of this new multivariate analysis of craniometric data have been used to examine the biological relationships of the inhabitants, prehistoric and modern, of Southeast Asia and neighbouring areas and some of the general archaeological and linguistic models that have been used to explain the settlement and colonisation of the region. Because of the recent explosion in molecular genetic research for this part of the world (e.g. Jin *et al.* 2001), very limited discussion of these studies is made in this chapter.

Materials and methods

Cranial series

Because multivariate statistical procedures do not allow for missing variables, only complete, or nearly complete, male crania are used in the present study. Comparable data for female specimens were not available for this study. The archaeological cranial series from Southeast Asia and East Asia (Table 3.1 and Figs. 1.1 (p. 5) and 1.2 (p. 6)) include four series from Thailand and crania from five different sites located in Laos and Vietnam. With the exception of Non Nok Tha, all data were recorded by Pietrusewsky.

Ban Chiang, northeast Thailand

Ban Chiang is a site located in northeast Thailand that spans the pre-metal (neolithic) to bronze/iron ages, or approximately 4,100 to 1,800 years BP (White 1986, Pietrusewsky and Douglas 2002). Male crania used in the present study are from two separate archaeological excavations directed by the University of Pennsylvania and the Thai Fine Arts Department within the village of Ban Chiang in 1974 and 1975. In addition to distinctive decorative pottery, ornaments and elaborate burial offerings, the archaeological sequence at Ban Chiang provides early evidence of agriculture and metallurgy, including domesticated rice and early bronze artefacts.

Ten of the twelve specimens are from the earliest phases of the Ban Chiang site (EPI–EPV, or 4,100 to 2,900 years BP) and two additional specimens are from the Middle Period (MPVI–MPVII, or 2,900–2,300 years BP). All specimens are currently located at the University of Hawaii, Honolulu.

Non Nok Tha, northeast Thailand

Non Nok Tha is a site that was primarily used as a cemetery. Located near the modern village of Ban Na Di, in northeast Thailand, the cranial specimens are from two seasons' excavation by the Thai Fine Arts Department and the University of Hawaii Archaeological Salvage Programme in 1965–66 and 1968 (Bayard 1971). Bayard (1996–97) has suggested that the site and its skeletal remains date from the early third to late first millennium BCE. A rich archaeological record accompanies the burials at Non Nok Tha, including stone tools, extensive faunal remains, items of personal adornment (including bronze artefacts) and evidence of domesticated fauna. The skeletons from the 1965–66 season of excavation

Table 3.1. *Archaeological cranial series from Southeast Asia and East Asia used in comparisons*

Cranial series	Location	No. of crania	Specimens used[a]	Dates (years BP)	References
Ban Chiang	Northeast Thailand	12	BC 20,23,43 BCES 22,31,35,45,47,50,51, 65,72	4,100–2,300	Pietrusewsky and Douglas (2002), White (1986)
Non Nok Tha	Northeast Thailand	17	SIR: Nos. 34,48,55,77,88 UNL: Nos. 9,10,32,33,47,55,61,62,64,71b,89,90	4,800–2,200	Bayard (1971), Douglas (1996)
Khok Phanom Di	Central Thailand	14	Nos.24,28,29,30,38,42,44,57,67,72, 74,93,129,132	4,000–3,500	Higham and Bannanurag (1990)
Early Holocene Indochina	Laos and Vietnam	5	Pho Binh Gia, Vietnam (No. 18504); Tam Hang, Laos (Nos. 20539,20540); Tam Pong, Laos (Nos. 120541,120542)	10,000–6,000	Pietrusewsky (1997), Pietrusewsky and Douglas (2002)
Mid Holocene Vietnam	Vietnam	5	Lang Cuom (Nos. 19416,19418,19455); Con Co Ngua (Nos. 3,4)	>6,500–5,000	Pietrusewsky (1997), Pietrusewsky and Douglas (2002)
Anyang	Northern China	15	Random selection of 15 of 56 specimens used in larger comparisons	3,385–3,112	Li (1977), Pietrusewsky (1988)
Jomon (Late–Latest)	Japan	15	Eleven crania are from the Ebishima site, Iwate Prefecture, Tohoku District; four are from the Tsukumo site, Okayama Prefecture, Chugoku District, Honshu Island	4,500–2,300	Akazawa (1983), Pietrusewsky (1999)

[a]Specimen sources: BC, 1974 excavations at Ban Chiang; BCES, 1975 excavations at Ban Chiang; SIR, 1965–66 excavation material in Siriraj Hospital, Bangkok; ULN, 1968 excavation material in University of Nevada, Las Vegas; Nos., catalogue numbers for specimens.

are housed in the Siriraj Hospital in Bangkok and the skeletons from the 1968 excavation are curated in the Department of Anthropology, University of Nevada, Las Vegas. The present study uses measurements recorded in 1993 by Michele Toomay Douglas from Non Nok Tha specimens of both series.

Khok Phanom Di, south central Thailand

Mortuary use of the Khok Phanom Di site, then located near the coast, in south central Thailand, has been dated to 4,000–3,500 years BP (Higham and Bannanurag 1990). The cultural sequence and subsistence orientation of this site differs substantially from the inland sites of Ban Chiang and Non Nok Tha. These differences include an adaptation to a rich tropical coastal ecosystem that contained marine and estuary resources, a clustering of graves, mortuary rites that included the sprinkling of red ochre over wrapped bodies, and a lack of domesticated animals. The prehistoric inhabitants of this site are also noteworthy for their high infant mortality, high levels of physiological stress, the presence of genetic anaemia (thalassaemia) and the practice of tooth ablation (Tayles 1999).

Early Holocene Indochina

Crania, now housed in the Musée de l'Homme, Paris, that date to the early Holocene (approximately 10,000–6,000 years BP) and were excavated in the early 1900s, are from three sites in Vietnam and Laos. Two of the crania are from the southern part of the rockshelter site of Tam Hang located in northern Laos. Two additional specimens are from Tam Pong rockshelter located in northern Laos and another specimen is from Pho Binh Gia Cave in northern Vietnam. As discussed in detail elsewhere (Pietrusewsky and Douglas 2002), these skeletal remains were excavated well before the introduction of chronometric dating methods and modern standards of archaeological investigation. The approximate cultural and chronological position for these skeletons has been estimated from associated material remains. Most of the skeletal remains from these sites were associated with ground or polished stone tools and have been attributed to the 'lower neolithic' or Bacsonian (Mansuy and Colani 1925). While the term neolithic, implying the presence of an agricultural lifestyle evidenced by ground or polished stone stools, is used to describe these assemblages in the older literature, this attribution is no longer applicable. These skeletal remains, for the most part, pre-date the introduction of an agricultural lifestyle and should be considered 'late lithic'.

Mid Holocene Vietnam

Crania from two sites in northern Vietnam, Lang Cuom and Con Co Ngua, are of mid Holocene dates, *c.* 5,000–3,500 years BP. Three crania are from the upper layers of the Lang Cuom rockshelter in the calcareous Bacson Massif. Although Mansuy and Colani (1925) attributed the lithic tool assemblages associated with these remains to the Bacsonian or 'lower neolithic', they are more likely attributable to the mid Holocene (Pietrusewsky and Douglas 2002, pp. 223–224). Two crania are from the open-air mound site Con Co Ngua. Con Co Ngua represents a late lithic assemblage from northern Vietnam. Most of the human skeletal remains from this site are from pit 2 (Parker 1998) and date to approximately 5,500 years BP (see Ch. 11 for additional details on this assemblage). Additional information on these sites and skeletal remains in the former Indochina is presented in Pietrusewsky and Douglas (2002).

Anyang, northern China

Bronze age (eleventh century BCE) male crania from 'sacrificial pits' excavated prior to World War II from Shang Dynasty tombs at Anyang, Henan Province, northern China (Li 1977) were examined at Academia Sinica, Taipei in 1983 and 1991. A random selection of 15 adult male crania is used in analysis I and a larger series of 56 is used in analysis II.

Late to final Jomon, Japan

The specimens used represent the late to final Jomon period (4,500–2,300 years BP) (Akazawa 1983) and derive primarily from two sites located on Honshu Island: Ebishima (11 specimens), Iwate Prefecture, Tohoku District, and Tsukumo (four specimens), Okayama Prefecture, Chugoku District.

Modern and near-modern comparative series

In addition to the prehistoric cranial series described above, the present study also included 71 cranial series representing modern and near-modern crania from all parts of Oceania (Polynesia, Micronesia, Melanesia, island Southeast Asia), Australia, Southeast Asia and East Asia. More detailed information on comparative series included in analysis II may be found in Pietrusewsky (1999, 2005) and Pietrusewsky and Ikehara-Quebral (2001).

Cranial measurements and multivariate statistical procedures

Multivariate statistical procedures

Two multivariate statistical procedures, stepwise discriminant function analysis and Mahalanobis' generalised distance statistic (Mahalanobis 1936), were applied to a total of 24 standard cranial measurements (see footnotes to Tables 3.3 and 3.5, below, for the exact measurements used) recorded for male crania in this study. A more detailed discussion of these methods is provided by Pietrusewsky (2000).

Stepwise discriminant function (canonical) analysis

The major purpose of discriminant function, or canonical, analysis is to maximise differences between groups by producing a linear array of weighted variables, referred to as discriminant functions or canonical variates, from the original measurements (Tatsuoka 1970). Typically, the first few functions, or canonical variates, account for most of the variation among groups. In this analysis, the original measurements were selected in a stepwise manner such that, at each step, the measurement that added most to the separation of the groups was the one entered into the discriminant function in advance of the others (Dixon and Brown 1979, p. 711). This procedure allows identification of those variables that are most responsible for the observed differentiation between individuals of the various groups. Interpretation of discriminant functions and the patterns of group separation is based on an inspection of standardised canonical coefficient values.

At the end of the stepping process, each individual specimen is classified into one of the original groups based on the discriminant scores it received through the calculation of posterior (regular classification) and/or typicality (jackknifed classification) probabilities (van Vark and Schaafsma 1992, pp. 244–255). Jackknifed classification represents a common cross-validation procedure in multiple discriminant analysis, where cases are classified without using misclassified individuals in computing the classification function. The 'correct' and 'incorrect' classification results provide a general guide for assessing the homogeneity or heterogeneity of the original series. Only the jackknifed classification results are presented in Table 3.2. Another useful feature of this procedure is that it allows group means to be plotted on the first few canonical variates, thus allowing visualisation of intergroup relationships. The computer program BMDP-7M (Dixon and Brown 1979) was used to perform the stepwise discriminant function analysis, while two-dimensional and three-dimensional plots were made using the SYGRAPH module of SYSTAT (Wilkinson 1992).

Table 3.2. *Number of cases reclassified in the jackknifed classification results for seven groups*[a]

Group	BC	NNT	KPD	ERH	MHL	JOM	ANY
Ban Chiang	1	2	4	2		2	1
Non Nok Tha	1	8		3		3	2
Khok Phanom Di	3		9		1		1
Early Holocene	3	2					
Mid Holocene	2		1		2		
Jomon	2	2	1	2		7	1
Anyang	3	1			2	2	7
Total	12	17	14	5	5	15	15
No. with correct assignment	1	8	9	0	2	7	7
Percentage with correct assignment	8.3	47.1	64.3	0.0	40.0	46.7	46.7

BC, Ban Chiang; NNT, Non Nok Tha; KPD, Khok Phanom Di; ERH, early Holocene Indochina; MHL, mid Holocene Vietnam; ANY, Anyang; JOM, Jomon (late–latest).
[a]Classification results for each group are read by scanning the rows across the page rather than down the page.

Mahalanobis' generalised distance

Mahalanobis' generalised distance, or the sum of squared differences, provides a single quantitative measure of dissimilarity (distance) between groups using several variables while removing the correlation between the variables (Mahalanobis 1936). The significance of these distances was determined using the method of Rao (1952, p. 245), a procedure recommended by Buranarugsa and Leach (1993, p. 17).

The average linkage within group-clustering algorithm (unweighted pair group method algorithm; Sneath and Sokal 1973) was the clustering procedure used to construct the diagrams of relationship, or dendrograms, using Mahalanobis' distances. This algorithm combines clusters so that the average distance among all cases in the resulting cluster is as small as possible and the distance between two clusters is taken to be the average among all possible pairs of cases in the cluster. The NTSYS-pc program was used to construct the dendrograms (Rohlf 1993).

Results

The results of two separate multivariate craniometric analyses are presented. In analysis I, 15 cranial measurements recorded in seven prehistoric male cranial series representing Southeast Asia and East Asia were used.

In analysis II, multivariate craniometric comparisons were made using 24 measurements recorded in 73 prehistoric, modern and near-modern male cranial series from mainland and island Southeast Asia, North and East Asia, the Pacific and Australia. Given that the multivariate procedures used in these analyses require complete data sets, the final number of measurements used represents the largest number facilitating intersample comparisons.

Analysis I: 7 male groups, 15 cranial measurements

In the first multivariate analysis, stepwise discriminant function analysis and Mahalanobis' generalised distance were applied to 15 cranial measurements common to 83 male crania representing seven series: Ban Chiang, Non Nok Tha, Khok Phanom Di, early Holocene Indochina, mid Holocene Vietnam, bronze age Chinese (Anyang) and late to final Jomon, Japan.

Stepwise discriminant function analysis

Overall, and the order in which they were entered into the stepwise discriminant function analysis, minimum frontal breadth, nasion-bregma chord, maximum cranial breadth and maximum cranial length were found to contribute most to the differences between these groups. The first three canonical variates, or discriminant functions, accounted for 82.7% of the total variation. Based on the inspection of standardised canonical coefficient values, maximum cranial and frontal breadths, followed by nasion–bregma chord and cheek height contributed most to the discrimination produced in the first canonical variate, while mastoid measurements and bistephanic breadth were most responsible for the differences in the second canonical variate. The two cranial length measurements, maximum cranial length and nasio-occipital length, were most responsible for the discrimination produced in the third canonical variate.

Jackknifed classification results (Table 3.2) were generally poor, for many of the cases were reassigned incorrectly. The early Holocene Indochina cases feature the poorest jackknifed classification, for not one was assigned to their original group. Three of the early Holocene specimens were reassigned to Ban Chiang and two more were classified as Non Nok Tha. Only one of the Ban Chiang specimens was correctly classified to that group. A total of eight cases originally assigned to Ban Chiang and Non Nok Tha were reassigned to Anyang (three) and Jomon (five). Of the

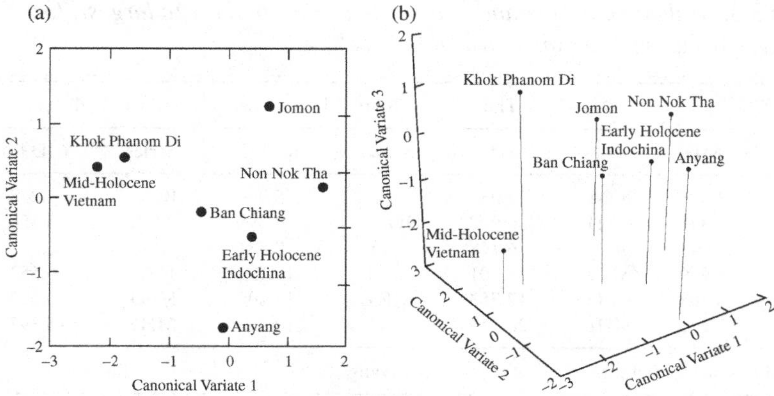

Figure 3.1. Plot of the seven group means on the first two (a) and first three (b) canonical variates using 15 cranial measurements.

30 cases originally assigned to Jomon and Anyang, eight were reclassified as Ban Chiang and Non Nok Tha. Three of the Khok Phanom Di cases were assigned to Ban Chiang. Khok Phanom Di had the best (64.3%) jackknifed classification results of all groups. Ban Chiang received more misclassifications (14) than any other group.

Plots of the seven male group means on the first two and then first three canonical variates (Fig. 3.1) demonstrated that the closest connection was between Ban Chiang and early Holocene Indochina. Non Nok Tha, Anyang and the Jomon series grouped with Ban Chiang and early Holocene Indochina in one sector of these diagrams. Khok Phanom Di and mid Holocene Vietnam assumed the most isolated positions in these representations.

Mahalanobis' generalised distance

Mahalanobis' distances, arranged from the smallest to the largest, for seven groups using 15 measurements are presented in Table 3.3. When tested according to the method of Rao (1952, p. 245) only two distance measures were found to be significant. The smallest distance, implying the greatest overall similarity, was between Ban Chiang and the early Holocene Indochina series. The largest distances, implying greatest dissimilarities, were generally between the mid Holocene Vietnam series and the remaining groups. Cluster analysis (Fig. 3.2) revealed that the Ban Chiang and early Holocene Indochina series formed a tight cluster to which Anyang joined at a more distant remove. Non Nok Tha and Jomon series exhibited moderate affinities to one another and more distant

Table 3.3. *Mahalanobis' distances, arranged from smallest to largest, for seven groups using 15 cranial measurements* [a]

Ban Chiang		Non Nok Tha		Khok Phanom Di		Early Holocene	
Series[b]	MD	Series[b]	MD	Series[b]	MD	Series[b]	MD
ERH	3.278	JOM	5.035	BC	5.969	BC	3.278
ANY	4.343	ERH	6.027	JOM	9.598	ANY	5.356
JOM	5.041	BC	6.168	ANY	9.891	NNT	6.027
KPD	5.969	ANY	7.901	ERH	10.392	JOM	6.552
NNT	6.168	KPD	12.753*	MHL	11.406	KPD	10.392
MHL	8.235	MHL	20.157	NNT	12.753*	MHL	15.397

Mid Holocene Vietnam		Anyang		Jomon	
Series[b]	MD	Series[b]	MD	Series[b]	MD
BC	8.235	BC	4.343	NNT	5.035
KPD	11.406	ERH	5.356	BC	5.041
ERH	15.397	NNT	7.901	ERH	6.552
JOM	15.792	KPD	9.891	KPD	9.598
ANY	16.078	JOM	10.110	ANY	10.110
NNT	20.157	MHL	16.078	MHL	15.792

[a] MD, Mahalanobis' distance. Measurements used: maximum cranial length (M-1); nasio-occipital length (M-1d); maximum cranial breadth (M-8); maximum frontal breadth (M-10); minimum frontal breadth (M-9); bistephanic breadth (H-STB); biauricular breadth (M-11b); biasterionic breadth (M-12); mastoid height (H-MDL); mastoid width (H-MDB); bifrontal breadth (M-43); cheek height (M-48(4)); nasion–bregma chord (M-29); bregma–lambda chord (M-30); lambda–opisthion chord (M-31). In these abbreviations M indicates Martin and Saller (1957) and H indicates Howells (1973).

[b] See Table 3.2 for cranial series abbreviations.

*Significance ($p < 0.05$) of the variance ratio for this distance.

affinities to series from Ban Chiang, Anyang and the early Holocene Indochina. Khok Phanom Di and mid Holocene Vietnamese series occupied isolated branches in this diagram, in which the mid Holocene Vietnam series occupied the most peripheral position in the dendrogram.

Discussion of analysis I

The generally poor jackknifed classification results suggest little variation across the groups in the comparison. The Ban Chiang and early Holocene Indochina series have the poorest classification results, indicating they are not well differentiated from the other groups. One exception is Khok Phanom Di, which has the best classification results, indicating that this cranial series is one of the most differentiated groups in this analysis.

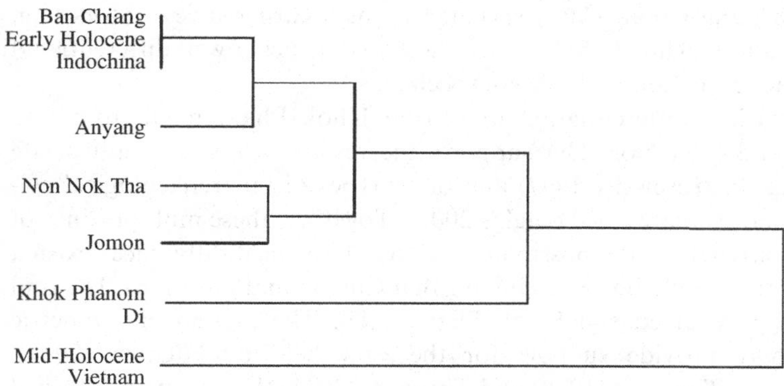

Figure 3.2. Diagram of relationship (dendrogram) based on a cluster analysis (unweighted pair group method algorithm) of Mahalanobis' generalised distances using 15 cranial measurements recorded in seven male groups.

The results of this first analysis reveal differences as well as similarities among cranial series from archaeological sites in Thailand, Laos and Vietnam. The cranial series from Ban Chiang, located in northeast Thailand, exhibits closest phenetic affinities to a cranial series of mostly early Holocene specimens from Laos, a connection not previously reported in analyses that combined the specimens from Laos and Vietnam (e.g. Pietrusewsky 1997). This similarity between Ban Chiang and the early Indochina cranial series could be interpreted as possible evidence for local continuity in these mostly inland mainland Southeast Asian series spanning the late lithic to neolithic/bronze age.

By way of contrast, the coastal, or near coastal, sites of Khok Phanom Di and the mid Holocene cranial series from Vietnam are in sharp contrast to other archaeological cranial series examined. The inferred isolation of the mid Holocene series from Vietnam is unexpected, given that two of the five specimens in this series are from Con Co Ngua, a site considered closest to Ban Chiang with respect to date and level of economic development (Higham 1996 (pp. 82–83), Pietrusewsky and Douglas 2002 (p. 225)). These results corroborate an earlier multivariate investigation (Pietrusewsky 1988) showing that Con Co Ngua was well differentiated from Ban Chiang and other cranial series from Thailand, Laos and Vietnam. A recent study of dental morphology by Matsumura *et al.* (2001) demonstrated a similar differentiation between Con Co Ngua and subsequent early metal age inhabitants of Vietnam. The remaining three specimens in the mid Holocene Vietnam series used in the present study are from the Lang Cuom Cave, specimens which earlier investigators (e.g. Mansuy and Colani 1925) have characterised as being well differentiated from other early Southeast Asian specimens.

The marked differentiation of coastal Khok Phanom Di from Ban Chiang and Non Nok Tha supports the results of earlier multivariate work (e.g. Pietrusewsky 1997) and other types of bioarchaeological evidence (Pietrusewsky and Douglas 2002). Together, these multiple lines of evidence strengthen the possibility that real biological differences existed between the people buried at inland Ban Chiang and Non Nok Tha and those interred at coastal Khok Phanom Di. The present craniometric results may provide support for the view held by Higham (2001), Higham and Thosarat (1998) and Thosarat (2001) that the people buried at Khok Phanom Di represent an earlier hunter–gatherer population distinct from the more newly arrived agriculturalists, who are held to be buried at sites such as Ban Chiang. However, the present craniometric results may also be interpreted to indicate that the inhabitants of Khok Phanom Di represent a settlement of intrusive people, well adapted to a marine subsistence base, who arrived in the region following a coastal route along the Gulf of Thailand. In a similar vein, Bulbeck and coworkers (Bulbeck 1999, Rayner and Bulbeck 2001; see Ch. 6) have suggested that certain aboriginal hunter–gatherer groups such as the Semang, Philippine Negritos and Andaman Islanders are the descendants of earlier migrations of people who arrived in island Southeast Asia following the southern coast of Asia. Since no Negrito cranial series are included in the present analyses, this hypothesis is not examined at this time.

As such, the present results not only reinforce an inland versus coastal distinction for cranial series within the modern political boundaries of Thailand but also suggest that a coastal–inland distinction may apply to Southeast Asia in general. Therefore, the biological separation between the early inland series for northeast Thailand and the mostly early series from Laos relative to coastal and near coastal series from Khok Phanom Di raise two potential implications. The first is that the coastal–inland distinction may have considerable antiquity in the region. The second is that people living along, or close to, the coast were generally more mobile than their inland counterparts.

These results also indicate a possible connection between bronze age Anyang (northern China) and Jomon Japanese with Ban Chiang, early Indochina and the Non Nok Tha series. This suggests broader associations for the early inhabitants of Southeast Asia, which may reflect movements of people in the past. Similar connections between some of the early Southeast Asian crania and those from northern and Northeast Asia have been observed in previous multivariate work (e.g. Pietrusewsky 1997), and other researchers (e.g. Hanihara 1993, 1994, 1996, Matsumura 1995) have reported biological connections between early Thai and Jomon series. Population expansion has been suggested by recent genetic data (Su *et al.* 1999). The craniometric study supports this or alternative models that posit a migration of agriculturalists into Southeast Asia from a northern source or separate origins for the ancestors of both southern and northern groups.

Analysis II: 73 male groups, 24 cranial measurements

In the second multivariate analysis, stepwise discriminant function analysis and Mahalanobis' generalised distance were applied to 24 cranial measurements (see Table 3.4) recorded in 3217 male crania from Ban Chiang, Khok Phanom Di, bronze age Chinese (Anyang), late to final Jomon from Japan, and 69 modern and near-modern cranial samples representing Southeast Asia, East/North Asia, the Pacific and Australia. This second analysis, which uses more measurements and larger sample sizes, should provide better resolution of group differences in a broader regional perspective than those obtained in the first analysis.

Stepwise discriminant function analysis

Those measurements that contributed most to the overall differences between these groups included three cranial breadth measurements (maximum cranial breadth, biorbital breadth and minimum cranial breadth) and basion–nasion length. The first three canonical variates, or discriminant functions, accounted for 61.4% of the total variation. The variables that contributed most to the differentiation in the first canonical variate were orbital breadth, maximum cranial length and minimum and maximum frontal breadths. The height of the nasal aperture, orbital breadth and minimum frontal breadth were the main discriminating variables for the second canonical variate, while orbital height, basion–nasion length and maximum cranial length contributed most to the third canonical variate.

Only some of the jackknifed classification results (Table 3.5) are summarised. Groups with the poorest classification results in this analysis include Solomon Islands, Ban Chiang, several Southeast Asian series (e.g. Lesser Sunda Islands, Sulawesi, Burma (Myanmar), Sumatra, Borneo, Philippines and Southern Moluccas), and some of the Chinese cranial series. Only 1 of the 12 Ban Chiang cases was correctly classified; three were misclassified to this group. Three of the Ban Chiang cases were

Table 3.4. *The smallest distances, and significance level, for 25 of the 73 groups using 24 cranial measurements*[a]

BC		KPD		SUM		JAV		BOR	
HAI	5.553*	THI	8.480*	LSN	1.976*	SLW	1.958*	LSN	1.459*
LSN	6.997*	BC	8.575	BOR	2.053*	THI	2.832	SUM	2.053*
EDO	7.401	BAC	9.992	SLW	2.697*	CML	2.994	SLW	2.516*
ANY	7.444	JAV	10.254	EDO	3.819	LSN	3.351	JAV	3.836
JAV	7.681	TGS	10.346*	JAV	3.916	SUL	3.700	SUL	3.935
KOR	7.701*	ANY	10.556	VTN	4.082	BOR	3.836	VTN	4.454
VTN	7.856	HAI	10.613	HAN	4.188	SUM	3.916	PHL	4.686
KAM	7.939	AMI	11.238	KAN	4.626	BAC	4.109	CML	5.125
HK	8.078	BUR	11.617*	YAY	4.642	PHL	4.260	SML	5.141
OKI	8.143	SLW	12.206	PHL	4.646	VTN	4.699	SOL	6.038

LSN		SML		SUL		PHL		VTN	
BOR	1.459*	BIK	3.250	SLW	2.605*	VTN	2.031*	PHL	2.031*
SUM	1.976*	NZ	3.322	CML	2.662*	LSN	2.983	LSN	3.280
SLW	2.621	NIR	3.353	JAV	3.700	SLW	3.054*	HK	3.741
PHL	2.983	SOL	4.083	BOR	3.935	JAV	4.260	THI	3.761
VTN	3.280	SEP	4.517	LSN	4.701	THI	4.320	KOR	3.805
JAV	3.351	LSN	4.526	BUR	4.827*	KOR	4.501*	SUM	4.082
EDO	3.703	BOR	5.141	SUM	4.836	HAI	4.520	HAI	4.188
SML	4.526	DCX	5.229	BAC	6.037	HK	4.540	OKI	4.247
KAM	4.546	VAN	5.301	PHL	6.355	SUM	4.646	EDO	4.375
SUL	4.701	CAR	5.314	THI	6.522	BOR	4.686	SLW	4.378

BAC		CML		THI		BUR		SLW	
THI	2.681	SLW	2.240*	BAC	2.681	THI	4.426*	JAV	1.958*
HAI	3.149	SUL	2.662*	JAV	2.832	BAC	4.430*	CML	2.240*
JAV	4.109	JAV	2.994	SLW	3.212	SLW	4.821*	BOR	2.516*
SLW	4.158	THI	5.072	HAI	3.387	SUL	4.827*	SUL	2.605*
BUR	4.430*	BOR	5.125	KOR	3.557	JAV	5.313*	LSN	2.621
VTN	4.754*	BAC	5.159	VTN	3.761	SUM	5.516*	SUM	2.697*
PHL	5.047	BUR	5.574*	PHL	4.320	CML	5.574*	PHL	3.054*
CML	5.159	LSN	5.634	BUR	4.426*	LSN	7.294	THI	3.212
KOR	5.559	PHL	5.675	HK	4.864	HAI	7.411	BAC	4.158
SUL	6.037	SUM	6.469	CML	5.072	KAN	7.549	KOR	4.276
TAI		**HAI**		**ATY**		**ANY**		**KOR**	
HAI	2.808	KOR	2.382*	KYU	4.781	HAI	3.020	EDO	2.173*
ANY	3.390	TAI	2.808	EDO	5.074	KOR	3.226	HAI	2.382*
KOR	3.613	ANY	3.020	KOR	5.174	AMI	3.383	KYU	2.544*
MAN	4.996	BAC	3.149	TOH	5.205	TAI	3.390	TOH	3.100
OKI	5.211	THI	3.387	LSN	5.267	HK	3.848	ANY	3.226
SAK	5.224	VTN	4.188	HAI	5.985	KYU	4.414	MAN	3.301
VTN	5.610	EDO	4.308	SAK	6.062	KAM	4.474	KAN	3.316
HK	5.708	HK	4.516	DAW	6.484	SAK	4.650	THI	3.557
EDO	5.805	PHL	4.520	VTN	6.613	EDO	4.673	TAI	3.613
KYU	5.867	KYU	4.697	PHL	6.698	MAN	4.959	VTN	3.805

Table 3.4. (*cont.*)

YAY		AIN		JOM		AMI		SAK	
KOF	1.232*	JOM	3.258	AIN	3.258	KAM	2.800*	OKI	2.864*
SAK	2.902*	EDO	3.950	YAY	4.135	YAY	3.306	YAY	2.902*
KAM	3.001	KAM	4.212	KOF	4.561	KOF	3.342	KAM	3.029*
AMI	3.306	KYU	4.289	KAM	4.817	SAK	3.346*	EDO	3.176*
EDO	3.402	TOH	4.376	OKI	5.221	ANY	3.383	AMI	3.346*
TOH	4.126	KAN	5.235	EDO	5.916	OKI	3.394	KOF	3.355
JOM	4.135	YAY	5.443	KYU	6.238	KOR	4.838	KYU	3.822
OKI	4.161	SUM	6.443	AMI	6.238	EDO	5.047	KOR	3.934*
KOR	4.491	KOF	6.481	KAN	6.348	VTN	5.093	TOH	4.638
SUM	4.642	SAK	7.012	TOH	6.410	KYU	5.448	ANY	4.650

ADR, Admiralty Is.; AIN, Ainu; AMI, Amami Islands, Ryukyu Is.; ANY, Anyang; ATY, Atayal, Taiwan; BAC, Bachuc Village, Vietnam; BC, Ban Chiang; BIK, Biak Is.; BOR, Borneo; BUR, Burma (Myanmar); CAR, Caroline Is.; CHG, Chengdu; CHT, Chatham Is.; CML, Cambodia/Laos; DCX, D'Entrecasteaux Is.; DAW, Dawson Strait Is.; EDO, Edo; FIJ, Fiji; FLY, Fly R.; GAM, Gambier Is.; GUA, Guam; HAI, Hainan Island; HAN, Hangzhou; HAW, Hawaii; HK, Hong Kong; JAV, Java; JOM, Jomon; KAM, Kamakura; KAN, Kanto Japanese; KOF, Kofun; KOR, Korea; KPD, Khok Phanom Di; KYU, Kyushu Japanese; LOY, Loyalty Is.; LSN, Lesser Sundas Is.; MAN, Manchuria; MOG, Mongolia; MRB, Murray R. Basin, Australia; MRG, Marshall/Gilbert Is.; MRQ, Marquesas Is.; NAN, Nanjing; NBR, New Britain; NCL, New Caledonia; NIR, New Ireland; NSW, New South Wales, Australia; NT, Northern Territory, Australia; NZ, New Zealand; OKI, Okinawa Is.; PHL, Philippines; QLD, Queensland, Australia; PUR, Purari Delta, Papua New Guinea; SAK, Sakishima Is., Ryukyu Is.; SAS, Swanport; South Australia; SCR, Santa Cruz Is.; SEP, Sepik R., Papua New Guinea; SLW, Sulawesi; SML, S. Moluccas Is.; SOC, Society Is.; SOL, Solomon Is.; SUL, Sulu Arch.; SUM, Sumatra; TAI, Taiwan Chinese; TAS, Tasmania; TGS, Tonga-Samoa; THI, Thailand; TOH, Tohoku Japanese; TUA, Tuamotu, Arch.; VAN, Vanuatu; VTN, Vietnam; WA, Western Australia; YAY, Yayoi.

[a]maximum cranial length (M-1); nasio-occipital length (M-1d); basion–nasion length (M-5); basion–bregma height (M-17); maximum cranial breadth (M-8); maximum frontal breadth (M-10); minimum frontal breadth (M-9); bistephanic breadth (H-STB); biauricular breadth (M-11b); minimum cranial breadth (M-14); biasterionic breadth (M-12); nasal height (H-NLH); nasal breadth (M-54); orbital height, left (M-52); orbital breadth, left (M-51a); alveolar breadth (M-61); mastoid height (H-MDL); mastoid breadth (H-MDB); bimaxillary breadth (M-46); bifrontal breadth (M-43); biorbital breadth (H-EKB); cheek height (H-WMH); nasion-bregma chord (M-29); bregma-lambda chord (M-30).

*Variance ratio not significant $p < 0.05$.

Table 3.5. *Number of cases reclassified for some of the jackknifed classification results for 24 of the 73 groups; numbers in parentheses represent the number of crania originally assigned to each group*

Ban Chiang (12)		Khok Phanom Di (14)		Thailand (50)		Vietnam (49)		L. Sundas (61)		Anyang (56)		Atayal (36)		Hainan (47)	
KPD	3	KPD	8	THI	10	VTN	13	BOR	6	ANY	13	ATY	14	TAI	6
BC	1	BC	2	BAC	5	ATY	4	VTN	6	HK	5	DAW	4	BC	4
BOR	1	TGS	2	JAV	4	HK	4	CML	4	AMA	5	SML	2	HAI	4
VTN	1	BUR	1	SLW	4	KOR	4	LSN	4	SAK	4	KOR	2	ANY	4
BUR	1	ANY	1	KOR	4	OKI	3	HAW	2	HAN	4	TOH	2	THI	3
AMA	1			BC	2	SLW	3	NZ	2	KYU	3	KAN	2	GUA	3
OKI	1			PHL	2	BC	2	JAV	2	KOR	3	KYU	2	VTN	2
HAI	1			HAI	2	TAI	2	PHL	2	NAN	2	LSN	1	BAC	2
NIR	1			KAN	2	SUM	1	HAI	2	TAI	2	TAI	1	BUR	2
FLY	1			AMA	2	BOR	1	ATY	2	ATY	2	HAI	1	SAK	2
Amami (31)		**Yayoi (62)**		**Ainu (50)**		**Jomon (51)**		**Java (50)**		**Cambodia/Laos (40)**		**Bachuc (51)**		**Burma (16)**	
AMA	12	YAY	9	AIN	23	JOM	18	JAV	12	CML	17	BAC	17	SLW	2
KAM	3	KOF	7	JOM	8	AIN	7	CML	4	SUL	4	THI	4	BAC	2
AIN	2	MOG	5	KAN	4	OKI	5	THI	4	JAV	2	BOR	3	SUM	1
BC	2	JOM	5	KAM	2	YAY	3	BUR	3	SML	2	CML	3	SML	1
YAY	2	BC	3	TOH	2	KOF	2	SLW	3	PHL	2	BUR	3	SUL	1
THI	2	NAN	3	KYU	2	BC	2	SUL	2	BAC	2	HK	3	CML	1
VTN	1	KAN	3	SUM	1	TGS	2	PHL	2	BUR	1	BC	2	BUR	1
HAI	1	OKI	3	HK	1	SUM	1	BAC	2	CHG	1	SLW	2	HAN	1
ATY	1	VTN	2	SAK	1	VTN	1	BUR	2	ATY	1	SUL	2	KPD	1
SAK	1	SAK	2	KOR	1	THI	1	HK	2	KPD	1	HAI	2	ANY	1

Table 3.5. (*cont.*)

Sumatra (55)		Sulu (38)		S. Moluccas (65)		Philippines (28)		Borneo (34)		Sulawesi (41)		Korea (32)		Taiwan (47)	
BOR	6	SUL	10	SML	10	PHL	4	BOR	4	CML	8	KOR	5	TAI	12
SUM	4	CML	7	NZ	8	VTN	4	SLW	2	JAV	3	HAI	4	ANY	6
JAV	2	JAV	4	NCL	4	MAN	2	SUM	2	SUL	3	BC	2	HAI	4
SML	2	BOR	3	NIR	4	BC	1	SUL	2	SUM	2	PHL	2	MAN	3
SUL	2	BUR	2	MRQ	3	SUM	1	VTN	2	SLW	2	HK	2	BAC	2
PHL	2	AIN	2	SCR	3	JAV	1	BAC	2	PHL	2	TOH	2	KAM	2
BUR	2	SLW	1	SEP	3	BOR	1	CML	2	VTN	2	HAW	2	TOH	2
CHG	2	SML	1	DAW	3	SLW	1	FLY	2	THI	2	SUM	1	BC	1
JOM	2	PHL	1	SLW	2	SUL	1	BC	1	SHA	2	BOR	1	VTN	1
LSN	1	BAC	1	SUL	2	BAC	1	JAV	1	BC	1	THI	1	THI	1

SHA, Shanghai, other abbreviations for group names are given in Tables 3.2 and 3.4.

misclassified to Khok Phanom Di, and one each was reclassified as Borneo, Vietnam, Burma (Myanmar), Hainan, Amami, Okinawa, Hainan Island, New Ireland and Fly River. Classification results for Khok Phanom Di were among the best, with 8 of 14 cases (57.1%) being correctly assigned. Two Khok Phanom Di individuals each were reassigned to Ban Chiang and Tonga-Samoa and one each was assigned to Burma (Myanmar) and Anyang. Among the cranial series that had two or more cases misclassified as Ban Chiang were Hainan (four), Yayoi (three), Jomon (two), Bachuc (two), Vietnam (two), Thailand (two), Shanghai (two), Tohoku (two), and Korea (two). Except for three Ban Chiang individuals, very few individuals from other cranial series were misclassified to the Khok Phanom Di cranial series.

A plot of the 73 group means on the first two canonical variates (Fig. 3.3) revealed three major constellations. All the Australian and Melanesian series fell into one of these major groupings while Polynesian and two Micronesia series occupied a second. The cranial series from Southeast Asia and East Asia formed a tight grouping, one that included Ban Chiang and Khok Phanom Di. Ban Chiang's position was peripheral but was nearest to modern cranial series from Thailand and Vietnam (Bachuc and Vietnam), while Khok Phanom Di was closest to several Chinese (e.g. Hong Kong, Hangzhou, Hainan and Shanghai) and Southeast Asian (e.g. Thailand, Vietnam) series.

Plots of 73 group means on the first three canonical variates are shown in Figs. 3.4 and 3.5. As in the previous plot of group means on the first two canonical variates, there was good separation between the Australian-Melanesian and Polynesian cranial series (Fig. 3.4). Ban Chiang's position (Fig. 3.5: No. 6) was closest to modern cranial series from Java, Vietnam and Sulawesi, while samples from Thailand, Cambodia-Laos, Philippines, Burma (Myanmar) and Hainan Island were not far removed. The spike for Khok Phanom Di (No. 17) was embedded among modern cranial series from Hong Kong, Vietnam, Philippines, Burma (Myanmar) and Sulawesi.

Mahalanobis' generalised distance

The ten smallest Mahalanobis' distances for some of the 73 groups (Table 3.4) are briefly summarised. Among the groups with the smallest distances to Ban Chiang were Hainan Island, Lesser Sundas, Edo Japanese, Anyang, Java, Korea and Vietnam. The remainder of these smallest distances generally included cranial series from Southeast Asia and East Asia. The groups closest to Khok Phanom Di included modern Thai, Ban Chiang, Bachuc and Java. Inspection of these distances indicated that the two archaeological series from Thailand (Ban Chiang and Khok Phanom Di)

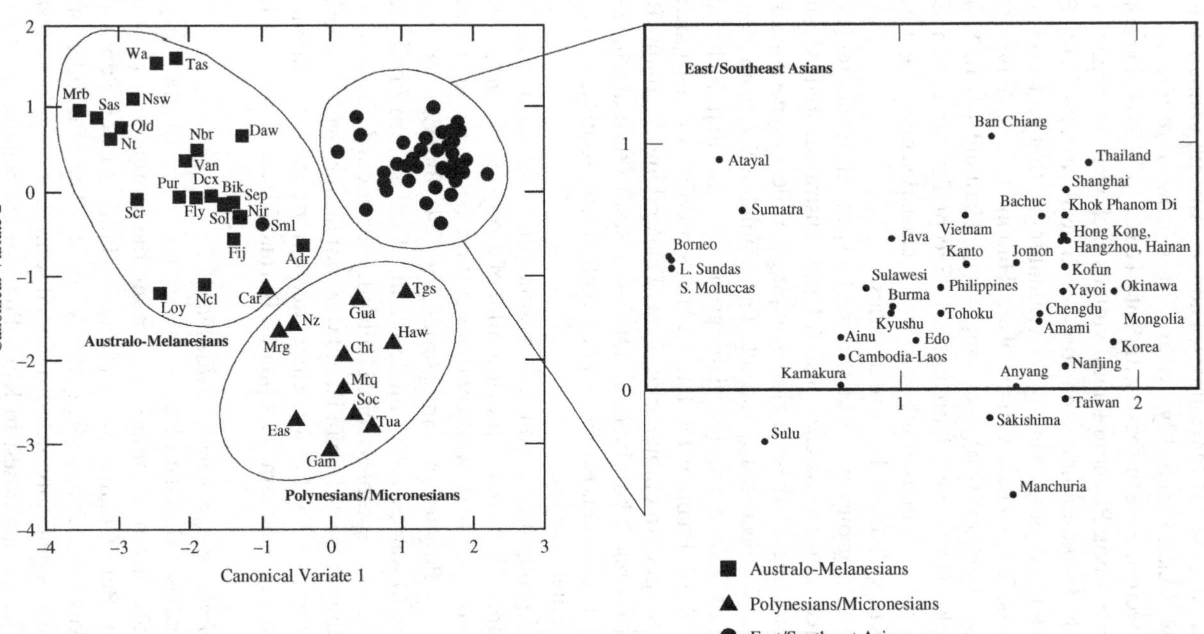

Figure 3.3. Plot of 73 group means in the first two canonical variates using 24 cranial measurements. (EAS, Easter Island; see Table 3.4 for other group abbreviations.)

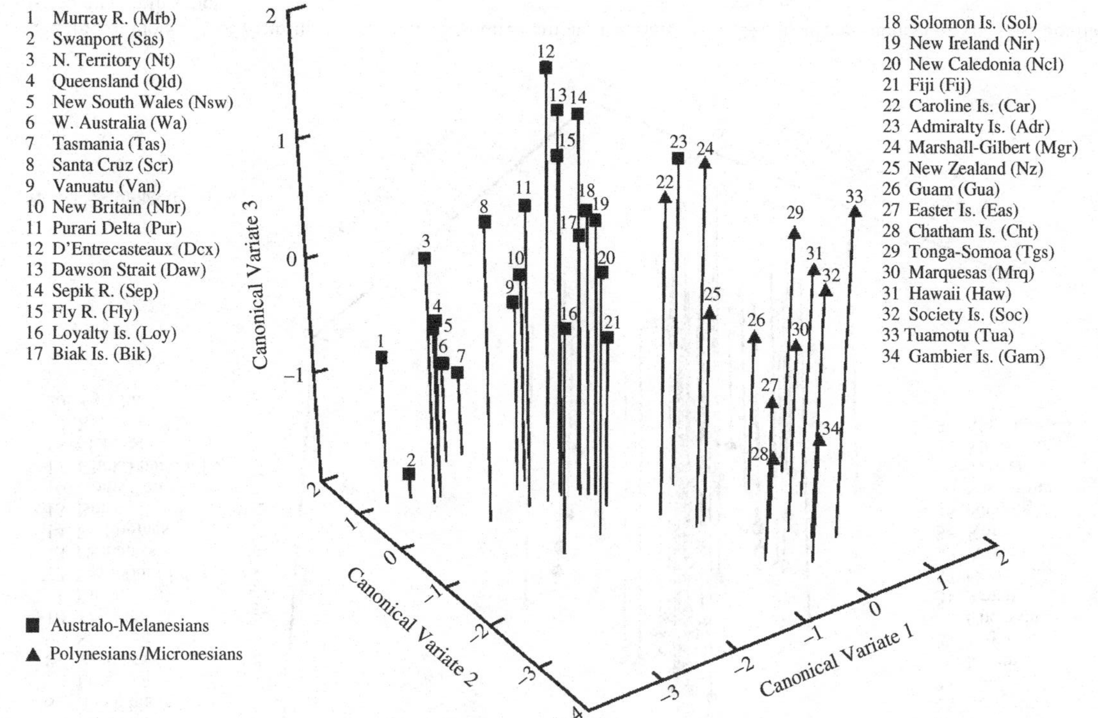

Figure 3.4. Plot of 34 of 73 group means on the first three canonical variates using 24 cranial measurements. The Australo-Melanesian and Polynesian/Micronesian series are shown in this plot.

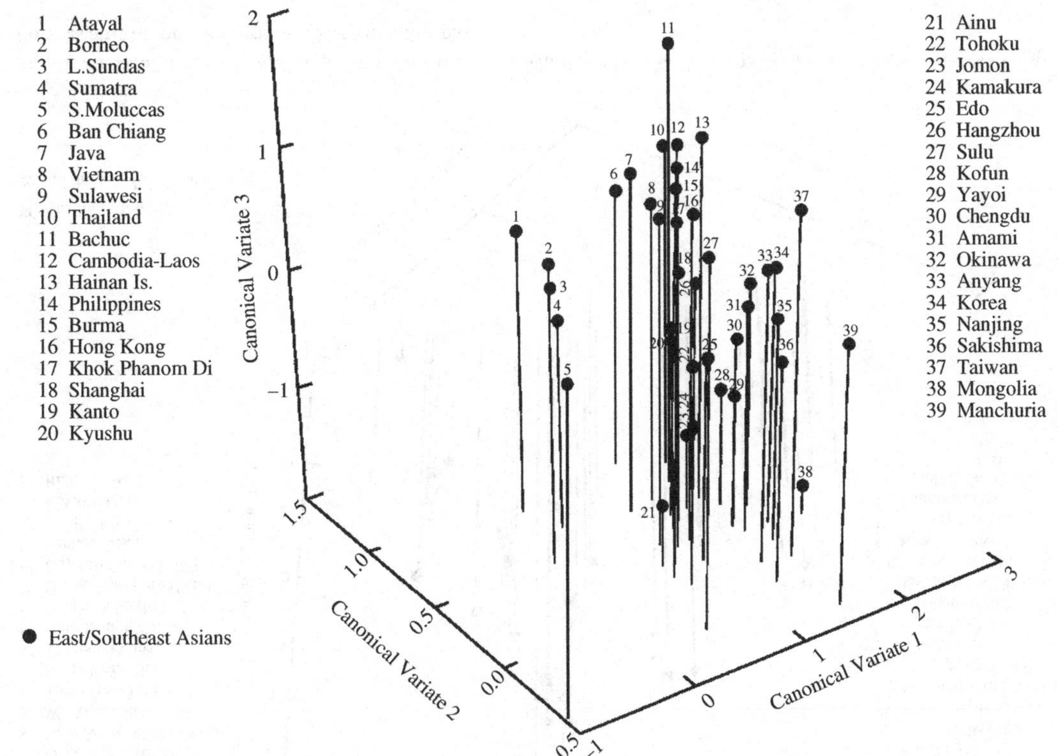

Figure 3.5. Plot of 39 of 73 group means on the first three canonical variates using 24 cranial measurements. The East/Southeast Asian series are shown in this plot.

were most similar to cranial series from Southeast Asia and East Asia and that these same series were very different from cranial series in Melanesia, Australia and Polynesia.

Applying the unweighted pair group method algorithm as a clustering technique to the distances for 73 groups resulted in the dendrogram in Fig. 3.6. Ban Chiang and Khok Phanom Di united to form an outlier to a large collection of East Asian and Southeast Asian series. Polynesian cranial series were the last to join this major Asiatic division. The second major grouping was one that contained all cranial series representing Australia, Tasmania and Melanesia.

Discussion of analysis II

The results of this second analysis reveal the existence of two major divisions among the inhabitants of eastern Asia and the Pacific. One of these divisions includes all cranial series from Australia, Tasmania, New Guinea and island Melanesia. The second major grouping is one that contains all cranial series from East Asia (mainland and island), Southeast Asia, Micronesia and Polynesia. The sharpness of the separation between these two divisions is highly suggestive of separate and ancient origins for these two major divisions of humanity. Series from Ban Chiang and Khok Phanom Di are members of the East/Southeast Asian division and demonstrate similarities to several modern and prehistoric cranial series from this greater geographical region. While these two prehistoric archaeological series appear to be isolated in the dendrogram in Fig. 3.6, inspection of the canonical plot in Fig. 3.5 indicates that the groups closest to Ban Chiang are primarily modern Southeast Asian cranial series. The groups closest to Khok Phanom Di, in the same plot, include several modern Southeast Asian series and Hong Kong. The classification results (Table 3.5) reiterate associations between Ban Chiang and Southeast Asian (e.g. Borneo, Vietnam, Burma (Myanmar)) series as well as connections to some of the Ryukyu Island cranial series. The possible connection between early Southeast Asian and Ryukyu Island cranial series, observed in these results, has been reported by others (e.g. Hanihara 1993, 1994). Closer inspection of the smallest distances for Ban Chiang and Khok Phanom Di (Table 3.4) indicate similarities between these groups, modern Southeast Asia and the bronze age Chinese series from Anyang. While possible connections between early cranial series from Thailand, bronze age Chinese and modern cranial series from the Ryukyu Islands hint at complex biological relationships, these same results convincingly and overwhelmingly demonstrate that there are very close connections between prehistoric and modern Southeast Asian cranial series.

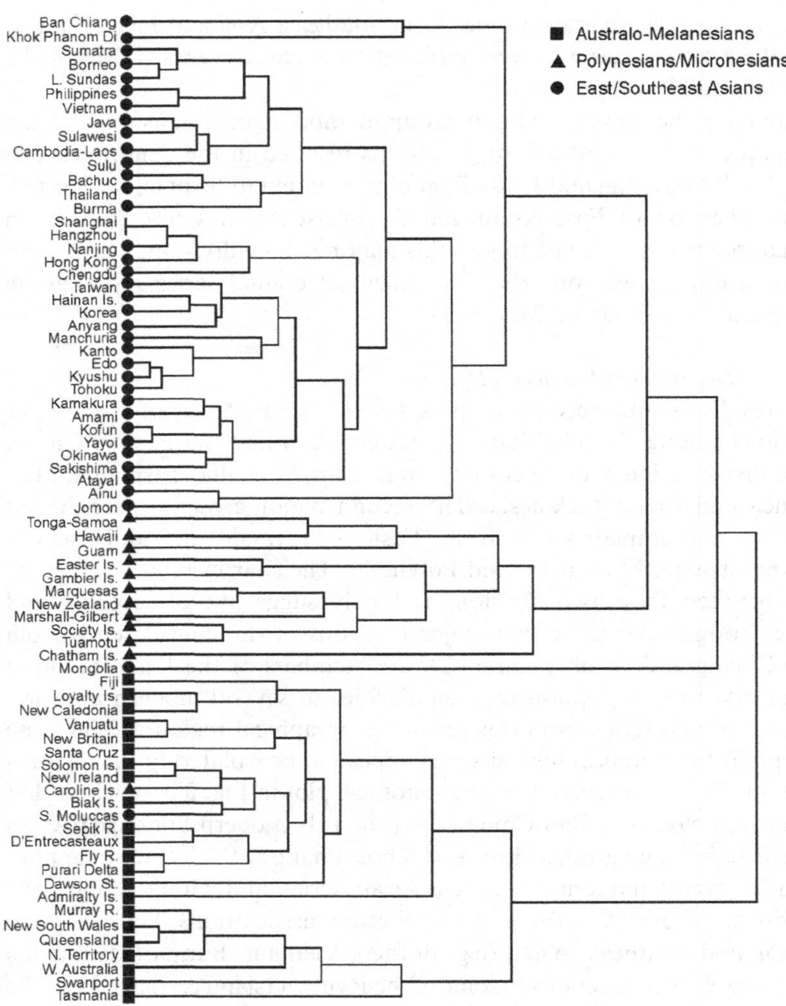

Figure 3.6. Diagram of relationship (dendrogram) based on a cluster analysis (unweighted pair group mean algorithm) of Mahalanobis' generalised distances using 24 cranial measurements recorded in 73 male cranial series.

The lack of strong morphological differences between earlier (preneolithic) and more recent inhabitants of northeastern Southeast Asia, an area that is very near one of the hypothesised zones of agricultural origins, may make it impossible to disentangle the demic expansion (e.g. Bellwood 1997, Higham 2001) versus continuity (e.g. Bulbeck 1982, Turner 1987) models that have been proposed to explain population history for this region. The group that grew demographically with agriculture could very well have displaced/replaced earlier groups in northeastern Southeast Asia without leaving a morphological signature. Likewise, the absence of morphological differences in the earlier and later inhabitants of this region could be used to support the Local Continuity Model.

The results of these broader comparisons also reveal a clear separation between East/North Asian and Southeast Asian cranial series and a clear connection between mainland and island Southeast Asia. Island Southeast Asian cranial series that are particularly close to mainland Southeast Asian series include Sulawesi, Java, Philippines and Sulu. When viewed in the context of current models that seek to account for the present distribution of peoples in Southeast Asia and East/North Asia, the results obtained in this study lend more support to models of long-term *in situ* evolution in these regions than to models that call for displacement to account for the modern inhabitants of Southeast Asia.

Connections between cranial series from East/Southeast Asia and Remote Oceania (including Polynesia), both well removed from a division that exclusively includes Melanesian and Australian cranial series, are also demonstrated in these results. This relationship is most consistent with an ancestral Polynesian homeland in East/Southeast Asia and not one within geographically adjacent Melanesia. Furthermore, although not shown in Table 3.4, several of the Polynesian cranial series (e.g. Hawaii, Marquesas, New Zealand, Chatham Island and Tonga–Samoa) reveal close ties with several island Southeast Asian series (e.g. Lesser Sunda Islands, Sulawesi, Southern Moluccas).

Finally, contrary to the view expressed by Brace and colleagues (Brace and Hunt 1990), a close biological connection between the Ainu and Polynesians is not supported by the present multivariate craniometric results. Rather, the Ainu are members (albeit marginal) of a greater East/North Asian division and do not connect directly with any of the Polynesian series. Similar conclusions have been reached by several other researchers using skeletal and dental evidence, as well as genetic evidence (e.g. Hanihara 1993, Ishida and Dodo 1993, Dodo *et al.* 1998, Ishida and Kondo 1999, Omoto and Saitou 1997).

Conclusions

The results of this new multivariate craniometric study, which includes archaeological and more modern human cranial series from Southeast

Asia and neighbouring regions, allow a number of conclusions regarding the biological relationships of the inhabitants of this region, ancient and modern. The relatively close biological relationship between the early pre-metal to bronze age inhabitants of Ban Chiang, a site located in northeast Thailand, and an early Holocene (before 6,000 years BP) cranial series from Laos provides support for local continuity in human groups from the late lithic to the neolithic/bronze age in mainland Southeast Asia. Earlier biological connections for the pre-metal age inhabitants of Southeast Asia are further implied in these results, with possible connections between archaeological series from northeast Thailand, bronze age Chinese (Anyang) and Jomon cranial series from Japan.

The people buried at Khok Phanom Di in south central Thailand and the mid Holocene crania from Vietnam are isolated from the remaining prehistoric cranial series examined, a separation that suggests a separate genetic heritage for these groups. The differentiation of Khok Phanom Di, a coastal site located in Thailand, and the two prehistoric cranial series from northeast Thailand is particularly interesting. A similar distinction between inland and coastal groups currently living in Southeast Asia may be further indicated.

The marked separation of Southeast Asian, including two prehistoric archaeological series from Thailand (Ban Chiang and Khok Phanom Di), and East/North Asian cranial series suggests long-term continuity in both regions. Possible connections between these two prehistoric archaeological series from Thailand and bronze age Chinese (Anyang) and the modern Ryukyu Island series are equally informative of past relationships.

Finally, the presence of two sharply contrasting divisions of humanity for eastern Asia and the Pacific, an Australo-Melanesian and an Asian complex, strongly suggests separate origins for the indigenous inhabitants of Oceania and Australasia and a homeland in (island) Southeast Asia for the Polynesians.

When additional, larger, well-provenanced skeletal series representing the early Holocene and later periods in Southeast Asia become available, our ability to decipher the human kaleidoscope, current and past, will be greatly advanced.

Acknowledgements

A number of individuals have helped in the analysis of data and preparation of this chapter. Mr Scott Reinke and Ms Rona Ikehara-Quebral assisted in various stages of data analyses and construction of the tables. Dr Michele Toomay Douglas generously allowed me to use the measurements she recorded in the Non Nok Tha specimens located in Las Vegas and Bangkok and gave helpful advice and comments on earlier drafts of this chapter. Ms Billie Ikeda is responsible for the figures. Finally, my thanks to Drs Nancy Tayles and Marc Oxenham and three anonymous reviewers for their constructive comments in preparing the final draft of this chapter.

References

Akazawa T. 1983. An outline of Japanese prehistory. In Kondo S., Hanihara K., Ikeda J. and Watanabe N., eds., *Recent Progress of Natural Sciences in Japan, Vol. 8: Anthropology*. Tokyo: Science Council of Japan, pp. 1–11.

Bayard D. T. 1971. *Otago University Studies in Prehistoric Anthropology*, Vol. 4: *Non Nok Tha: The 1968 Excavation. Procedure, Stratigraphy, and Summary of the Evidence*. Dunedin: University of Otago Press.

1996–97. Bones of contention: the Non Nok Tha burials and the chronology and context of early Southeast Asian bronze. In Bulbeck F. D., ed., *Ancient Chinese and Southeast Asian Bronze Age Cultures*, Vol. 2. Taipei: SMC, pp. 889–940.

Bellwood P. 1996. Early agriculture and the dispersal of the southern Mongoloids. In Akazawa T. and Szathmáry E. J. E., eds., *Prehistoric Mongoloid Dispersals*. Oxford: Oxford University Press, pp. 287–302.

1997. *Prehistory of the Indo-Malaysian Archipelago*, revised edn. Honolulu: University of Hawaii Press.

2000. Some thoughts on understanding the human colonization of the Pacific. *People and Culture in Oceania* **16**: 5–17.

Blust R. 1996. Beyond the Austronesian homeland: the Austric hypothesis and its implications for archaeology. In Goodenough W., ed., *Prehistoric Settlement of the Pacific*. Philadelphia, PA: Transactions of the American Philosophical Society, pp. 117–140.

Bowles G. T. 1977. *The People of Asia*. New York: Charles Scribner.

Brace C. L. and Hunt K. D. 1990. A nonracial craniofacial perspective on human variation: A(ustralia) to Z(uni). *American Journal of Physical Anthropology* **82**: 341–360.

Buikstra J. E., Frankenberg S. R. and Konigsberg L. W. 1990. Skeletal biological distance studies in American physical anthropology: recent trends. *American Journal of Physical Anthropology* **82**: 1–7.

Bulbeck D. 1982. A re-evaluation of possible evolutionary processes in Southeast Asia since the late Pleistocene. *Bulletin of the Indo-Pacific Prehistory Association* **3**: 1–21.

1999. Current biological anthropological research on Southeast Asia's Negritos. *SPAFA Journal* **9**: 14–22.

Buranarugsa M. and Leach F. 1993. Coordinate geometry of Moriori crania and comparisons with Maori. *Man and Culture in Oceania* **9**: 1–43.

Dixon W. J. and Brown M. B. (eds.). 1979. *BMDP-79. Biomedical Computer Programs P-series*. Berkeley, CA: University of California Press.

Dodo Y., Doi N. and Kondo O. 1998. Ainu and Ryukyuan cranial non-metric variation: evidence which disputes the Ainu–Ryukyu common origin theory. *Anthropological Science* **106**: 99–120.

Douglas M. T. 1996. Paleopathology in human skeletal remains from the pre-metal, bronze, and iron ages Northeastern Thailand. Ph.D. thesis, University of Hawaii. [Ann Arbor, MI: University Microfilms.]

Glover I. C. and Higham C. F. W. 1996. New evidence for early rice cultivation in South, Southeast, and East Asia. In Harris D. R., ed., *The Origins and Spread of Agriculture and Pastoralism in Eurasia*. Washington, DC: Smithsonian Institution Press, pp. 413–441.

Hanihara T. 1993. Population prehistory of East Asia and the Pacific as viewed from craniofacial morphology: the basic populations in East Asia, VII. *American Journal of Physical Anthropology* **91**: 173–187.

1994. Craniofacial continuity and discontinuity of far easterners in the late Pleistocene and Holocene. *Journal of Human Evolution* **27**: 417–441.

1996. Comparison of craniofacial features of major human groups. *American Journal of Physical Anthropology* **99**: 389–412.

Higham C. F. W. 1996. *The Bronze Age of Southeast Asia*. Cambridge, UK: Cambridge University Press.

2001. Prehistory, language and human biology: is there a consensus in East and Southeast Asia? In Jin L., Seielstad M. and Xiao C., eds., *Genetic, Linguistic and Archaeological Perspectives on Human Diversity in Southeast Asia*. Singapore: World Scientific, pp. 3–16.

Higham C. F. W. and Bannanurag R. 1990. *Research Report* XLVII. *The Excavation of Khok Phanom Di, A Prehistoric Site in Central Thailand*, Vol. 1: *The Excavation, Chronology and Human Burials*. London: Society of Antiquaries of London.

Higham C. F. W. and Thosarat R. 1998. *Prehistory of Thailand*. Bangkok: River Books.

Howells W. W. 1973. *Papers of the Peabody Museum of Archaeology and Ethnology* Vol. 67, *Cranial Variation in Man*. Cambridge, MA: Harvard University Press.

Ishida H. and Dodo Y. 1993. Non-metric cranial variation and the populational affinities of the Pacific peoples. *American Journal of Physical Anthropology* **90**: 49–57.

Ishida H. and Kondo O. 1999. Non-metric cranial variation of the Ainu and neighbouring human populations. *Perspectives in Human Biology* **4**: 127–138.

Jin L., Seielstad M. and Xiao C. (eds.). 2001. *Genetic, Linguistic and Archaeological Perspectives on Human Diversity in Southeast Asia*. Singapore: World Scientific.

Li C. 1977. *Anyang*. Seattle, WA: University of Washington Press.

Mahalanobis P. C. 1936. On the generalized distance in statistics. *Proceedings of the National Institute of Sciences Calcutta* **2**: 49–55.

Mansuy H. and Colani M. 1925. Contribution à l'étude de la préhistoire de l'Indochine. VII. Néolithique inférieur (Bac-Sonien) et néolithique supérieur dans le Haut-Tonkin (dernières recherches) avec la description des crânes du gisement de Lang-Cuom. *Mémoires du Service Géologique de'Indochine* **12**: 1–45.

Martin R. and Saller K. 1957. *Lehrbuch der Anthropologie*. Stuttgart: Gustav Fischer Verlag.

Matsumura H. 1995. Dental characteristics affinities of the prehistoric to modern Japanese with the East Asians, American natives and Australo-Melanesians. *Anthropological Science* **103**: 235–261.

Matsumura H., Cuong N. L., Thuy N. K. and Anezaki T. Y. 2001. Dental morphology of the early Hoabinian, the Neolithic Da But and the Metal Age Dong Son civilized peoples in Vietnam. *Zeitschrift für Morphologie und Anthropologie* **83**: 59–73.

Omoto K. and Saitou N. 1997. Genetic origins of the Japanese: a partial support of the dual structure hypothesis. *American Journal of Physical Anthropology* **102**: 437–446.

Parker K. 1998. The early Neolithic cemetery at Con Co Ngua, Northern Vietnam: non-metric cranial variation and relationships. B.A. (Hons.) thesis, Department of Anthropology, Northern Territory University, Darwin.

Pietrusewsky M. 1974. Neolithic populations of Southeast Asia studied by multivariate craniometric analysis. *Homo* **25**: 207–230.

1981. Cranial variation in early metal age Thailand and Southeast Asia studied by multivariate procedures. *Homo* **32**: 1–26.

1988. Multivariate comparisons of recently excavated Neolithic human crania from Thanh Hoa Province, Socialist Republic of Vietnam. *International Journal of Anthropology* **3**: 267–283.

1990. Craniofacial variation in Australasian and Pacific populations. *American Journal of Physical Anthropology* **82**: 319–340.

1994. Pacific–Asian relationships: a physical anthropological perspective. *Oceanic Linguistics* **33**: 407–430.

1997. The people of Ban Chiang: an early bronze-age site in Northeast Thailand. *Bulletin of the Indo-Pacific Prehistory Association* **16**: 119–148.

1999. A multivariate craniometric investigation of the inhabitants of the Ryukyu Islands and comparisons with cranial series from Japan, Asia, and the Pacific. *Anthropological Science* **107**: 255–281.

2000. Metric analysis of skeletal remains: methods and applications. In Katzenberg M. A. and Saunders S. R., eds., *Biological Anthropology of the Human Skeleton*. New York: Wiley-Liss, pp. 375–415.

2005. The physical anthropology of the Pacific, East Asia, and Southeast Asia: a multivariate craniometric analysis. In Sagart L., Blench R. and Sanchez-Mazas A., eds., *The Peopling of East Asia: Putting Together Archaeology, Linguistics, and Genetics*. London: Routledge Curzon, pp. 203–231.

Pietrusewsky M. and Douglas M. T. 2002. *University Monograph 111: Ban Chiang, A Prehistoric Site in Northeast Thailand. I: The Human Skeletal Remains*. Philadelphia, PA: University of Pennsylvania Museum of Archaeology and Anthropology.

Pietrusewsky M. and Ikehara-Quebral R. 2001. Multivariate comparisons of Rapa Nui (Easter Island), Polynesian, and circum-Polynesian crania. In

Stevenson C. M., Lee G. and Morin F. J., eds., *Pacific 2000. Proceedings of the Fifth International Conference on Easter Island and the Pacific, Los Osos.* California: Easter Island Foundation, pp. 457–494.

Rao R. C. 1952. *Advanced Statistical Methods in Biomedical Research*. New York: Wiley.

Rayner D. R. T. and Bulbeck D. 2001. Dental morphology of the 'Orang Asli' aborigines of the Malay peninsula. In Henneberg M., ed., *Causes and Effects of Human Variation*. Adelaide: Australasian Society for Human Biology, pp. 19–41.

Rohlf F. J. 1993. *NTSYS-pc. Numerical Taxonomy and Multivariate Analysis System*, version 1.80. Setauket, NY: Exeter Software.

Sjøvold T. 1984. A report on the heritability of some cranial measurements and non-metric traits. In van Vark G. N. and Howells W. W., eds., *Multivariate Statistics in Physical Anthropology*. Dordrecht: Reidel, pp. 223–246.

Sneath P. H. A. and Sokal R. R. 1973. *Numerical Taxonomy*. San Francisco, CA: Freeman.

Su B., Xiao J. H., Underhill P. *et al.* 1999. Y-chromosome evidence for a northward migration of modern humans into East Asia during the last ice age. *American Journal of Human Genetics* **65**: 1718–1724.

Tatsuoka M. M. 1970. *Discriminant Analysis*. Champaign, IL: Illinois Institute of Personality and Ability Testing.

Tayles N. G. 1999. *Report of the Research Committee LXI. The Excavation of Khok Phanom Di, A Prehistoric Site in central Thailand*, Vol. V: *The People*. London: Society of Antiquaries.

Thosarat R. 2001. Before the Neolithic: hunter–gatherer societies in central Thailand. In Jin L., Seielstad M. and Xiao C., eds., *Genetic, Linguistic and Archaeological Perspectives on Human Diversity in Southeast Asia*. Singapore: World Scientific, pp. 35–39.

Turner C. G. II. 1987. Late Pleistocene and Holocene population history of East Asia based on dental variation. *American Journal of Physical Anthropology* **73**: 305–321.

1990. Major features of sundadonty and sinodonty including suggestions about East Asian microevolution, population history, and late Pleistocene relationships with Australian Aborigines. *American Journal of Physical Anthropology* **82**: 295–317.

van Vark G. N. and Howells W. W. (eds.). 1984. *Multivariate Statistics in Physical Anthropology*. Dordrecht: Reidel.

van Vark G. N. and Schaafsma M. 1992. Advances in the quantitative analysis of skeletal morphology. In Saunders S. R. and Katzenberg M. A., eds., *Skeletal Biology of Past Peoples: Research Methods*, New York: Wiley-Liss, pp. 225–257.

White J. 1986. A revision of the chronology of Ban Chiang and its implications for the prehistory of Northeast Thailand. Ph.D. thesis, University of Pennsylvania, Philadelphia, PA.

Wilkinson L. 1992. *Systat for Windows*, version 5. Evanston, IL: Systat.

4 *Interpretation of craniofacial variation and diversification of East and Southeast Asians*

TSUNEHIKO HANIHARA
Saga Medical School, Saga, Japan

Introduction

The origins and affinities of Southeast Asians have been much debated over the years. The classic view held that the indigenous inhabitants of Southeast Asia were of Australian and Melanesian lineages. Then a wave of migrants from somewhere in southern China displaced and absorbed the majority of the indigenous population at some point between 4,000 and 5,000 years BP (Brothwell 1960, Howells 1976, 1977, Birdsell 1977, Bowles 1977, Glinka 1981, Bellwood 1985, 1987). In stark contrast to the population replacement model described above, other researchers espouse a local evolution model in which there was no significant gene flow into Southeast Asia from populations of East Asia (Bulbeck 1981, 1982, Omoto 1984, Turner 1987, 1990, 1992, Pietrusewsky 1988, Pietrusewsky and Chang 2003). Recently, Lahr (1996) and Underhill *et al.* (2001) have suggested that the colonisation of Southeast Asia by modern humans may have taken place through multiple dispersals. The first may have been an eastward movement of coastal peoples from East Africa to Arabia via the Horn, along the Indian Ocean rim towards Southeast Asia and finally into Australia. The second phase of modern occupation of Southeast Asia might have resulted from an expansion of the range of modern humans in North Africa and West Asia (Lahr 1996).

Turner (1987, 1990, 1992), Ballinger *et al.* (1992) and Chu *et al.* (1998) proposed that the human occupation of East and Northeast Asia resulted from the expansion of late Pleistocene Southeast Asians. The East and Northeast Asians would have subsequently differentiated from early sundadont populations, evolving specialised cranial features and the sinodont dental pattern. Turner's (1990) main argument for a Southeast Asian origin of East and Northeast Asians was based on the facts that the

Bioarchaeology of Southeast Asia. Marc Oxenham and Nancy Tayles.
Published by Cambridge University Press.

sinodont dental pattern shows more derived features and that the sundadont dental pattern can be found in the range of sinodont populations such as the prehistoric Jomon in Japan, and to certain extent in the modern Ainu. By comparison, Underhill *et al.* (2001) proposed that ancestors of Northeast Asians may be traced back to a lineage from the Levant moving towards Eurasia and reaching the northern part of East Asia through southern Siberia. The relationship between East and Northeast Asians and western Eurasians is also indicated by genetic studies (Cavalli-Sforza *et al.* 1988, 1994). Several lines of genetic evidence suggest a closer association between indigenous inhabitants of Japan, Jomon and their possible descendants, the Ainu, with northeastern Asian continental populations (Nei 1995, Omoto 1995, Omoto and Saitou 1997). Recently, moreover, morphological similarities between Ainu and Palaeoindians of the New World have been suggested (Brace *et al.* 2001, Jantz and Owsley 2003).

A conventional framework for understanding the affinities of East/Southeast Asians and Pacific populations was provided by Pietrusewsky's (1984, 1988, 1990, 1997, 1999, Pietrusewsky and Ikehara-Quebral 2001) extensive work on craniofacial morphology, suggesting which lineages, and of what origin, made the prehistoric and present patterns of variation within East/Southeast Asia and the Pacific. Lahr (1995) emphasised that analyses based on a greater temporal and geographical scale are essential for understanding the differentiation of local groups. However, with the notable exception of the cranial studies of Howells (1973, 1989, 1995), Brace (Brace and Hunt 1990; Brace *et al.* 2001) and Lahr (1995, 1996; Lahr and Wright 1996) and the dental studies of Scott and Turner (1997), research into morphological variation and diversification of East/Southeast Asians has seldom been addressed in a broader context, including western neighbours as well as Arctic and New World populations.

The study described here examined the morphological relationship between East/Southeast Asians and neighbouring populations using 103 population groups, including eastern neighbours of Southeast Asia (e.g. Australians and Melanesians), Northeast Asian-related groups (e.g. Arctic and New World populations), South Asians and samples from the Middle East.

Materials and methods

Quantitative morphological traits in the form of craniometric data were used in this study. Sample names, sizes and related information are provided in Appendix 4.1 (p. 107). All samples were composed of non-deformed

Table 4.1. *Craniometric variables used*[a]

General measurements
Maximum cranial length (H-GOL)
Nasion-opisthocranion (H-NOL)
Cranial base length (H-BNL)
Maximum cranial breadth (H-XCB)
Minimum frontal breadth (M9)
Maximum frontal breadth (H-XFB)
Biauricular breadth (H-AUB)
Maximum occipital breadth (H-ASB)
Basion–bregma height (H-BBH)
Sagittal frontal arc (M26)
Sagittal parietal arc (M27)
Sagittal occipital arc (M28)
Sagittal frontal chord (H-FRC)
Sagittal parietal chord (H-PAC)
Sagittal occipital chord (H-OCC)
Facial length (H-BPL)
Upper facial breadth (M43)
Bizygomatic breadth (H-ZYB)
Middle facial breadth (M46)
Upper facial height (H-NPH)
Interorbital breadth (H-DKB)
Orbital breadth (M51)
Orbital height (H-OBH)
Nasal breadth (H-NLB)
Nasal height (H-NLH)
Maxilloalveolar breadth (H-MAB)
Mastoid height (H-MDH)
Mastoid width (H-MDB)
Facial flatness measurements[b]
Frontal chord between the frontomalaria orbitalia (M43(1))
Subtense of the nasion from the frontal chord (No. 43c)
Minimum horizontal breadth of the nasalia (H-WNB)
Simotic subtense (H-SIS)
Zygomaxillary chord between the zygomaxillaria anteriora (H-ZMB)
Zygomaxillary subtense (H-SSS)

[a]The sources for the variables were: M, Martin and Saller (1957); H, Howells (1973); No., Bräuer (1988).

[b]Facial flatness measurements described by Yamaguchi (1973) and Hanihara *et al.* (1999).

adult crania. For each specimen, 34 standard craniofacial measurements were recorded, including six facial flatness measurements defined by Howells (1973), Martin and Saller (1957) and Bräuer (1988) (Table 4.1). With the exception of the prehistoric archaeological samples, complete or substantially complete male crania were used as male sample sizes were larger than those of females in many cases. I recorded all of the data myself and details of the samples used are provided elsewhere (Hanihara 2000, Hanihara and Ishida 2001, Hanihara et al. 1998, 2003).

Biological distance analysis based on *C*-scores was employed to represent the craniofacial affinities of the samples. Raw measurement data were converted into Z-scores, which in turn were used to generate *C*-scores as defined by Howells (1989). First, *C*-scores were calculated for all individuals using the mean of the sample means compared, to avoid weighting by different sample numbers, and standard deviations based on pooled variance of the same samples. Second, Mahalanobis' generalised distances were calculated from the *C*-score dataset. After this two-step procedure, the effects of both cranial size and intercorrelation of variables could be removed from the final distance matrix.

To estimate the relative degree of variation in each regional group, part of the **R**-matrix methods (Relethford and Blangero 1990, Relethford 1991, 1994, 1996, Relethford and Harpending 1994, 1995) were applied. Essentially the **R**-matrix is given as

$$\mathbf{R} = \mathbf{C}(1 - F_{ST})/2t$$

F_{ST} is defined as

$$F_{ST} = (\sum w_i C_{ii})/(2t + \sum w_i C_{ii}),$$

where t is the number of traits and w_i is the weighting factors, the relative size of population i, defined as

$$w_i = N_i / \sum N_j.$$

N_j is the effective size of population j. C_{ii} is the diagonal element of codivergence matrix (**C**), computed as

$$\mathbf{C} = \Delta \mathbf{G}^{-1} \Delta',$$

where $\mathbf{G}^{-1}$ is the inverse of the pooled within-group additive genetic variance–covariance matrix, and the elements of Δ are the deviations of group means from the total means averaged over all populations weighted by population size. Calculation of **G** from the phenotypic variance–covariance matrix (**V**) requires an estimate of average heritability for phenotypic traits. According to Relethford and Harpending (1994), for any given trait, $\mathbf{G} = h^2\mathbf{V}$, where h^2 is the heritability of the trait. An average estimate of $h^2 = 0.55$ obtained from Devor (1987) was used.

The expected average phenotypic variation in population *i* is $E[\bar{\nu}_i]$ and is defined as

$$E[\bar{\nu}_i] = \bar{\nu}_w(1 - r_{ii})/(1 - F_{ST}),$$

where $\bar{\nu}_w$ is the pooled average within-group phenotypic variation over all populations. The residual variance ($R[E_i]$) is defined as

$$R[E_i] = \bar{\nu}_w - E[\bar{\nu}_i].$$

The standard errors of the residual variance were approximated by the jackknife method (Miller 1974). The residual variance divided by its standard error is distributed approximately as a *t*-statistic with $t-1$ degrees of freedom. Using this approximation, significance testing of residual variance was performed.

Results

The first two principal coordinates of the Mahalanobis' generalised distance matrix based on the *C*-scores of 44 East/Southeast Asian and Australian/Melanesian samples are shown in Fig. 4.1. A clear separation among Australian/Melanesian samples, East/Southeast Asian samples, and Northeast Asian samples is evident. The internal organisation of the Southeast Asian branch suggests the separation of the early Southeast Asian, Nicobar and Andaman samples. The early Southeast Asian sample is situated between the predominantly Southeast Asian samples and the Australian/Melanesian cluster. The typical East Asian samples, such as Chinese, Koreans and Japanese, form a satellite removed from the Southeast Asian samples. The Ainu and Jomon samples are plotted separately. The Tibetan sample is included in the Southeast Asian branch, while the Nepalese sample is closest to the Nicobar sample.

Figure 4.2 shows the two dimensional scattergram resulting from principal coordinate analysis of the *d*-squared results based on the *C*-score datasets in which the series from the western neighbours of Southeast Asia, the Indian subcontinent and Middle East, are added. The intergroup relationships of the East/Southeast Asian and Australian/Melanesian samples shown in Fig. 4.1 are retained. The early Southeast Asian sample is situated at an intermediate position between Southeast Asian and Australian/Melanesian samples. The Nicobar sample is plotted at the centre of the Southeast Asian, Australian/Melanesian and South Asian triangle. The Andaman sample may be identified as the most distinctive sample in Southeast Asia, showing loose affinity to Middle Eastern

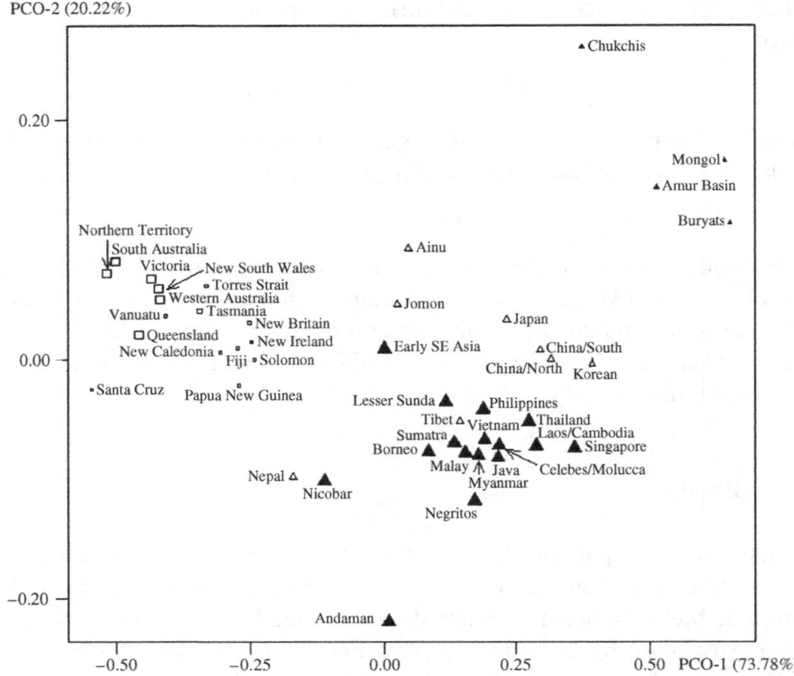

Figure 4.1. The first and second principal coordinates from the Mahalanobis' generalised distance matrix obtained using *C*-scores. The samples included are from Southeast Asia, East Asia, Northeast Asia, Australia and Melanesia. Using the first two dimensions, 94.0% of the total variance is expressed.

samples. The Jomon and Ainu samples show more or less different morphological patterns from other East/Southeast Asian samples, suggesting their distinctiveness.

Figure 4.3 is a plot of 85 Asian and New World samples based on the principal coordinate analysis applied to Mahalanobis' generalised distance resulting from *C*-score datasets. The Ainu sample aligns as a peripheral member of the Native American series and shows close similarity to the northeastern American samples. The Jomon sample occupies a position between the early Southeast Asian and Native American series. The Northeast Asian series form a separate cluster, showing closer affinities to the Kodiak and eastern Aleut series than to East Asian series. Except for the Fuegian/Patagonian sample, the series from Central and South America are closest to East/Southeast Asian samples.

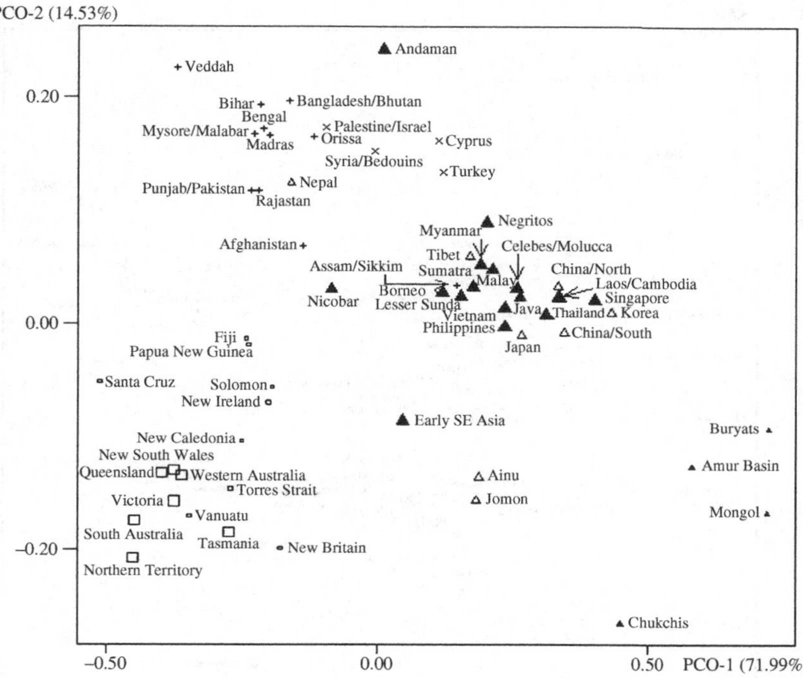

Figure 4.2. The first and second principal coordinates from the Mahalanobis' generalised distance matrix obtained using *C*-scores. In addition to the samples used in Fig. 4.1, the South and West Asian samples are included. Using the first two dimensions, 86.5% of the total variance is expressed.

Figure 4.4 shows what happens when the series from both South/West Asian and Arctic/New World samples are included in the analyses. The intergroup relationships shown in Figs. 4.1, 4.2 and 4.3 are retained. The Southeast Asian series are positioned close to the Central/South American series on the one hand, and the Middle East series on the other. The Nicobar sample is included in the South/West Asian cluster. The Andaman sample is the most differentiated of the Southeast Asian groups.

Table 4.2 gives the results of Relethford and Blangero's (1990) method of estimating intraregional variation under the assumption of equal effective population size. In this analysis, the Tibetan and Nepalese samples and the Chukchi sample are excluded from the East Asian and Northeast Asian datasets, respectively. The Arctic and New World series are also excluded since they represent newly diverged populations (Relethford and Harpending 1994). The Northeast Asian, East Asian, Southeast Asian

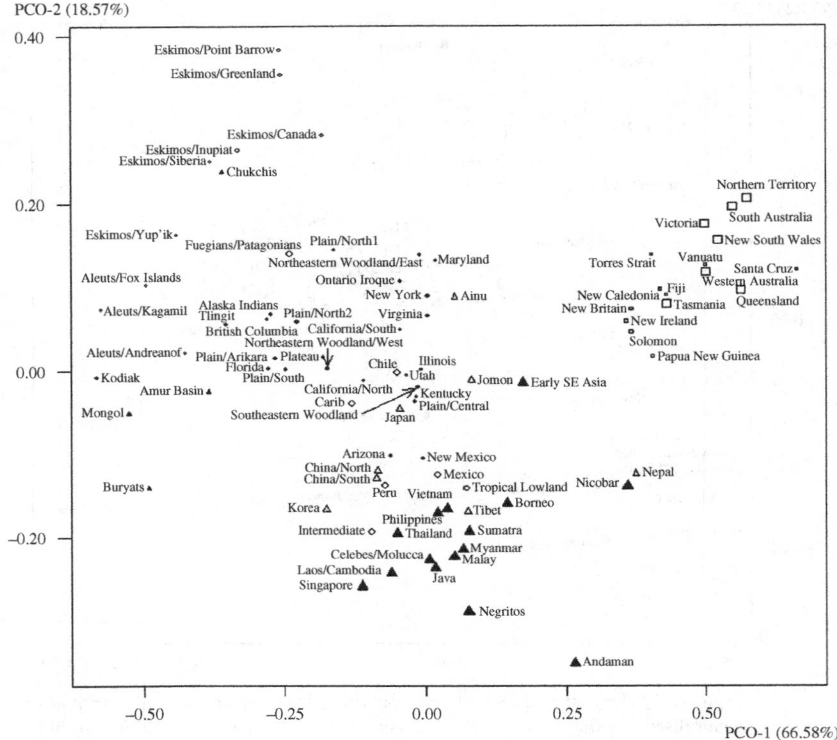

Figure 4.3. The first and second principal coordinates from the Mahalanobis' generalised distance matrix obtained using *C*-scores. In addition to the samples used in Fig. 4.1, the samples from New World are included. Using the first two dimensions, 85.2% of the total variance is expressed.

and Middle Eastern series show relatively large intraregional variance, followed by the Indian subcontinent and Australian series. Among these series, three Asian series and the Australian series show positive residual variances. According to Relethford and Harpending (1994), significance testing of the difference between observed and expected variance can be performed by testing the null hypothesis that the residual variance in a given region is equal to zero. The results of the test using the jackknife method are reported in Table 4.2. The three Asian and the Australian series show significantly greater average phenotypic variation. By comparison, the Jomon/Ainu, Andaman and Nicobar series show less intraregional variance, although the Andaman series show significant differences between observed and expected variance.

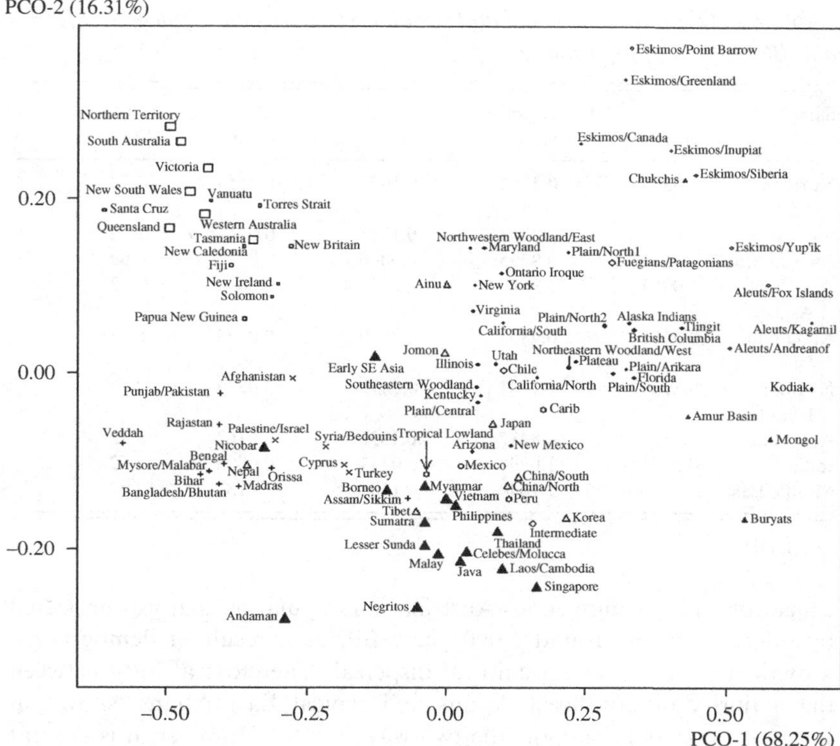

Figure 4.4. The first and second principal coordinates from the Mahalanobis' generalised distance matrix obtained using *C*-scores. A total of 103 samples are included. Using the first two dimensions, 84.4% of the total variance is expressed.

Discussion

Southeast Asia has been thought to be the population hearth of eastern Asia and Oceania. It is obvious that any dispersal into Australia had to go through Southeast Asia. The palaeolithic human fossil record in Southeast Asia from such sites as Niah Cave, Wajak and Tabon support this hypothesis (Brothwell 1960, Jacob 1975, Macintosh 1978, Kennedy 1979). This remote association of Australians and early Southeast Asians has been extended to contemporary Negritos of present day Southeast Asia (Coon 1962, Garn 1965, Jacob 1967, Brues 1977, Birdsell 1977). Bellwood (1978, 1985, 1987) and many others offered a two-wave or dual structure model for the formation of contemporary Southeast Asian physical features that

Table 4.2. *Observed, expected and residual variances based on Relethford and Blangero's (1990) method*

Sample	Observed variance	Expected variance	Residual	Standard error	t (jackknife) d.f. = 33
Northeast Asia	1.0376	0.5354	0.5022	0.0255	19.70*
East Asia	1.0569	0.9358	0.1211	0.0212	5.72*
Jomon/Ainu	0.9661	0.9595	0.0066	0.0736	0.09
Southeast Asia	1.0057	0.9840	0.0217	0.0058	3.73*
Andaman Islands	0.8463	0.4853	0.3610	0.0673	5.37*
Nicobar Islands	0.8959	1.1381	−0.2422	0.0889	2.73
Australia	0.9823	0.8487	0.1336	0.0162	8.22*
India	0.9851	1.0295	−0.0444	0.0096	4.62
Middle East	1.0143	1.1844	−0.1701	0.0137	12.42

*$p < 0.001$.

called for an early indigenous Australasian population that was impacted by migrants from around 5,000 years BP, as a result of demographic growth related to the agricultural dispersal. The close affinity between the majority of Southeast Asians and typical East Asians, shown in Fig. 4.1, appears to support the two-wave model. However, it is evident that the early Southeast Asian series from approximately 5,000 to 10,000 years BP are situated at an intermediate position between present-day Southeast Asians and Australians/Melanesians. The Philippine Negritos with phenotypic affinities with the Australian/Melanesian group are located on the periphery of the Southeast Asian cluster. Moreover, the Central and South Americans are plotted between Southeast Asians and East Asians. This indicates that, morphologically, Southeast Asians and East Asians are not necessarily the closest groups to each other. Such findings argue against the complete or nearly complete replacement hypothesis for the formation of present-day Southeast Asian physical characteristics.

Recent genetic and morphological studies indicate that an important factor behind the unique physical makeup of contemporary Southeast Asians may be gene flow from the west in the distant past (Nei and Roychoudhury 1993, Lahr 1996, Underhill *et al.* 2001). In this study, Andamanese and Nicobarese tend to cluster with South and West Asians. Many years ago, Howells (1959) pointed out a possible association between Southeast Asian Negritos and the Hadramaut of southern Arabia and India. The language of the inhabitants of the Andaman Islands is, moreover, not related to any language family of Southeast Asia (Greenberg 1971, Lahr 1996), indicating long-term isolation. This may allow us to suppose that the western-like features of Andamanese and Nicobarese are not necessarily attributed to recent gene flow from India. The affinities of Andamanese and Nicobarese with South/West Asians, together with a clinal variation from west to east as shown in Figs. 4.2 and 4.4, indicate that the characteristics of Southeast Asian craniofacial features change from the west of the Malay Peninsula, Andaman/Nicobar island chain and of the Myanmar-Assam/Sikkim region, Bangladesh/Bhutan. However, from the present materials, based mainly on recent samples, it is still far from obvious whether the results are consistent with recent genetic evidence suggesting multiple dispersals from the west to Southeast Asia (Lahr 1996, Underhill *et al.* 2001).

Recent metric and non-metric cranial morphological studies as well as genetic studies found no direct relationship between Ainu and populations in southern East Asia (Cavalli-Sforza *et al.* 1988, Nei 1995, Omoto 1995, Omoto and Saitou 1997, Pietrusewsky 1997, 1999, Ishida and Kondo 1999, Hanihara *et al.* 2003). It is a matter of interest that the Ainu show some association with the Native American groups. Recently, morphological affinities between the Ainu and Palaeoindians in the New World have been suggested (Brace *et al.* 2001). The present findings, however, do not necessarily support the Ainu–Palaeoindian connection because an ancestor–descendent relationship between early and late Holocene American series is still tentative (Steele and Powell 1992, Lahr 1995, Jantz and Owsley 2001, 2003, van Vark *et al.* 2003). However, if we accept the hypothesis that Native Americans are derived from prehistoric Northeast Asians with more generalised craniofacial features (Howells 1959, Lahr 1995), the morphological shift of Ainu to the Native Americans may throw light on the relationships between prehistoric Northeast Asia and Hokkaido.

However, prehistoric Jomonese shared craniofacial similarities with not only Ainu but also both early Southeast Asians and some Native Americans. These findings are in agreement with Turner's (1987, 1990, 1992) Sundaland origin hypothesis for the peopling of East/Northeast Asia and ultimately the New World. However, the present study suggests that intraregional variation within East and Northeast Asia is relatively large, comparable to that of Southeast Asia. Under Relethford's model, positive residual variances reflect (a) higher levels of long-range gene flow from an outside source, producing greater heterozygosity; (b) long-term population history or a larger effective population size; and (c) a higher mutation or drift rate than other groups (Relethford and Blangero 1990, Relethford and Harpending 1994, 1995, Powell and Neves 1999). Under the assumption of the first hypothesis, we can suppose multiple migrations to the East/Northeast Asian region. This has been supported by a genetic study (Li and Bing 2000) suggesting that the entry of anatomically modern humans into the southern part of East Asia occurred in the late Pleistocene, followed by a northward migration. There might have been a contribution from central Asians who arrived in East Asia at a later time (Li and Bing 2000, Uinuk-Ool *et al.* 2003). If we accept the second hypothesis, the present findings may support genetic evidence suggesting that the East/Northeast Asians and Southeast Asians are two separate phylogenetic units (Cavalli-Sforza *et al.* 1988, Omoto 1995, Underhill et al. 2001). The third hypothesis is also an undeniable possibility, because Northeast Asian physical characteristics are strongly influenced by the harsh environment.

Regardless of which hypothesis is correct, the present findings, such as Jomon–early Southeast Asian connection on the one hand, and Jomon–Ainu–Native American connection on the other hand, suggest that (a) the Jomon/Ainu group retained some morphological characteristics attributed to those of the upper palaeolithic inhabitants of the eastern Asian region, although their origin still remains to be resolved; and (b) the morphological diversity of the early prehistoric populations distributed in the relatively vast region from Southeast Asia to the northern part of the Japanese archipelago might not have been so large.

Acknowledgements

I am deeply grateful to M. F. Oxenham and N. Tayles for inviting me as a contributor to this book. I wish to express my sincere thanks to the following for the kind permissions to study materials under their care: T. Molleson, R. Kruszynski, L. T. Humphrey and C. Stringer of the Natural History Museum, London; R. Foley, M. M. Lahr, and M. Bellatti of the Department of Biological Anthropology, University of Cambridge; A. Langaney and M. A. Pereira da Silva of Laboratoire d'Anthropologie biologieque, Musée de l'Homme, Paris; D. Hunt, D. Owsley, S. Ousley, R. Potts, M. London and D. H. Ubelaker of the Department of Anthropology, National Museum of Natural History, Smithsonian Institution, Washington DC; I. Tattersall, K. Mowbray and G. Sawyer of the Department of Anthropology, American Museum of Natural History, New York; J. Specht, P. Gordon, L. Bonshek and

N. Goodsell of the Department of Anthropology, Australian Museum, Sydney; J. Stone and D. Donlon of the Department of Anatomy and Histology, University of Sydney; D. Henley of the New South Wales Aboriginal Land Council, Sydney; M. Chow, a dentist in Sydney; M. Hanihara of the School of Languages, Macquarie University, Sydney; C. Pardoe and G. L. Pretty of the Department of Anthropology, South Australian Museum, Adelaide; Y. Dodo of the Department of Anatomy and Anthropology, Tohoku University; and G. Murakami of the Department of Anatomy, Sapporo Medical University.

This study was supported in part by Grant-in-Aid for Scientific Research (Nos. 1440521 and 14540659) from the Ministry of Education, Science and Culture in Japan; a Japan Fellowship for Research in United Kingdom from the Japan Society for the Promotion of Science; and Smithsonian Opportunities for Research and Study: Smithsonian Institution Fellowship Program (Senior Fellow in 2001–2002).

References

Ballinger S. W., Schurr T. G., Torroni A. *et al.* 1992. Southeast Asian mitochondrial DNA analysis reveals genetic continuity of ancient Mongoloid migration. *Genetics*, **130**: 139–152.

Bellwood P. 1978. *Man's Conquest of the Pacific: the Prehistory of Southeast Asia and Oceania.* New York: Oxford University Press.

1985. *Prehistory of the Indo-Malaysian Archipelago.* Sydney: Academic Press.

1987. The prehistory of island Southeast Asia: a multidisciplinary review of recent research. *Journal of World Prehistory* **1**: 171–224.

Birdsell J. B. 1977. The recalibration of paradigm for the first peopling of greater Australia. In Allen J., Golson J. and Jones R., eds., *Sunda and Sahul: Prehistoric Studies in Southeast Asia, Melanesia, and Australia.* London: Academic Press, pp. 113–167.

Bowles G. T. 1977. *The People of Asia.* New York: Charles Scribner.

Brace C. L. and Hunt K. D. 1990. A nonracial craniofacial perspective on human variation A(ustralia) to Z(uni). *American Journal of Physical Anthropology*, **82**: 341–360.

Brace C. L., Nelson A. R., Seguchi N. *et al.* 2001. Old World sources of the first New World Human inhabitants: a comparative craniofacial view. *Proceedings of the National Academy of Sciences of the USA*, **98**: 10017–10022.

Bräuer G. 1988. Osteometrie: a. Kraniometrie. In Knußmann R., ed., *Anthropologie: Handbuch der Vergleichenden Biologie des Menschen*, Vol. I. Stuttgart: Gustav Fischer, pp. 160–193.

Brothwell D. R. 1960. Upper Pleistocene human skull from Niah Caves, Sarawak. *Sarawak Museum Journal*, **9**: 323–349.

Brues A. M. 1977. *The Macmillan Series in Physical* Anthropology: *People and Race*. New York: Macmillan.

Bulbeck D. 1981. *Continuities in Southeast Asian evolution since the late Pleistocene, some new material described and some old questions reviewed.* M.A. thesis, Australian National University, Canberra.

1982. A re-evaluation of possible evolutionary processes in Southeast Asia since the late Pleistocene. *Bulletin of the Indo-Pacific Prehistory Association* **3**: 1–21.

Cavalli-Sforza L. L., Piazza A., Menozzi P. and Mountain J. 1988. Reconstruction of human evolution: bringing together genetic, archaeological, and linguistic data. *Proceedings of the National Academy of Sciences of the USA* **85**: 6002–6006.

Cavalli-Sforza L. L., Menozzi P. and Piazza A. 1994. *The History and Geography of Human Genes*. Princeton, NJ: Princeton University Press.

Chu J. Y., Huang W., Kuang S. Q. *et al.* 1998. Genetic relationship of populations in China. *Proceedings of the National Academy of Sciences of the USA* **95**: 11763–11768.

Coon C. S. 1962. *The Origin of Races*. New York: Alfred A. Knopf.

Devor E. J. 1987. Transmission of human craniofacial dimensions. *Journal of Craniofacial, Genetic, and Developmental Biology* **7**: 95–106.

Garn S. M. 1965. *Human Races*. Springfield, IL: Charles C. Thomas.

Glinka J. 1981. Racial history of Indonesia. In Schwidetzky I, ed., *Rassengeschichte der menschheit*. Munich: R. Lodenbourg, pp. 97–113.

Greenberg J. H. 1971. The Indo-Pacific hypothesis. [In Sebeok T. A., ed., *Linguistics in Oceania*.] *Current Trends in Linguistics* **8**: 807–871.

Hanihara T. 2000. Frontal and facial flatness of major human populations. *American Journal of Physical Anthropology* **111**: 105–134.

Hanihara T. and Ishida H. 2001. Os incae: variation in frequency in major human population groups. *Journal of Anatomy* **198**: 137–152.

Hanihara T., Ishida H. and Dodo Y. 1998. Os zygomaticum bipartitum: frequency distribution in major human populations. *Journal of Anatomy* **192**: 539–555.

Hanihara T., Dodo Y., Kondo O. *et al.* 1999. Intra- and interobserver errors in facial flatness measurements. *Anthropological Science* **107**: 25–39.

Hanihara T., Ishida H. and Dodo Y. 2003. Characterization of biological diversity through analysis of discrete cranial traits. *American Journal of Physical Anthropology* **121**: 241–251.

Howells W. W. 1959. *Mankind in the Making*. New York: Doubleday.

1973. *Papers of the Peabody Museum of Archaeology and Ethnology*, Vol. 67, *Cranial Variation in Man*. Cambridge, MA: Harvard University Press.

1976. Physical variation and history in Melanesia and Australia. *American Journal of Physical Anthropology* **45**: 641–650.

1977. The sources of human variation in Melanesia and Australia. In Allen J., Golson J. and Jones R., eds., *Sunda and Sahul: Prehistoric Studies in Southeast Asia, Melanesia, and Australia*. London: Academic Press, pp. 169–186.

1989. *Papers of the Peabody Museum of Archaeology and Ethnology*, Vol. 79, *Skull Shapes and the Map: Craniometric Analyses in the Dispersion of Modern* Homo. Cambridge, MA: Harvard University Press.

1995. *Papers of the Peabody Museum of Archaeology and Ethnology*, Vol. 82, *Who's Who in Skulls: Ethnic Identification of Crania from Measurements* Cambridge, MA: Harvard University Press.

Ishida H. and Kondo O. 1999. Nonmetric cranial variation of the Ainu and neighbouring human populations. *Perspectives in Human Biology* **4**: 127–138.

Jacob T. 1967. Racial identification of bronze age human dentitions from Bali, Indonesia. *Journal of Dental Research* **46**: 903–910.

1975. Morphology and palaeoecology of early man in Java. In Tuttle R. H., ed., *Palaeoanthropology: Morphology and Palaeoecology*. The Hague: Mouton, pp. 311–324.

Jantz R. L. and Owsley D. W. 2001. Variation among early North American crania. *American Journal of Physical Anthropology* **114**: 146–155.

2003. Reply to van Vark *et al.*: is European upper Paleolithic cranial morphology a useful analogy for early Americans? *American Journal of Physical Anthropology* **121**: 185–188.

Kennedy K. A. R. 1979. The deep skull of Niah: an assessment of twenty years of speculation concerning its evolutionary significance. *Asian Perspectives* **20**: 32–50.

Lahr M. M. 1995. Patterns of modern human diversification: implications for Amerindian origins. *Yearbook of Physical Anthropology* **38**: 163–198.

1996. *The Evolution of Modern Human Diversity: a Study of Cranial Variation.* Cambridge, UK: Cambridge University Press.

Lahr M. M. and Wright R. V. S. 1996. The question of robusticity and the relationships between cranial size and shape in *Homo sapiens*. *Journal of Human Evolution* **31**: 157–191.

Li J. and Bing S. 2000. Natives or immigrants: modern human origin in East Asia. *Nature Genetics* **1**: 126–133.

Macintosh N. W. G. 1978. The Tabon cave mandible. *Archaeology and Physical Anthropology in Oceania* **13**: 143–159.

Martin R. and Saller K. 1957. *Lehrbuch der Anthropologie*. Stuttgart: Fischer Verlag.

Miller R. G. 1974. The jackknife: a review. *Biometrika* **61**: 1–15.

Nei M. 1995. The origins of human populations: genetic, linguistic, and archeological data. In Brenner S. and Hanihara K., eds., *The Origin and Past of Modern Humans as Viewed from DNA*. Singapore: World Scientific, pp. 71–91.

Nei M. and Roychoudhury A. K. 1993. Evolutionary relationships of human populations on a global scale. *Molecular Biology and Evolution* **10**: 927–943.

Omoto K. 1984. The Negritos: genetic origins and microevolution. *Acta Anthropogenetica* **8**: 137–147.

1995. Genetic diversity and the origins of the 'Mongoloids'. In Brenner S. and Hanihara K., eds., *The Origin and Past of Modern Humans as Viewed from DNA*. Singapore: World Scientific, pp. 92–109.

Omoto K. and Saitou N. 1997. Genetic origins of the Japanese: a partial support for the dual structure hypothesis. *American Journal of Physical Anthropology* **102**: 437–446.

Pietrusewsky M. 1984. Metric and non-metric cranial variation in Australian Aboriginal populations compared with populations from the Pacific and Asia. *Occasional Papers in Human Biology* **3**: 1–113.

1988. Multivariate comparisons of recently excavated Neolithic human crania from the Socialist Republic of Vietnam. *International Journal of Anthropology* **3**: 267–283.

1990. Craniofacial variation in Australasian and Pacific populations. *American Journal of Physical Anthropology* **82**: 319–340.

1997. The people of Ban Chiang: an early Bronze site in Northeast Thailand. *Bulletin of the Indo-Pacific Prehistory Association* **16**: 119–148.

1999. A multivariate craniometric study of the inhabitants of the Ryukyu Islands and comparisons with cranial series from Japan, Asia, and the Pacific. *Anthropological Science* **107**: 255–281.

Pietrusewsky M. and Chang C.-F. 2003. Taiwan Aboriginals and peoples of the Pacific–Asia region: multivariate craniometric comparisons. *Anthropological Science* **111**: 293–332.

Pietrusewsky M. and Ikehara-Quebral R. 2001. Multivariate comparisons of Rapa Nui (Easter Island), Polynesian and circum-Polynesian crania. In Stevenson C. M., Lee G. and Morin J. F., eds., *Proceedings of the Fifth International Conference on Easter Island and the Pacific*. Los Osos, CA: Easter Island Foundation, pp. 457–494.

Powell J. F. and Neves W. A. 1999. Craniofacial morphology of the first Americans: pattern and process in the peopling of the New World. *Yearbook of Physical Anthropology* **42**: 153–188.

Relethford J. H. 1991. Genetic drift and anthropometric variation in Ireland. *Human Biology* **63**: 155–165.

1994. Craniometric variation among modern human populations. *American Journal of Physical Anthropology* **95**: 53–62.

1996. Genetic drift can obscure population history: problem and solution. *Human Biology* **68**: 29–44.

Relethford J. H. and Blangero J. 1990. Detection of differential gene flow from patterns of quantitative variation. *Human Biology* **62**: 5–25.

Relethford J. H. and Harpending H. C. 1994. Craniometric variation, genetic theory, and modern human origins. *American Journal of Physical Anthropology* **95**: 249–270.

1995. Ancient differences in population size can mimic a recent African origin of modern humans. *Current Anthropology* **36**: 667–674.

Scott G. R. and Turner C. G. II 1997. *The Anthropology of Modern Human Teeth: Dental Morphology and its Variation in Recent Human Populations*. Cambridge, UK: Cambridge University Press.

Steele D. G. and Powell J. F. 1992. Peopling of the Americas: palaeobiological evidence. *Human Biology* **64**: 303–336.

Turner C. G. II. 1987. Late Pleistocene and Holocene population history of East Asia based on dental variation. *American Journal of Physical Anthropology* **73**: 305–321.

1990. Major features of sundadonty and sinodonty, including suggestions about East Asian microevolution, population history and late Pleistocene relationships with Australian Aboriginals. *American Journal of Physical Anthropology* **82**: 295–317.

1992. The dental bridge between Australia and Asia: following Macintosh into the East Asian hearth of humanity. *Perspectives in Human Biology* **2**: 143–152.

Uinuk-Ool T. S., Takezaki N. and Klein J. 2003. Ancestry and kinships of native Siberian populations: the HLA evidence. *Evolutionary Anthropology* **12**: 231–245.

Underhill P. A., Passarino G., Lin A. A. *et al.* 2001. The phylogeography of Y-chromosome binary haplotypes and the origins of modern human populations. *Annals of Human Genetics* **65**: 43–62.

van Vark G. N., Kuizenga D. and Williams F. L. 2003. Kennewick and Luzia: lessons from the European upper Paleolithic. *American Journal of Physical Anthropology* **121**: 181–184.

Yamaguchi B. 1973. Facial flatness measurements of the Ainu and Japanese crania. *Bulletin of National Science Museum* **16**: 161–171.

Appendix

Appendix 4.1. *Names, sizes of samples and related information for the adult crania used in this study*

Sample name (sample size)	Details
Mainland Southeast Asia	
1. Laos/Cambodia (43)	Recent people from Laos and Cambodia
2. Thailand (47)	Bangkok, Thailand (recent)
3. Vietnam (24)	Recent Vietnamese
4. Burma (Myanmar) (103)	Recent Burmese (now called Myanmar)
5. Malay (53)	Recent inhabitants from Malay Peninsula
6. Singapore (43)	Recent Singaporeans
7. Early Southeast Asia (32)	Laos: Abri-sous-roche de Tam-Nang Anh/Neolithic; Vietnam: Lang-Cuom, Dong Thûoc, Keo Phay, Pho Binh Gia/Neolithic; Malaysia: mesolithic and neolithic Malay from Gua Cha site (*c.* 5,000–10,000 years BP)
Island Southeast Asia	
8. Andaman Islands (53)	From both Great and Little Andaman Islands
9. Nicobar Islands (21)	Recent native people from Nicobar Islands
10. Sumatra (27)	Recent inhabitants of Sumatra Island
11. Java (62)	Recent inhabitants of Java Island
12. Borneo (95)	Mainly the so-called Land Dayaks
13. Philippines (100)	Tagalog, Bisaya, Bilan, etc., Filipinos
14. Philippine Negritos (22)	Aeta, Agta and other tribes from Luzon Island

Appendix 4.1. (*cont.*)

Sample name (sample size)	Details
15. Lesser Sunda (15)	Timor, Sumbawa, Flores and Bali Islands
16. Celebes (19)	Recent Celebes Islanders
17. Molucca (22)	Several Islands of the Molucca chain
East Asia	
18. Northern China (113)	North of Cheng River, mainly from Liaoning Prefecture
19. Southern China (66)	South of Cheng River
20. Japan (150)	Tokyo and Tohoku (northern part of Honshu Island)
21. Korea (33)	Recent Koreans, mainly from South Korea
22. Tibet (58)	Recent Tibetans, mainly from Tibetan soldiers of the nineteenth century
23. Nepal (36)	Sunwar and other regions: low lander
24. Jomon (50)	Neolithic people in Japan, *c.* 5,300–2,300 years BP
25. Ainu (35)	Recent Ainu from Hokkaido
Northeast Asia	
26. Amur basin (15)	Ulchs, Negidals, Nivkhs and other tribes
27. Mongol (151)	Route au nord d'Ourga, Environs d'Ouliassoutai, Environs de Kobdo
28. Buryats (27)	Sables de l'Angara, Troiskasavsk, North Kiachta
29. Chukchis (21)	Arctic region of northeast Siberia
Australia	
30. Queensland (22)	Australians (recent) from Queensland
31. New South Wales (66)	Coastal region of New South Wales
32. Victoria (8)	Australians (recent) from Victoria
33. Northern Territory (12)	Arnhem Land, Northern Territory
34. Murray River (51)	Mainly from Roonka site, including a recent sample from Murray River basin
35. South Australia (143)	From near Adelaide
36. Western Australia (28)	Shark's Bay and other regions
37. Tasmania (16)	Recent Tasmanian Aborigines
Melanesia	
38. Papua New Guinea (152)	Sepik River basin, Fly River basin, Purari River delta, etc.
39. Torres Strait (61)	Thursday, Prince of Wales, Banks, Darnley, Hammond, Mer Islands
40. New Britain (74)	Gazelle, Ralum, Matupi, Branch Bay
41. New Ireland (40)	Recent inhabitants of New Ireland
42. Solomon (67)	New Georgia, Guadalcanal, San Cristobal Islands
43. Santa Cruz (16)	Recent inhabitants of Santa Cruz Islands
44. Vanuatu (28)	Recent inhabitants of Vanuatu
45. New Caledonia (42)	Recent inhabitants of New Caledonia
46. Fiji (46)	Rotumah, Viti Levu, Cikobia and other Islands

Appendix 4.1. (*cont.*)

Sample name (sample size)	Details
South Asia	
47. Assam/Sikkim (40)	Mishme, Naga, Thado, Kuki, Singho, Lepcha, Darjeeling, Assam and Sikkim Districts
48. Bangladesh/Bhutan (27)	Recent inhabitants of Bangladesh and Bhutan
49. Bengal (51)	West Bengal District
50. Bihar (21)	Patna and other regions of the Province of Bihar
51. Madras (100)	Native Indians from around Madras, mainly Dravidians
52. Malabar/Mysore (28)	Malabar coast of India, Province of Karanataka
53. Orissa (21)	Koa and others from Cuttack, Province of Orissa
54. Rajastan (38)	Recent people from Rajastan Province
55. Punjab/Pakistan (73)	Punjab and Kashmir Districts, northwest India and Pakistan
56. Veddah (17)	From Ceylon Island
Middle East	
57. Afghanistan (40)	Kelati, Pecheen Valley, Kandahar/native of Afghanistan
58. Palestine/Israel (91)	Tell Duweir (Lachish), bronze and iron age, *c.* 5,000–3,000 years BP
59. Syria/Bedouins (36)	Montefik, Arab, Hurbat, Damascus, Palmyra
60. Turkey (55)	Adalia, Aintab, Kurd, Armenian
61. Cyprus (22)	Bronze and early iron age/including a few Hellenistic, Roman and recent period
Arctic	
62. Aleuts/Fox (82)	Fox Islands, Amaknak, Unalaska, Unga, Umnak, Wislow and Shiplock Islands
63. Aleuts/Kagamil (45)	Samalga Island Quad, Kagamil Island
64. Aleuts/Andreanof (36)	Kanaga, Amlia, Adak and Atka Islands, Andreanof Islands including a few specimens from Rat and Near Islands chain
65. Eskimos/Point Barrow (70)	Utkiavik, Nixerak (near Barrow): northernmost part of Alaska
66. Eskimos/Yup'ik (135)	Togiak, Mumtrak, Tanunak, Bethel, Hooper Bay, Iliamna Lake, Yukon, Kwi
67. Eskimos/Inupiat (107)	Seward Peninsula, Shishmaref, Wales, Postolik, Nome, Little Diomede Island, Point Hope
68. Eskimos/Siberia (22)	Port Providence, Plover Bay, Puoten, Indian Point/ northeast Siberia
69. Eskimos/Canada (30)	Pond's Inlet, Sculpin Island, Labrador, Manico Point, Southampton Island, Cumberland Gulf, Baffin Island
70. Eskimos/Greenland (107)	Smith Sound, Port Foulke, Western District, Nuussuaq, Ikertok Fiord, North Star Bay, Saunders Island

Appendix 4.1. (*cont.*)

Sample name (sample size)	Details
North America	
71. Alaska (39)	Yukon River basin, Bonasila, North Shageluk, Ghost Creek, Holy Cross/Near Fort McPherson, Mackenzie River Delta, Wrangell, etc.
72. Kodiak (15)	Chief's Point, Uyak Bay, Mouth Uyak Bay, Koniag, Kiavak, Kodiak Island, Hog Island, Afongak Quad
74. Tlingit (35)	Southeast Alaska and northeast coast of Canada
75. British Columbia (78)	Columbia River, east end of Peel Island, Kwakiutl, Fort Rupert, Eburne, Nimpkish River, Port Hammond, Clayoquot, Vancouver Island, Nootka, Lytton, Kamloops, Bella Bella, Nicola, Thompson
76. Plateau (15)	Oregon and Washington
77. Utah (69)	Native Americans from Great Basin
78. California/North (61)	Sacramento; Alameda, Calaveras and Marin Counties Angel Island, San Jose, Mares Island
79. California/South (163)	San Nicolas Island, Santa Cruz Island, Angeles Bay, Santa Rosa Island, Santa Barbara County.
80. Arizona (41)	Apache Indians, Canyon del Muerto, Canyon de Chelly
81. New Mexico (63)	Mainly from Ketchipauan: Santa Fe Country, Mescalera, Bosque Rodondo Reservation
82. Plains/north 1 (30)	Plains Indians from North Dakota
83. Plains/north 2 (30)	Plains Indians from South Dakota and Nebraska
84. Plains/Arikara (119)	Arikara Indians
85. Plains/central (39)	Plains Indians from Missouri and Arkansas
86. Plains/south (27)	Plains Indians from Texas and Kansas
87. Northeastern woodland/ east (33)	Michigan, Wisconsin, Indiana, Ohio
88. Northeastern woodland/ west (23)	Maine, Pennsylvania, New Jersey, Massachusetts
89. Illinois (118)	Jersey, Henderson, Randolph, Schuyler, Cass, Madison Counties
90. Ontario Iroque (37)	Iroque Atlantic and adjoining states
91. New York (27)	Pelham Bay Park, Bronx, Staten Island, Throg's Neck, Westchester County, Aqueduct, Port Washington, Shinnecock Hills, Long Island
92. Maryland (54)	Port Tobacco; Ossuary; Choptank; Poolesville; Accokeek Prince George's, Charles and Montgomery Counties
93. Kentucky (37)	Indian Knoll; Green River; Ohio, Union, Warren and Ballard Counties; Calvin Lake
94. Virginia (59)	Werren, Botatourt, Shenandoah Counties
95. Southeastern woodland (47)	Tennessee, North/South Carolina, Mississippi, Georgia, Alabama, Louisiana
96. Florida (96)	St Augustine, Cedar Keys, Seminole, Canoveral, Perico Island, Cativa Island, Belle Grade mound site, Palm Beach county

Appendix 4.1. (*cont.*)

Sample name (sample size)	Details
Central/South America	
97. Mexico (73)	Sierra Madre of Durargo, Tombat Xico, Jarasco, Tarasco, Yucatan, Duraugo, Pueblo, Chihuahua
98. Intermediate (28)	Colombia, Ecuador, Panama
99. Carib (25)	From Carib Islands
100. Tropical lowland (14)	Guyana, Bolivia
101. Peru (223)	Cerro del Oro, Arica, Quichua, Chincha
102. Chile (20)	Guaytecas, Guayaneco, Chiloe, Alacatuf, Nacaluf, Orillas del Lago, Tehuelche, Cura Cautin, Inoian
103. Fuegians/Patagonians (57)	Terra del Fuego and Patagonia region

5 *New perspectives on the peopling of Southeast and East Asia during the late upper Pleistocene*

FABRICE DEMETER
Collège de France, Paris, France

Introduction

Palaeogeographical studies show constant temporal changes of the environment, particularly during the Pleistocene, and emphasise the important effects of such variability on floral and faunal development and adaptation (Fairbridge 1961, van Heekeren 1972, Shackleton and Opdyke 1973, Farrell and Clark 1976, Dunn and Dunn 1977, Chappell and Shackleton 1986, Gibbons and Clunie 1986, Kershawa 1988, Ferguson 1993, Robert and Wright 1993). The close of the Pleistocene was characterised by several cold period peaks at approximately 60,000, 30,000 and 18,000 years BP. During these cold periods, sea levels dropped at least 125 m from current levels (Shackleton and Opdyke 1973, Kershawa 1988, Robert and Wright 1993, Stringer 1993), exposing large areas of land including the Sunda shelf in Southeast Asia (Figs. 1.1, p. 5 and 1.2, p. 6). During the last glaciation, the Japanese archipelago, Taiwan, Malaysia, parts of the Philippine islands and the Indonesian archipelago were connected to the Asian continent (Fairbridge 1961, Shackleton and Opdyke 1973, Dunn and Dunn 1977, Chappell and Shackleton 1986, Demeter 2000, Robert and Wright 1993), providing an unbroken land route from Japan to Indonesia. Climate changes associated with periods of peak glaciation had marked effects on the flora and fauna. For instance, palynological analyses (Frenzel 1968, Kershawa 1988, Robert and Wright 1993) show that drops in temperature were associated with the replacement of species by others better adapted to cold and aridity.

The drop in temperature at each glaciation was progressive and played a pivotal role in the history of the modern human settlement of the world and the Southeast and East Asian region. It is reasonable to suppose that north to south migrations following the sea coast (Demeter 2000) were

Bioarchaeology of Southeast Asia. Marc Oxenham and Nancy Tayles.
Published by Cambridge University Press.

heavily influenced by people searching out warmer regions. While humans could certainly have adapted to cold conditions in the north, this does not appear to have been the case in East Asia at least. The tendency for early humans to live near the coast may have provided two advantages for southerly migration. The first is that movement may have been facilitated by way of the coast. The second, related, reason is that populations with a marine subsistence focus would tend to follow the coast south in search of warmer conditions, while maintaining their subsistence orientation, until reaching the Indonesian archipelago and Java. In practical terms, to cover the distance from Japan to Indonesia in less than 10,000 years, each human generation would only have to travel 50 km (Allen *et al.* 1977). As fast as this seems, it needs to be remembered that southerly human migrations would have been dependent on eustatic variations.

Currently, two major theoretical positions are being debated concerning the modern human peopling of the world: the Multiregional Evolution hypothesis and the Unique and Recent Origin hypothesis. Both of these explanatory models are based, for the most part, on morphological analyses of fossil hominid material recovered from various geographical areas, and/or reconstructions of ancestral lineages based on modern genetic data. The Multiregional model was developed from Franz Weidenreich's (1946) early polycentric theory of human origins, which saw the evolutionary process as gradual and continuous. Weidenreich (1946) argued for the evolution of modern humans from archaic *Homo sapiens*, which, in turn, derived from *Homo erectus* in each of the main regions of the old world. The early version of this model saw limited amounts of genetic exchange between the early regional populations, while it is now believed by scholars such as Milford Wolpoff (1984, 1996) that interpopulational gene flow is of pivotal importance to the model. With respect to Southeast and East Asia, multiregionalists such as Wu Xin Zhi (1995) believe that two modern groups, a Chinese and an Indonesian, developed from *in situ H. erectus* populations in each region.

The alternative to the Multiregional model was first developed by Louis Leakey in the 1960s (Leakey *et al.* 1995) and suggests that the origin of modern humans is to be found only within Africa (the Out-of-Africa model). By the end of the middle Pleistocene, modern humans had evolved directly from archaic *H. sapiens* on the African continent. These archaic *H. sapiens* groups are then proposed to have migrated into Eurasia (Stringer 1993), replacing the first (*H. erectus*) human groups already established there before continuing to spread to the rest of the world. With respect to developments in Southeast and East Asia, some palaeoanthropologists, namely Tsunehiko Hanihara (cf. Ch. 4) and Hirofumi Matsumura (Matsumura 1998), have proposed that these modern humans would initially have reached the northern portions of East Asia, including the Japanese archipelago, before moving south.

The aim of this chapter is to examine the modern human fossil record for Southeast and East Asia from the earliest known specimen (Liujiang dated to 67,000 years BP) to more modern forms (such as Samrong Sen, formerly Som Ron Sen, dated to 1,200 years BP) in an attempt to characterise prehistoric human settlement in the area. Within the context of the Multiregional Continuity and Out-of-Africa models, was human settlement in Southeast and East Asia continuous or not and was it from one or several regional centres? Further, if more than one centre is implicated, is it probable that one of these is to be found in Northeast Asia? The analysis of prehistoric human cranial morphology described here uses a range of multivariate techniques to offer new perspectives on prehistoric human settlement and movement in Southeast and East Asia.

Materials and methods

The reference sample

The reference sample for this study was restricted to modern human cranial material from Southeast and East Asia. An attempt was made to develop a reference sample with similar proportions of individuals from each region represented by the fossil sample. The reference sample (housed in the Anthropology Laboratory/Musée de l'Homme, Paris) was composed of 100 individuals of known provenance (represented by cranial and mandibular remains) albeit of unknown sex. No effort was made to attribute sex to any specimens, as the statistical operations used in this study control for sexual dimorphism. For each specimen, I personally took 118 measurements. The sample comprised Japan ($n = 6$, Hiogo-Kobé, Steenackers Collection registered in 1886), China ($n = 9$, Soller Collection registered in 1886), Vietnam ($n = 47$, Canoville Collection registered in 1884), Laos ($n = 23$, Noël Bernard Collection registered in 1920), Thailand ($n = 10$, Bel Collection registered in 1896), Cambodia ($n = 1$, Pannetier Collection registered in 1920), Philippines ($n = 1$, Philippines, Eugène Salomon Collection registered in 1897) and Indonesia ($n = 3$, Dumoutier Collection registered in 1874).

The fossil sample

The fossil sample comprised 44 fossil modern human crania discovered in Southeast and East Asia from Japan to Indonesia (Table 5.1 and Figs. 1.1 (p. 5) and 1.2 (p. 6)) (Demeter 2000). All measurements were taken on the original fossils except for the Chinese material where only casts are now available (Upper Cave from Zhoukoudian, the originals of which disappeared in the 1940s, and the Hong Kong fossil (Jacob 1968) housed in the Senckenberg Museum in Frankfurt). The other Chinese fossils, Liujiang, Mapa and Salawusu, are stored in the Institute of Vertebrate Palaeontology and Palaeoanthropology in Beijing (IVPP). The Japanese Minatogawa, Hamakita and Mikkabi fossils are housed in the University of Tokyo. The Vietnamese fossils Mai Da Dieu, Mai Da Nuoc and Dong Can are housed in the Institute of Archaeology in Hanoi. The Indonesian fossils are housed in the National Museum of Natural History of Leiden in the Netherlands. Fossils from Cambodia, Laos and Vietnam were brought to France at the beginning of the twentieth century and in the 1930s by researchers working for the Ecole Française d'Extrême Orient, the National Museum of Natural History of Paris and even the Indochina Geological Service. All of these fossils are currently housed in the Biological Anthropology Laboratory of the National Museum of Natural History in Paris (Musée de l'Homme). Recently, several of these fossils have been carbon-14 dated (Demeter and Long 2002). It is worthwhile recounting the circumstances in which some of these fossils were rediscovered in France. While engaged in doctoral research in 1998, I was inventorying all the Southeast Asian fossils in the Biological Anthropology Laboratory collections, with the help of Ph. Mennecier and M. Chech. We were fortunate in identifying fossils discovered 70 years previously in Cambodia (Samrong Sen (Demeter *et al.* 1999)), Vietnam (Cau Giat (Colani 1930, Demeter *et al.* 2000), Pho Binh Gia (Mansuy 1924)) and Laos (Tam Hang (Fromaget 1940)). These fossils had either never or only incompletely been published and some of them are very fragmentary. Nevertheless, they were stored in several wooden boxes in 1936 when the old Ethnographic Museum was replaced by the Palais de Chaillot, where the Musée de l'Homme is currently located. In such a manner, the scientific community forgot about these boxes and their contents for more than 70 years.

Evolutionary considerations

The geological age of the fossils was not considered to constitute 'same age' groups of individuals. The temporal range of the fossil sample was

Table 5.1. *Southeast and East Asian fossil sample summary*

Specimen name	Abbreviation	Location (country)	Location (museum)	Geographic location	Date (years BP)
Dong Thûoc 19424	Dthuoc	Vietnam	Musée de l'Homme, Paris, France	Coastal	Neolithic
Jinchuan	Jinchuan	China	IVPP, Beijing, China	Inland	Neolithic
Than Hoa	THOA DB	Vietnam	Musée de l'Homme, Paris, France	Coastal	Neolithic
Pho Binh Gia 23096	PBG 3	Vietnam	Musée de l'Homme, Paris, France	Coastal	Neolithic
Tam Nang An Tta 20543	TH TTA	Laos	Musée de l'Homme, Paris, France	Inland	Neolithic
Samrong Sen	Srom sen	Cambodia	Musée de l'Homme, Paris, France	Inland	1,200
Tam Pong 20541	TP 1	Laos	Musée de l'Homme, Paris, France	Inland	5,380 ± 60
Ban Kao BKInconnu	BKI.INCONNU	Thailand	Sood Sangvichian Prehistory Museum, Bangkok, Thailand	Inland	6,000
Ban Kao BKKBSP	BKKBSP	Thailand	Sood Sangvichian Prehistory Museum, Bangkok, Thailand	Inland	6,000
Ban Kao BKVII	BKI.BVII	Thailand	Sood Sangvichian Prehistory Museum, Bangkok, Thailand	Inland	6,000
Lang Cuom 19409	Lcuom 1	Vietnam	Musée de l'Homme, Paris, France	Coastal	6,440 ± 50
Lang Cuom 19411	Lcuom 4	Vietnam	Musée de l'Homme, Paris, France	Coastal	6,440 ± 50
Lang Cuom 19412	Lcuom 5	Vietnam	Musée de l'Homme, Paris, France	Coastal	6,440 ± 50
Lang Cuom 19415	Lcuom 8	Vietnam	Musée de l'Homme, Paris, France	Coastal	6,440 ± 50
Lang Cuom 19416	Lcuom 9	Vietnam	Musée de l'Homme, Paris, France	Coastal	6,440 ± 50
Lang Cuom 19417	Lcuom 10	Vietnam	Musée de l'Homme, Paris, France	Coastal	6,440 ± 50
Lang Cuom 19418	Lcuom 11	Vietnam	Musée de l'Homme, Paris, France	Coastal	6,440 ± 50
Lang Cuom 19420	Lcuom 13	Vietnam	Musée de l'Homme, Paris, France	Coastal	6,440 ± 50
Wadjak 1 WH24	Wadjak 1	Indonesia	Museum of Natural History, Leiden, the Netherlands	Coastal	7,000
Wadjak 2 WH22–23–4–5	Wadjak 2	Indonesia	Museum of Natural History, Leiden, the Netherlands	Coastal	7,000

Pho Binh Gia 18504	PBG 1	Vietnam	Musée de l'Homme, Paris, France	Coastal	7,470 ± 50
Cau Giat A4 23105	Cgiat A4	Vietnam	Musée de l'Homme, Paris, France	Coastal	7,520 ± 50
Cau Giat 2	Cgiat 2	Vietnam	Musée de l'Homme, Paris, France	Coastal	7,520 ± 50
Cau Giat 1	Cgiat 1	Vietnam	Musée de l'Homme, Paris, France	Coastal	7,520 ± 50
Mai Da Dieu 84MDDM1	MDD	Vietnam	Institute of Archaeology, Hanoi, Vietnam	Coastal	8,200 ± 70
Mai Da Nuoc 84MDNM1	MDN	Vietnam	Institute of Archaeology, Hanoi, Vietnam	Coastal	8,200 ± 70
Dong Can	Dong Can	Vietnam	Institute of Archaeology, Hanoi, Vietnam	Coastal	11,600 ± 90
Tam Hang 20533	THN 3	Laos	Musée de l'Homme, Paris, France	Inland	15,740 ± 80
Tam Hang 20534	THS 13	Laos	Musée de l'Homme, Paris, France	Inland	15,740 ± 80
Tam Hang 20535	THS 2	Laos	Musée de l'Homme, Paris, France	Inland	15,740 ± 80
Tam Hang 20537	THS 3	Laos	Musée de l'Homme, Paris, France	Inland	15,740 ± 80
Tam Hang 20538	THS 14	Laos	Musée de l'Homme, Paris, France	Inland	15,740 ± 80
Tam Hang 20539	THS 10	Laos	Musée de l'Homme, Paris, France	Inland	15,740 ± 80
Tam Hang 20540	THS 11	Laos	Musée de l'Homme, Paris, France	Inland	15,740 ± 80
Tam Hang 20542	THS 4	Laos	Musée de l'Homme, Paris, France	Inland	15,740 ± 80
Tam Hang 20550	THS 22	Laos	Musée de l'Homme, Paris, France	Inland	15,740 ± 80
Minatogawa 1	Minat 1	Japan	University of Tokyo, Tokyo, Japan	Coastal	18,250 ± 650
Minatogawa 2	Minat 2	Japan	University of Tokyo, Tokyo, Japan	Coastal	18,250 ± 650
Minatogawa 4	Minat 4	Japan	University of Tokyo, Tokyo, Japan	Coastal	18,250 ± 650
Upper Cave 1	Ucave 1	China	IVPP, Beijing, China	Inland	33,000
Upper Cave 2	Ucave 2	China	IVPP, Beijing, China	Inland	33,000
Upper Cave 3	Ucave 3	China	IVPP, Beijing, China	Inland	33,000
Zhiyang	Zhiyang	China	IVPP, Beijing, China	Inland	37,400 ± 3,000
Liujiang	Liujiang	China	IVPP, Beijing, China	Coastal	67,500 ± 5,000

IVPP, Institute of Vertebrate Palaeontology and Palaeoanthropology.

considerable in that the earliest fossil is the Liujiang cranium from China where associated faunal remains were carbon-14 dated to 67,000 years BP (Wu 1995), and the youngest fossil is the Samrong Sen cranium from Cambodia, which was directly carbon-14 dated to 1,200 years BP (Demeter 2000). Clearly, the fossil sample cannot be considered to constitute a 'same age' group, but it can be assumed that evolutionary processes had a minimal impact on the sample in that no speciation events are observable. In other words, only micro-evolutionary changes are likely to have occurred. This assumption that major evolutionary changes have not occurred is confirmed by the results of the multivariate analyses carried out in this study, which do not isolate the Lijiuang specimen. Both the resultant dendrogram and the multidimensional scaling plot grouped Liujiang with the Wadjak specimens which are believed to be approximately 9,000 years old (John de Vos personal communication; but see Storm (1995) who dated it to 6,500 years BP).

Statistical techniques

Hierarchical cluster analysis

A well-established methodology was followed (Martin 1928, Twiesselmann and Brabant 1960, Menin 1977, Peyre 1977) and morphological characteristics were highlighted on the 45 fossil crania using sigmoid profiles upon 13 variables with constant presence (see Appendix 5.1 (p. 131) for the original measurements). As with the reference sample, the fossil crania were not analysed by sex. Here, the sigmoid profiles (standardised deviation curves) showed the morphological features of individuals by comparing fossil cranial variables with the reference sample in terms of sigma (Olivier 1960). The ordinate axis, or sigmoid scale, was divided into standard deviations (σ). Only the standard deviation for each character was used as a reference. The reference sample was then represented by a baseline on the ordinate axis at the zero point (0). Starting from this point, individuals, or groups of individuals, were compared with this baseline. If the mean value of some characters in the selected sample is $m1$ and the corresponding mean value in some other population is $m2$, then the quantity is represented by:

$$\frac{m1 - m2}{\sigma}$$

In a second step, multivariate analyses were run to characterise the fossil sample. It was necessary to check if the fossil database was structured and if the specimens could be grouped by similar morphology. Consequently, hierarchical cluster analysis (HCA) was used. As demonstrated by Howells (1989), this method shows, by way of a dendrogram, the different main clusters according to an aggregation criterion. For this analysis, the Ward criterion (also called the Ward association coefficient), which groups individuals with the smallest loss of information, was selected (Ward 1963).

To control for the potential problems introduced by sexual dimorphism, the 'size' factor was controlled for and the 'shape' factor preserved by normalising the database using Howells' recommendations (1989). First, the individual raw measurements were rendered into a standard form (Z-scores). The deviation of each score from the general mean was divided by the general standard deviation estimated from the pooled data. Then the mean of each individual's Z-scores was calculated, which is called 'pensize': $(Z1 + Z2 + \cdots + Z_g)/G$, where G is the number of measurements. Next, the Z-scores were centred once again by subtracting from each of them that individual's pensize, so that the sum of these deviated scores was zero. These newly centred figures are called C-scores.

Once the C-score database matrix was calculated, the HCA using the Ward association coefficient was performed to study the relations between individuals. This method, which only considers the cases rather than the variables, is also called Q-mode analysis. The results of the HCA can be represented by a dendrogram, as shown in Fig. 5.1. The meaning of the different branches is as follows. The individuals grouped on the same side have a similar morphology. The hierarchy is indexed, which means that each cut corresponds to a numeric value indicating at what level the clustering happens. This index, called the aggregation level, shows the distance between clusters. The higher this aggregation level is, the less similar is the morphology. When describing the different morphological groups, the set of traits that characterises them was used, avoiding any racial classification.

Multidimensional scaling analysis and distance matrices

Multidimensional scaling analysis (MDS; Torgerson 1958, Seber 1984) was applied to the fossil database to identify graphically possible patterns of spatial variation. This multivariate method takes a set of dissimilarities (as in a distance matrix) and returns a set of points such that distances among the points in the plot are approximately equal to the dissimilarities. The MDS method is not so much an exact procedure as

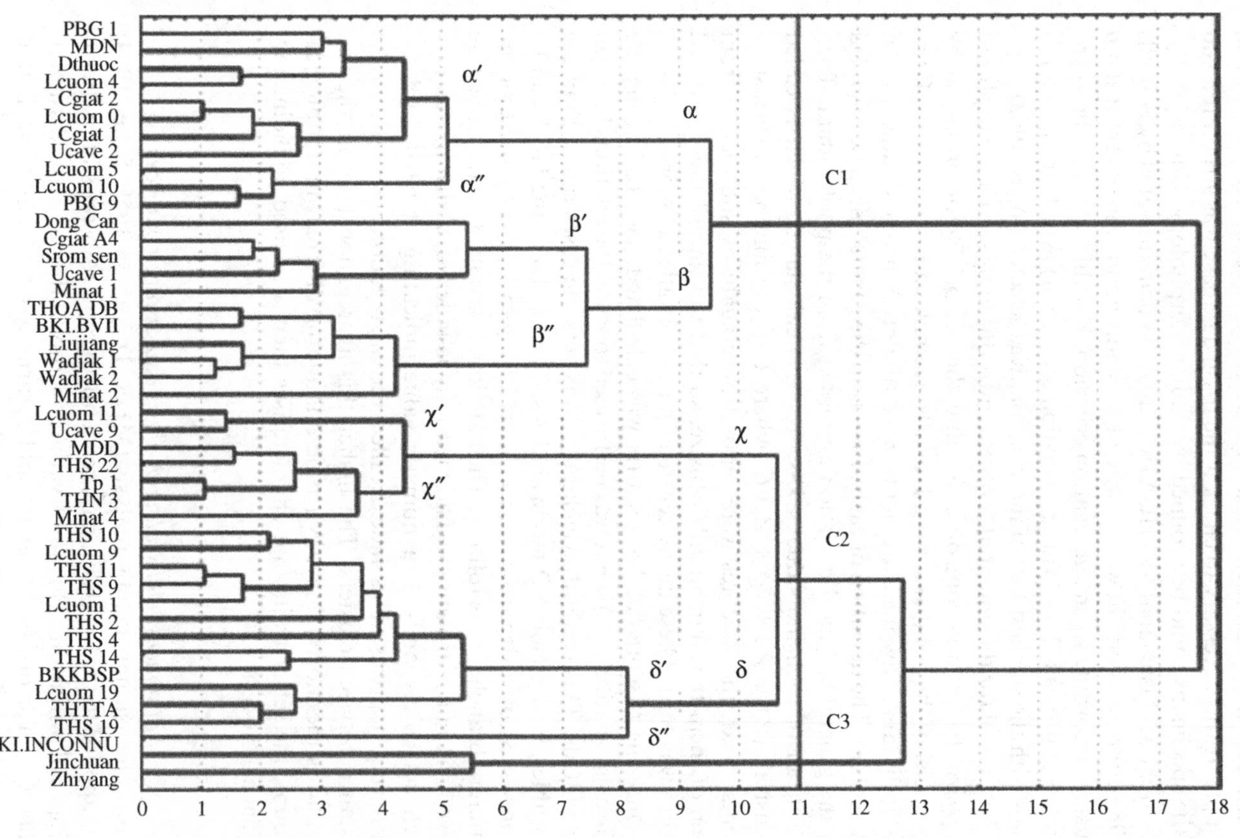

Figure. 5.1. Dendrogram using 16 variables of the calvarium according the Ward aggregation criterion. Abbreviations are defined in Table 5.1.

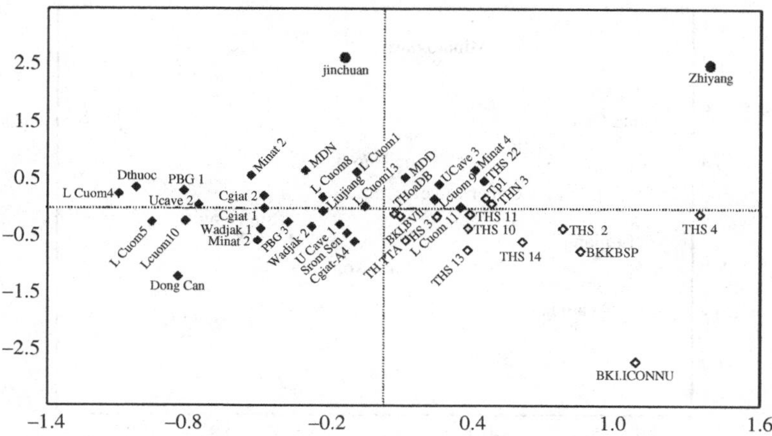

Figure 5.2. Multidimensional scaling scatterplot (individual) of the fossils considered in this study. The plot was computed on a distance matrix between all 45 individuals and accounts for 13 measures of length, width, height and mastoid height of the crania. Stress: 0.16; (●) Chinese samples (◆) coastal and insular specimens (◇) inland specimens Abbreviations are defined in Table 5.1.

rather a way to 'rearrange' objects in an efficient manner so as to arrive at a configuration that best approximates the observed distances. It actually moves objects around in the space defined by the requested number of dimensions and checks how well the distances between objects can be reproduced by the new configuration. In effect, it uses a function minimisation algorithm that evaluates different configurations with the goal of maximising the goodness of fit (or minimising 'lack of fit'). The most common measure that is used to evaluate how well (or poorly) a particular configuration reproduces the observed distance matrix is the stress measure. The raw stress value phi of a configuration is defined by:

$$\sum [d_{ij} - f(\delta_{ij})]^2$$

In this formula, d_{ij} stands for the reproduced distances, given the respective number of dimensions, and δ_{ij} (*delta*$_{ij}$) stands for the input data (i.e. observed distances). The expression $f(\delta_{ij})$ indicates a *non-metric*, monotone transformation of the observed input data (distances). Thus, it will attempt to reproduce the general rank ordering of distances between the objects in the analysis. The MDS analysis (Figs. 5.2 and 5.3), as well as the dendrogram (Fig. 5.1), were computed using Statistica 5.0 on the same standardised data matrix (Torgerson 1958, Howells 1989).

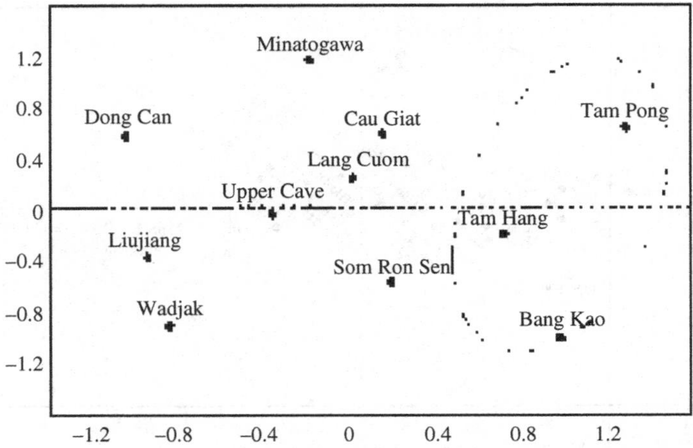

Figure 5.3. Multidimensional scaling scatterplot (sample) of the fossils considered in this study grouped in populations. The plot was computed on a distance matrix between 11 populations (Chinese outliers Jinchuan and Zhiyang excluded). Stress: 0.14.

Results

Intersample and interspecimen variability

Figure 5.1 shows the three main clusters of specimens or groups labelled C1, C2 and C3. The first group (C1) consists of specimens characterised by a high, long and broad neurocranium with a high, long and fairly broad face (Fig. 5.4). The second group (C2, Fig. 5.5) consists of specimens themselves divided into two other distinct subgroups (χ and δ). The first sub-group χ consists of specimens with a relatively high, short and narrow neurocranium and a low and long face that is very wide at the level of the orbits with a narrow frontal bone. The second subgroup δ consists of specimens with a rather short and narrow neurocranium, narrow frontal, rather broad face and very broad interorbital region. The third group (C3, Fig. 5.6) is represented by two specimens that show the same morphological characteristics as the previous C2 subgroup δ, but with a very long face.

MDS analysis of interspecimen differences and intersample differences are shown in Fig. 5.2 and 5.3, respectively. Both analyses show a clear differentiation between coastal and inland excavated sites. At an individual level (Fig. 5.2), the two groups can be distinguished, roughly on each side of the *y* axis, as follows. The first group, composed of specimens from

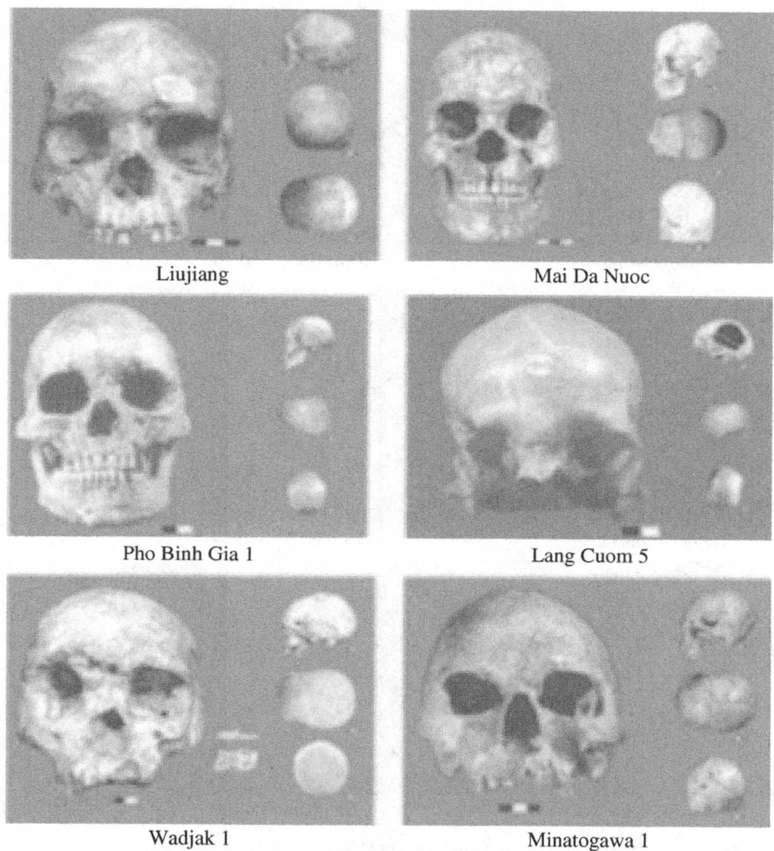

Figure 5.4. Examples of well-preserved C1 morphology.

coastal sites with a C1 morphology, include Lang Cuom 4 5 10 13 1 8; Dong Thûoc; Pho Binh Gia 1 and 2; Mai Da Nuoc; Minatogawa 1 and 2; Liujiang; Wadjak 1 and 2; Upper Cave 1 and 2; Samrong Sen; Cau Giat 1, 2, A4; and Dong Can. The second group is composed of the following specimens: Tam Hang 3, 13, 11, 10, 14, 4, 22, 2, THN3, TH TTA; A; Than Hoa; Tam Pong 1; Ban Kao BKVII, BKLSII, BKKBSP, BKInconnu; Mai Da Dieu; Lang Cuom C9 and 11; Upper Cave 3; and Minatogawa 4. Besides the Jinchuan and Zhiyang crania, which are clearly outliers, there is an absence of specimen (Fig. 5.2) or sample (Fig. 5.3) clustering. Further, these specimens can all be characterised as having a C2 morphology and the majority of them (with the exception of Upper Cave 3, Minatogawa 4, Lang Cuom 9 and 11) Mai Da Dieu, derive from inland sites.

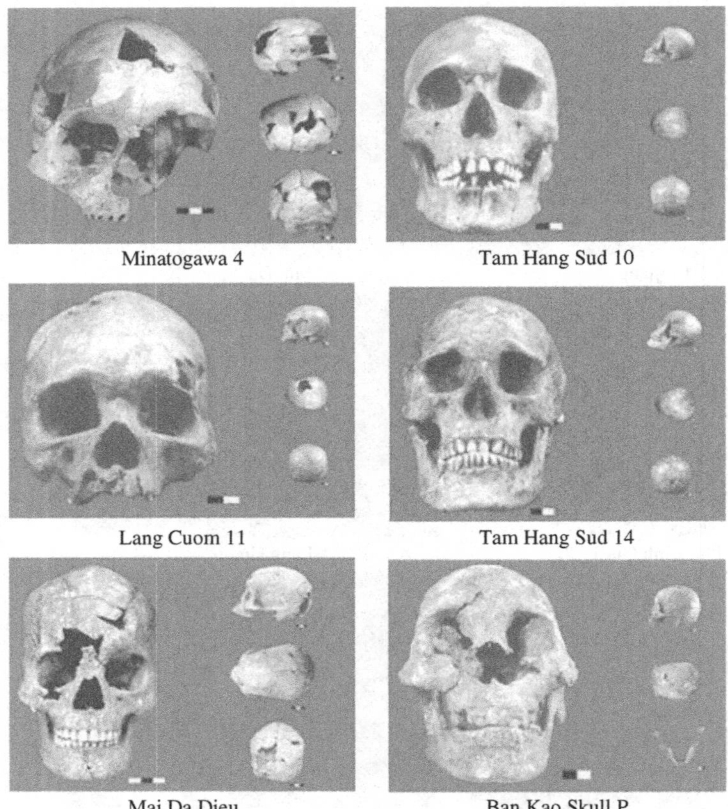

Figure 5.5. Examples of well-preserved C2 morphology.

Discussion

The two multivariate analyses (HCA and the multidimensional scaling) revealed a tripartite division of the fossil sample characterised by the C1, C2 and C3 morphologies. The clear differentiation of the two Chinese fossils, Jinchuan and Zhiyang, is evident from the multivariate analyses at both specimen and sample levels (Figs. 5.2 and 5.3). These fossils are outliers belonging to two separate samples (since they differ considerably from one another) that are distinct from the other samples from the region. The remaining samples are more homogeneous and equally spaced on the multidimensional scaling plots, (Figs. 5.2 and 5.3), suggesting the absence of clustering. This configuration does not mean that neighbouring populations on the plots are geographically close. The total lack of correlation

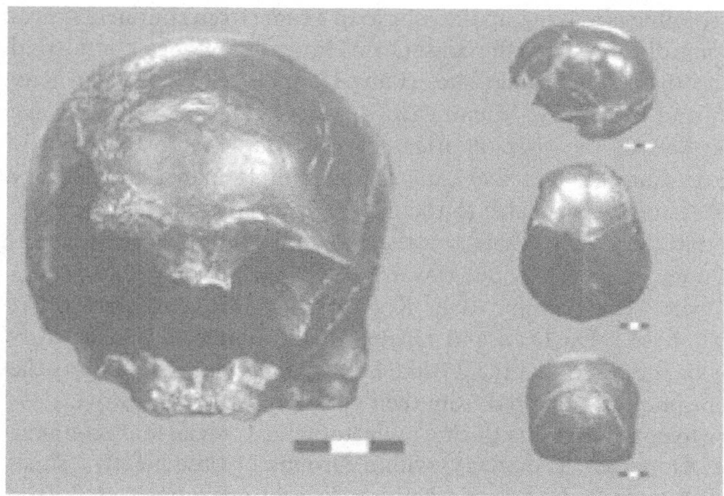

Figure 5.6. An example of well-preserved C3 morphology (Zhiyang).

between the distance matrices based on biometrical measures and those based on the geographic distances indicates the absence of any latitudinal cline in the pattern of morphological change. This means that the considerable variability between fossils could be the result of different migration patterns in the early peopling of the area. This observation has to be related to strong differences among the Chinese samples and to the differentiation between coastal and inland samples, as suggested by the analysis of fossils both at the specimen (Fig. 5.2) and at the sample (Fig. 5.3) level.

Concerning the first group (C1), one specimen (Liujiang) was dated to 67,000 ± 5,000 years BP, making it the oldest member of this group. The other specimens in this group were dated to between 33,000 years BP (Upper Cave) and 1,200 years BP (Samrong Sen). The C1 group demonstrates that a similar morphology was present from at least 67,000 years BP to at least 1,200 years BP. The sites where the fossils forming the C1 group were found are either insular or continental sites (Figs. 1.1, p. 5 and 1.2, p. 6), although the latter are close to the current coast. Furthermore, these sites are distributed from north to south along the East Asian coast. This very particular pattern of human settlement close to the sea, with specimens sharing the same morphology, suggests a history of continuous coastal peopling.

The geological ages of the fossils from the second (C2) and third (C3) groups all fell between 33,000 years BP (Upper Cave 3) and 6,000 years BP (Ban Kao). The C3 group consisted of two individuals (Jinchuan and Zhiyang) from continental sites, far from the present coast. Within the C2 group, the specimens making up the subgroup δ came from continental sites (Fig. 1.1) some distance from the coast (Tam Hang, Tam Pong), continental sites close to the sea or insular sites (Lang Cuom 1, 9, 11, 13, Ban Kao BKInconnu and BKKBXV, Minatogawa 4, Mai Da Dieu and Upper Cave 3). The location of the sites of origin of subgroup δ is geographically variable. For example, half the Lang Cuom sample had a morphology similar to C1, and in particular to the 67,000-year-old Liujiang specimen, suggesting that the C1 morphology is the ancestral condition. The remaining Lang Cuom specimens tended toward the C2 morphology, thus grouping them with Tam Hang, Ban Kao BKInconnu and BKKBXV, Minatogawa 4, Mai Da Dieu and Upper Cave 3 (dated to 33,000 years BP). It may be proposed that the C1-like Lang Cuom specimens display the ancestral morphology inherited from their coastally settled ancestors.

It may be hypothesised that during prehistoric times social and economic constraints supported endogamy within groups. Consequently, small groups would have had a genetic heritage largely unchanged, if drift is assumed to be minimal, until the arrival of new immigrants. If this hypothesis is correct, the more-derived C2-like morphology of the other half of the Lang Cuom sample can be interpreted in two ways. The first interpretation suggests that *in situ* micro-evolutionary change would have led to the emergence of new morphological characters. The second interpretation suggests the arrival of immigrants coming from mainland continental sites, with either replacement of original populations or intermixing. The Tam Hang site could attest to the presence of such immigrants moving toward the coast as early as 33,000 years BP (Upper Cave 3). This scenario seems more probable given the observation that insular and continental sites close to the coast yield numerous C2-like specimens (Lang Cuom, Ban Kao, Mai Da Dieu, Upper Cave, Minatogawa). The observation that the C1 specimens were confined to the East Asian coast suggests the possibility of early maritime navigation. Some very early evidence for maritime travel comes from Flores Island, where 800,000-year-old lithic artefacts were recovered from Mata Menge (Morwood *et al.* 1999). Nonetheless, such examples are uncommon and it is unlikely that maritime travel was a necessarily significant factor in prehistoric migration throughout this region.

Conclusions

The main objective of this work was to characterise the population history, including migration, settlement and microevolutionary change, of anatomically modern humans in Southeast and East Asia from approximately 67,000 years BP until 1,200 years BP. This population history was examined within the context of the competing Multiregional model and the Unique and Recent Origin (Out-of-Africa) hypothesis. The results of this study indicate that during the last 67,000 years there were at least two major regional centres from which humans migrated to settle Southeast and East Asia. One group, identified here as C1 (characterised by a high, long and broad neurocranium, very broad face and particularly thick cranial bones), was located in northern East Asia, including the Japanese archipelago. Late Pleistocene climate changes affected the plant and animal distributions and periods of lowered temperature initiated southward human migrations, using coastal routes for the most part, toward the Sunda shelf. This group reached Southeast Asia around 30,000 years ago and extensive admixture with existing local populations, characterised by the C2 morphology (short, broad and fairly high neurocranium, narrow frontal bone, rather broad face and very wide interorbital breadth), occurred. The two main morphologies (C1 and C2) described in this study demonstrate the coexistence and deep antiquity of two quite different types of anatomically modern human living and interacting in this region. The findings in this study lend some support to the Out-of-Africa model whereby anatomically modern humans recently migrated into Southeast and East Asia and replaced all those populations that had descended from *H. erectus*.

Acknowledgements

The author would like to thank Professor Yves Coppens from the Collège de France whose support enabled me to accomplished this work. I would like also to thank Professor André Langaney, Director of the Biological Anthropology Laboratory of the Musée de l'Homme in Paris. He gave me the best environment in which to study the ex-Indochinese material housed in Paris. I would also like to thank Dr Franz Manni, who performed the statitical analyses of this study.

Reference

Allen J., Golson J. and Jones R. 1977. *Sunda and Sahul: Prehistoric Studies in Southeast Asia, Melanesia and Australia.* London: Academic Press.

Broca P. 1875. Instructions crâniologiques, *Mémoire de la Société d'Anthropologie Paris, 2nd series* **2**: 203.

Chappell J. and Shackleton N. 1986. Oxygen isotopes and sea level. *Nature* **324**: 137–140.

Colani M. 1930. Recherches sur le préhistorique Indochinois. *Bulletin de l'Ecole Française d'Extrême Orient*, **XXX**: 299–422.

Demeter F. 2000. Histoire du peuplement humain de l'Asie extrême-orientale depuis le Pléistocène supérieur récent. Second thesis (Ph.D.), University of Paris-1, Paris.

Demeter F. and Vu The Long. 2002. New absolute dates of three early *Homo sapiens* sites in Vietnam, Pho Binh Gia, Lang Cuom and Cau Giat. Paper Presented at the *XIV Congress of the UISPP*, September 2001, Liege.

Demeter F., Peyre E. and Coppens Y. 1999. Le crâne humain préhistorique de Som Ron Sen (Cambodge), *Compte Rendu de l'Académie des Sciences Paris Series IIa* **328**: 125–132.

2000. Présence probable de formes de type Wadjak dans la baie fossile de Quyhn Luu au Nord Viêt-Nam sur le site de Cau Gia. *Compte Rendu de l'Académie des Sciences, Paris Series IIa* **328**: 451–456.

Dunn F. L. and Dunn F. D. 1977. Maritime adaptations and exploitation of marine resources in Sundaic Southeast Asian prehistory. *Modern Quaternary Research in Southeast Asia* **3**: 1–28, 243–272.

Fairbridge R. W. 1961. Eustatic changes in sea level. *Physics and Chemistry on Earth* **4**: 99–164.

Farrell W. E. and Clark J. A. 1976. Postglacial sea level. *Geophysical Journal of the Royal Astronomy Society* **46**: 647–667.

Ferguson D. K. 1993. The impact of late Cenozoic environmental changes in East Asia on the distribution of terrestrial plants and animals. In Jablonski N. G., ed., *Evolving Landscapes and Evolving Biotas of East Asia Since the mid-Tertiary*. Hong Kong: Center of Asian studies, University of Hong Kong, pp. 45–196.

Frenzel B. 1968. The Pleistocene vegetation of northern Eurasia. *Science* **161**: 637–639.

Fromaget J. 1940. Les récentes découvertes anthropologiques dans les formations préhistoriques de la chaîne annamitique. *Proceedings of the Third Far Eastern Prehistory Congress* Singapore, pp. 60–70.

Gibbons J. R. and Clunie F. G. A. U. 1986. Sea level changes and Pacific prehistory. *Journal of Pacific History* **21**: 58–82.

Howells W. W. 1989. *Papers of the Peabody Museum of Archaeology and Ethnology*, Vol. 79, *Skull Shape and the Map: Craniometric Analysis in the Dispersion of Modern* Homo. Cambridge, MA: Harvard University Press.

Jacob T. 1968. A human Wadjakoid maxillary fragment from China. *Proceedings of the Koninklijke Nederlandse Akademie van Wetenschappen, Serie B, Physical Sciences* **71**: 232–235.

Kershawa P. 1988. Australasia, vegetation history. In Huntley B. and Webb T. III, eds., *Vegetation History*. London: Kluwer Academic, pp. 237–306.

Leakey M. G., Feibel C. S., McDougall I. and Walker A. 1995. New 4 million years old hominid species from Kanapoi and Allia Bay, Kenya. *Nature* **376**: 565–571.

Mansuy H. 1924. Stations préhistoriques dans les cavernes du massif calcaire de Bac Son. *Mémoires du Service Géologique de l'Indochine* **11**: 2.

Martin R. 1928. *Lehrbuch der Anthropologie in systematischer Darstellung, Vol. 2, Kraniologie, Osteologie*. Jena: Gustav Fischer.

Matsumura H. 1998. Native or migrant lineage: The Aeneolithic Yayoi people in western and eastern Japan. *Anthropological Science* **106** (supplement): 17–25.

Menin C. 1977. La population gallo-romaine de la nécropole de Maule (Yvelines): étude anthropologique. Third thesis (Ph.D.) University of Pierre and Marie Curie, Paris.

Morwood M. J, Aziz F, O'Sullivan P *et al.* 1999. Archaeological and palaeontological research in central Flores, East Indonesia: results of fieldwork 1997–98. *Antiquity* **73**: 273–286.

Olivier G. 1960. *Pratique Anthropologique*. Paris: Vigot.

Peyre E. 1977. Etude anthropologique qualitative et quantitative de la population mérovingienne de la nécropole de Maule (France, Yvelines). Third thesis (Ph.D.), University of Pierre and Marie Curie, Paris.

Robert N. and Wright H. E. Jr. 1993. Vegetational, lake-level and climatic history of the Near East and Southwest Asia. In Wright H. E. Jr, ed., *Global Climates Since the Last Glacial Maximum*. Minneapolis, MN: University of Minnesota Press, pp. 194–220.

Seber G. A. F. 1984. *Multivariate Analysis*. New York: Wiley.

Shackleton N. J. and Opdyke N. D. 1973. Oxygen isotope and paleomagnetic stratigraphy of equatorial Pacific core V28–238: oxygen isotope temperature and ice volume over a 105 year and 106 year scale. *Quaternary Research* **3**: 39–55.

Storm P. 1995. The evolutionary significance of the Wajak skulls. *Scripta Geologica* **110**: 1–247.

Stringer C. B. 1993. Reconstruction of recent human evolution. In Aitken M. J., Stringer C. B. and Mellars P. A, eds., *The Origin of Modern Humans and the Impact of Chronometric Dating*. Princeton, NJ: Princeton University Press, pp. 179–195.

Torgerson W. S. 1958. *Theory and Methods of Scaling*. New York: Wiley.

Twiesselmann F. and Brabant H. 1960. Observations sur les dents et les maxillaires d'une population ancienne d'âge Franc de Coxyde (Belgique). *Bulletin du Groupe International de Recherche Scientifique en Stomatologie* **10**: 5–180.

van Heekeren H. R. 1972. *The Stone Age of Indonesia*. The Hague: verhandelingen van Het Kononklijk Instituut Voor Taal, Land End Volkenkunde, Martinus Nijhoff.

Ward J. H. Jr. 1963. Hierarchical grouping to optimize an objective function. *Journal of the American Statistical Association* **58**: 236–244.

Weidenreich F. 1946. *Apes, Giants and Man*. Chicago: Chicago University Press.

Wolpoff M. H. 1984. Modern *Homo sapiens* origins: a general theory of hominid evolution involving the fossil evidence from East Asia. In Smith F. H. and Spencer F., eds., *Origins of Modern Humans: a World Survey of the Fossil Evidence*. New York: Liss, pp. 411–483.

1996. Interpretation of multiregional evolution. *Science* **274**: 704–707.

Wu X. 1995. *Human Evolution in China. A Metric Description of the Fossils and a Review of the Sites*. Oxford: Oxford University Press.

Appendix 5.1. *Original measurements on the 45 Fossil crania used in this study*

Specimen[a]	Measurements[b] (to nearest mm)												
	LMX	LNB	WMX	WFS	WFI	HBB	MZG	MZD	LFT	WBO	WIO	WBZ	HFS
PBG 1	196		143	124	99	155	26	26		96	25	138	67
Dthuoc	195		132	115	95		24	24		106	20	135	
Cgiat 2	193		133	118	103								
Cgiat 1	192		135	123	102						27	143	
Dong Can	192	140	150	127	116	127	34	34	97	116	29	140	74
Lcuom 4	192		130	113	86		20	20			22		
Cgiat A4	190	103	152	123	101	129	30	30	103	96	24	144	71
Lcuom 5	189		140	114	92		17	17			27		
Lcuom 1	188	109	126	106	96	135	26	26			24	116	
MDN	187		132	117	96	145	25	34	116	97	21	135	65
Lcuom 8	186		130	115	101						23		
Lcuom 10	185		134	115	95		19	19				148	
THOA DB	184	90	127	117	94	134	24	24	88	91	28	131	59
Lcuom 9	181	103	130	116	97	140	33	33	99	104	26	136	68
PBG 3	180		141	115	93		21	21			24	140	
Lcuom 11	179	102	141	118	100	138	34	34	98	102	26	134	63
MDD	177	93	127	102	89	131	27	27	98	91	23	120	59
Lcuom 13	177	99	133	117	94	143	24	24	93		24	120	72
THS 10	185	105	144	121	94	145	35	35	98	102	28	142	77
THS 14	181	95	153	125	100	140	36	36	94	98	27	144	74
THS 11	181	91	135	116	99	134	30	30	87	91	27	126	65
TP 1	177	105	137	110	97	136		34	105	112	27	130	73
THS 3	176	96	137	116	95	134	28	28	86	91	27	126	64
TH TTA	175	93	140	121	95	140	22	22	86	91		128	70
THS 4	171	99	142	115	94	135	37	36	94	96	34	129	65

Appendix 5.1. (*cont.*)

Specimen[a]	Measurements[b] (to nearest mm)												
	LMX	LNB	WMX	WFS	WFI	HBB	MZG	MZD	LFT	WBO	WIO	WBZ	HFS
THS 22	167	95	120	101	86	120	25		88	82	22	114	58
THN 3	164	91	129	105	89	125	26	26	91	91	25	120	65
THS 13	161	91	146	120	93	135	25	25	92	96	26	128	73
THS 2	156	86	135	116	104	130	25	26					
BKI.BVII	188	104	145	126	99	145	30	30	106	96	30	136	65
BKKBSP	180	106	142	121	101	147	38	38	83	100	26	145	80
BKI.Inconnu	165		147	115	87		23	23		96		138	72
Minat 1	183	102	149	115	94	135	25	29	103	105	20	144	66
Minat 4	177	101	137	112	89	126		38	99	95	22	122	60
Minat 2	162		131	114	90	123	18	19	100	95	21	121	49
U Cave 1	202	110	144	124	108	134	31	31	105	102	27	143	76
UCave 3	196		132	112	103		32	32		96	27	123	68
Liujiang	190	104	140	123	97	134	23	25	101	95	28	124	64
UCave 2	183	108	130	120	102	144	20	20		100	26	137	67
Jinchuan	182	118	120	120	101	157		40					
Zhiyang	170		131	107	91	123	35		125	34	22	125	
Srom Sen	182		148	118	100						24		73
Wadjak 1	200	109	152	126	99	138			107	100	30	140	68
Wadjak 2					104			31			30		

[a]Abbreviations for specimens given in Table 5.1.

[b]Measurements defined by Martin (1928) include: LMX, 1; LNB, 5; WMX, 8; WFS, 10; WFI, 9; HBB, 17; HFS, 48; HAS, 48(1); WIO, 50; WBZ, 45. Measurements defined by Twiesselmann (1960) include LFT, 12; Measurement of mastoid length (MZ) according to Broca (1875).

6 *Human variation and evolution in Holocene Peninsular Malaysia*

DAVID BULBECK
Australian National University, Canberra, Australia

ADAM LAUER
University of Hawaii at Manoa, USA

Introduction

Peninsular Malaysia is home to four groups recognised by the Malaysian government as *bumi putra* or 'children of the soil'. These are the mainstream Melayu Malays, who became the dominant occupants after the declaration of the sultanate of Melaka in the early fifteenth century, and three 'Orang Asli' aboriginal groups, which each include several tribes (Benjamin 2002). Experts agree on the classification of the Orang Asli into three groups, although opinion varies on how these groups arose and why their distinctions can be ambiguous (Rashid 1995, Fix 2002). Best known but least numerous are the Semang Negritos, rainforest foragers characterised by dark skin, woolly hair and, at least until recent decades, short stature (Bulbeck 2004). They have all now been resettled except for one group of Batek, who traverse the Taman Negara National Park and combine paid work as guides with traditional foraging and desultory agriculture in the form of replanting tubers (Tuck Po 2000). The most numerous Orang Asli are the Senoi, especially the Semai and Temiar, who, until recent times, subsisted through swidden farming in remote forest tracts. Skin colour tends to be light; hair form is highly variable, while stature and body build are slight (Fix 2002). The so-called Aboriginal Malays distinguish themselves from Melayu Malays mainly through their tardy uptake of Islam and other Malay cultural trademarks, and their aptitude in collecting forest produce for trade (Benjamin 1985). Many Aboriginal Malays speak dialects of Malay whereas the mother tongues of all Semang and Senoi groups belong to the Aslian branch of Austroasiatic (Fig. 6.1).

Early explanations for this diversity, epitomised by Carey (1976), took migration as the sole relevant factor and posited four successive waves of

Bioarchaeology of Southeast Asia. Marc Oxenham and Nancy Tayles.
Published by Cambridge University Press.

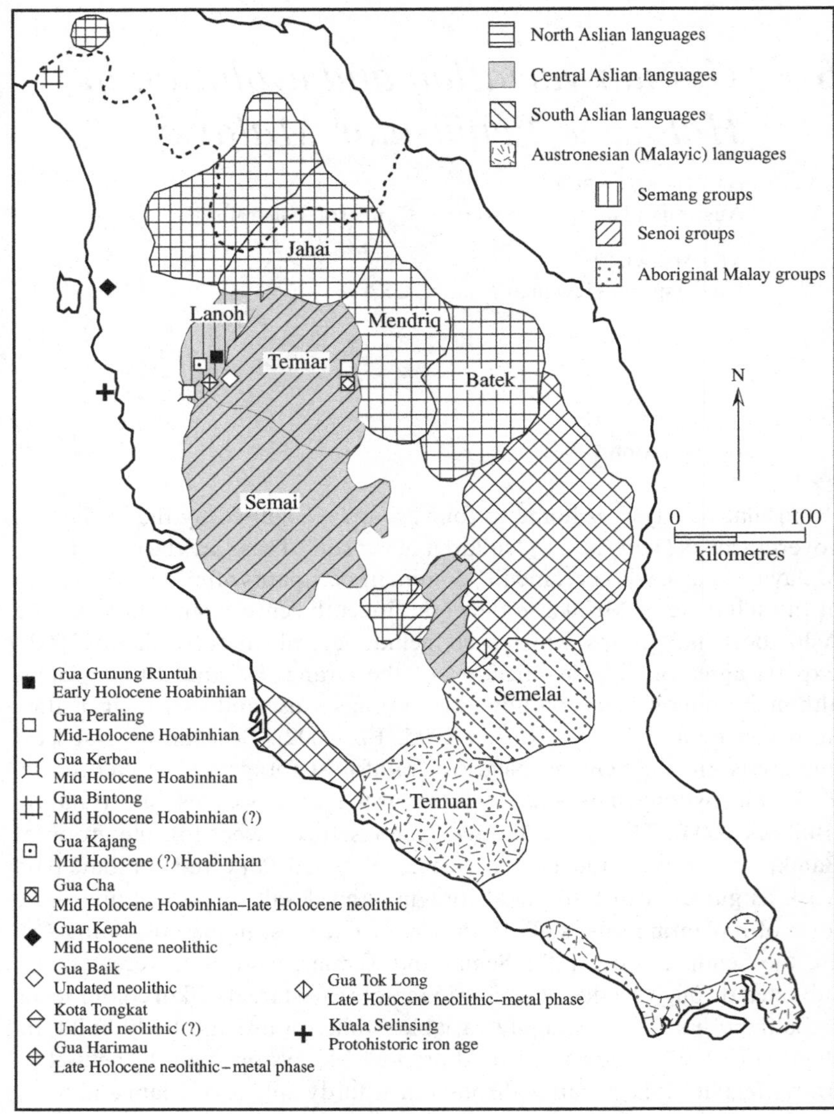

Figure 6.1. Orang Asli groups in this chapter, distribution of Orang Asli ethnological and language groups (after Rashid (1995) and Benjamin (2004)) and sites in West Malaysia with human remains referred to in this study.

Negritos from Africa, Senoi ancestors from South Asia, and Proto-Malays (Aboriginal Malays) and then Deutero-Malays (Melayu Malays) from Southeast Asian homelands. Local evolutionary accounts gained credence as an alternative in the 1980s (Benjamin 1987). These included a late Holocene Gracilisation model to explain postneolithic osteological change in Southeast Asia, including Malaya (Bulbeck 1982, Storm 1995), and Rambo's (1988) Human Ecology model relating Orang Asli phenotypic variation to niche differentiation. At the same time, Bellwood (1985, 1993) crystallised his Demic Diffusion model in which, prior to the widespread adoption of Malay in the Peninsula, 'Mongoloid' farmers speaking proto-Aslian spread south from Thailand and began absorbing or replacing Negrito hunter–gatherers.

Recent research in dental morphology (Rayner and Bulbeck 2002) and mitochondrial DNA (Hill *et al.* 2002) has shown that the Temuan and Semelai, Aboriginal Malay groups speaking Malay and southern Aslian, respectively (Fig. 6.1), are very similar to each other and to Southeast Asians generally. The Temiar Senoi are slightly unusual by 'Mongoloid' Southeast Asian standards and the Semang are more distinct. Mitochondrial DNA analysis finds great diversity amongst the Semang and suggests that one group, the Jahai, may be closer to the Temiar than to other sampled Semang (Mendriq and Batek). Intra-Semang diversity was not tested in the dental morphology study, which pooled its Jahai and Batek data to boost sample size, and then found these Semang to be intermediate between other Orang Asli and populations living as far away as North Africa and New Guinea. However, dental morphology and mitochondrial DNA studies agree that (a) the signature of an indigenous Pleistocene origin appears strongest among the Semang; (b) the Temiar combine this local ancestry with exogenous genes, most probably of mainland Southeast Asian origin, perhaps related to a mid-Holocene arrival of proto-Aslian in the Peninsula; and (c) the distinguishing feature of the Aboriginal Malays is enhanced genetic input from island Southeast Asia, most probably Sumatra.

These results support Bellwood's (1993, 1997) linguistically driven interpretation of the Peninsula's archaeological record more than Rambo's (1988) local differentiation approach. However, as we shall show, the different Orang Asli groups resemble each other in their stature, tooth size and craniometrics, which all show significant change in Peninsula Malaysia since the mid Holocene. Gene flow between the groups, which Bellwood's model allows, can hardly explain this convergence in morphology when the differences in external phenotype, dental morphology and mitochondria DNA imply restricted gene exchange. Also, Bellwood's model would expect an increase over time in Mongoloid features, but this expectation often fails with the skeletal samples assigned to the neolithic and metal phase. The crania retain a preneolithic resemblance to Pacific and African crania when subjected to discriminant function analysis, and the teeth show morphological expressions more similar to African and Australian Aboriginal tendencies than is true of either the preneolithic or Orang Asli samples (Lauer 2002).

As detailed below, when we relate the Orang Asli to their prehistoric counterparts represented in the Peninsula's archaeological record, we need to implicate both local evolution, whereby the Orang Asli would have adapted biologically to selection pressures in force during recent millennia, and migration from extra-Peninsula sources. Further, a simple composite model invoking both local evolution and immigration as complementary processes would predict a smoother osteological transition than we observe on certain features. Consequently, we shall develop a more complex model that takes into account the social and economic repercussions of immigration, resultant changes in indigenous population structure, and genetic isolation as a micro-evolutionary force that generated variation amongst late Holocene populations.

Materials and methods

Bulbeck (1996, 2003) and Rayner and Bulbeck (2002), respectively, provided the provenances for the Orang Asli skeletal specimens and the dental casts taken from living subjects employed in this study. Lauer (2002) and Bulbeck (2004) documented the most likely dating of the Hoabinhian, neolithic and metal phase human remains used in our contribution (Fig. 6.1), and the sample sizes per site available for each suite of characters. The likely subsistence orientation of the different sites' occupants is not always a straightforward matter and this calls for some background discussion.

Cultural evolutionary theory based on Darwinian processes can explain both the ubiquitous shift to agriculture in temperate and tropical zones during the Holocene and the variability in how and when this shift occurred. More intensive land-utilisation strategies eventually prevail, but a frontier between farmer- and forager-dominated regions can remain static for millennia, and cases are known of foragers using land more intensively than farmers and replacing the latter (Richerson *et al.* 2001). This perspective allows for localised infiltration of farming groups, dynamic interactions between groups, and even the creation of novel hunter–gatherer niches. It thus addresses Malaya's late Holocene archaeology (Bulbeck 2003) better than the paradigm of monolithic hunter–gatherer and food-producing blocs assumed in Demic Diffusion models (e.g. Ammerman and Cavalli-Sforza 1984). Also, the uneven spread of agriculture across Malaya evidently allowed foragers to become 'neolithic', when this term – instead of being interpreted as some vague synonym for agricultural – is defined to denote material culture that lacks metal yet includes polished stone artefacts and, as is usually the case, pottery (Bulbeck 2003, 2004). The analogy may be drawn with Andaman Islanders, who have used whetstones and pottery for 2000 years and so should be classified as neolithic – or perhaps iron age, when they started to salvage iron from local shipwrecks – notwithstanding their forager economy (Cooper 2002). Further, intensity of land use is not simply a forager–farmer distinction but also distinguishes between forager economies (Richerson *et al.* 2001), an important point when we contemplate the selection pressures responsible for local evolutionary changes observed amongst Malayan populations.

Lack of clear evidence for Hoabinhian agriculture in Malaya suggests that the Hoabinhians relied on foraging, tempered perhaps with some desultory agriculture, while direct archaeological evidence for neolithic agriculture in Malaya is equally elusive (Bellwood 1997, pp. 158–173, 203, 260–264). Hence, we accept our Hoabinhian specimens represent foragers but suspect the same for the Guar Kepah remains deposited in shell middens where waisted axes and pot sherds co-occur with Hoabinhian pebble tools. The extended supine burials at Gua Cha and Gua Baik, which reflect the Ban Kao mortuary tradition (otherwise found at neolithic sites in south and central Thailand; see Bellwood 1997) in terms of their age (*c.* 3,000 years BP), interment method and grave goods of ceramics and polished stone, could also represent a foraging population. Opinion varies on which subsistence economy to assign to these people (see Bellwood 1997, pp. 261, 265) and, even if a case were established for the same type of economy as practised by some given Orang Asli group, this would not justify tracing an ancestor–descendant relationship. Metal phase groups may have altered their procurement activities as they responded to the trade networks that increasingly tied the fortunes of local communities to the expanding economies of coastally based polities. However, when we come to the late neolithic and metal phase burials at Gua Tok Long, Gua Harimau and Kuala Selinsing, we would be on firmer grounds in relating them to forager–trader, farmer–manufacturer, and coastal–trader orientations, respectively (Bulbeck 2004).

The limb bone lengths from some of the burials have been measured in terms of Martin and Saller's (1959) definitions or estimated from fragments using standard regression formulae (Bulbeck 2003). *In situ* measurements of extended skeletons, and Australasian regression formulae that relate limb bone lengths to stature, provide some prehistoric stature estimates (Bulbeck 1996, 2004). To record tooth size, mesiodistal lengths (except in cases of excessive interstitial wear) and buccolingual breadths were taken according to Martin and Saller's (1959) prescriptions, and the averages from the left and right sides calculated. The data for prehistoric Malayan teeth were computed by Lauer (2002) from measurements made by Bulbeck; for the Orang Asli, Lauer measured the Semang and Senoi dental casts while Bulbeck measured the Semelai dental casts. Dental morphology was recorded by Bulbeck (Lauer 2002: Appendix A) on the prehistoric Malayan specimens, and by Rayner (2000) on the Orang Asli dental casts, according to the Arizona State University System (Scott and Turner 1997). Most of the cranial measurements employed here were collated from the literature on the Orang Asli (Bulbeck 1996) or made by Bulbeck on prehistoric skulls (Lauer 2002) using equivalent definitions in the Biometric (von Bonin 1931) and German systems (Martin and Saller 1959). Many Orang Asli cranial measurements were taken before either system was formally defined, so it is not always certain which exact measurement prescription had been followed (Bulbeck 1996). The statistical methods employed in our analyses are explained in the following text.

Results

Stature and tooth size reduction

Bulbeck (2004) showed that average stature apparently reduced by 10–15 cm between the middle Holocene and the early twentieth century in the hinterland, and by 5–10 cm on the coast during the same period. Limb bones shortened steadily over time, and stature estimates based either on prehistoric limb bone lengths or on the skeleton lengths of Gua Cha neolithic males also showed a gradual, temporal decrease toward the Orang Asli averages (Fig. 6.2). The only prehistoric limb bone that is shorter than the recorded Orang Asli average is the radius of the Gua Baik neolithic male (Bulbeck 2003: Table 4.8). Consequently, we regard this evidence for a *c.* 10 cm reduction in stature over the last five millennia as reliable and, as we shall see, a critical guide to our interpretation of the dental and cranial evidence.

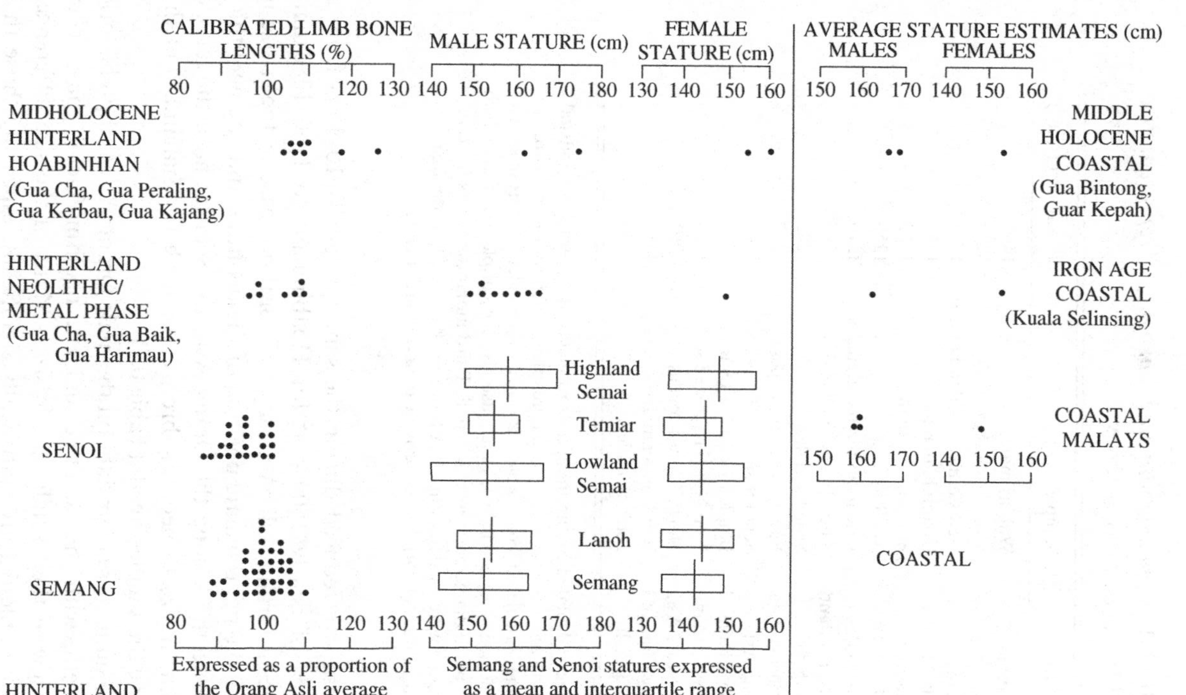

Figure 6.2. Body-size reduction since the mid Holocene in West Malaysia. Hinterland limb-bone lengths are given as a proportion of the recorded Orang Asli average (from Bulbeck 2003: Table 6.8); prehistoric stature estimates and recent stature data are from Bulbeck (1996, 2004).

Table 6.1. *Summed tooth areas (in mm^2) of representative prehistoric and recent series*

Affinity	Sample	Tooth area (mm^2)[a]
Australia	Walpiri males	1450
Australia	Murray Valley males	1444
Melanesia	New Guinea highland males	1443
Early Malaya	Hoabinhian males	1391
Melanesia	Bougainville males	1362
Early Malaya	Neolithic	1350
Early Malaya	Late neolithic/metal males	1310
Early Thailand	Khok Phanom Di males	1295
Orang Asli	Semelai males	1282
Southeast Asia	Batawi Javanese males	1272
Orang Asli	Senoi males	1249
Early Thailand	Ban Kao/Ban Na Di males	1237
Early Thailand	Non Nok Tha mixed sex	1224
Northeast Asia	North Chinese males	1201
Early Thailand	Ban Chiang males	1186
Orang Asli	Semang males	1164

[a] Areas calculated by summing the product of the average mesiodistal length and average buccolingual breadth of each tooth. Malayan values calculated from Table 6.2; other values calculated from sources indicated in Appendix 6.1, which also give the sample sizes from which the mean diameters were calculated. Third molar diameters for Ban Chiang were used in estimating Ban Kao/Ban Na Di tooth area as Matsumura (1994) did not provide data on third molars.

Male tooth size reduced during the same period by 10–15% from a summed area of nearly 1400 mm^2 for Hoabinhians to 1160–1280 mm^2 amongst recent Orang Asli. A steady temporal decrease in tooth size is suggested by the intermediate position of the neolithic and late neolithic/metal phase teeth. Among the Orang Asli, the Semelai have the largest teeth, similar in size to neolithic/bronze age teeth in Thailand, and the Semang have the smallest teeth (Table 6.1).

Table 6.2 summarises our data on the male Orang Asli and prehistoric Malayan teeth, with the neolithic and late neolithic/metal phase teeth pooled in view of their small sample sizes. Many of the size differences suggested by Table 6.2 are statistically significant, especially those that reflect the Hoabinhians' large posterior molars and the small size of the Semang teeth (Table 6.3). When analysed with Penrose's (1954) size statistic (Table 6.4) in conjunction with benchmark Australasian samples, the data in Table 6.2 yield results that parallel observations made on summed

Table 6.2. *Malayan male tooth measurements*

Measurement	Hoabinhian		Neolithic/metal age		Semelai		Temiar Senoi		Semang	
	No.	Mean (mm (SD))	No.	Mean (mm (SD))	No.	Mean (mm (SD))	No.	Mean (mm (SD))	No.	Mean (mm (SD))
I^1 mesiodistal	4	8.45 (0.51)	8	8.35 (0.95)	22	8.79 (0.36)	25	8.75 (0.45)	29	8.22 (0.66)
I^2 mesiodistal	5	7.05 (0.89)	8	6.83 (1.06)	22	6.83 (0.76)	25	6.93 (0.62)	29	6.73 (0.70)
$\underline{C}$ mesiodistal	5	7.90 (0.42)	7	7.81 (0.48)	21	7.95 (0.79)	25	7.84 (0.64)	29	7.74 (0.58)
P^1 mesiodistal	6	7.39 (0.45)	7	7.71 (0.27)	22	7.47 (0.39)	24	7.34 (0.41)	28	7.11 (0.59)
P^2 mesiodistal	8	7.05 (0.44)	8	7.29 (0.71)	22	7.08 (0.53)	23	7.16 (0.67)	28	6.67 (0.59)
M^1 mesiodistal	7	10.99 (0.88)	11	10.61 (0.63)	22	10.78 (0.58)	24	10.78 (0.51)	25	10.20 (1.18)
M^2 mesiodistal	7	10.67 (0.58)	13	10.39 (0.48)	22	10.09 (0.72)	22	9.28 (1.06)	27	9.02 (0.88)
M^3 mesiodistal	6	9.50 (0.41)	11	9.41 (1.01)	14	9.08 (1.09)	8	8.82 (0.82)	19	8.18 (0.98)
I^1 buccolingual	5	7.08 (1.86)	10	7.17 (1.42)	22	7.34 (0.47)	22	7.77 (0.78)	29	7.20 (0.59)
I^2 buccolingual	6	7.81 (1.24)	9	6.89 (0.58)	22	6.56 (0.79)	25	6.57 (0.80)	27	6.41 (0.81)
$\underline{C}$ buccolingual	6	9.05 (0.49)	9	8.65 (0.89)	21	8.08 (0.89)	25	8.09 (0.72)	29	8.29 (0.63)
P^1 buccolingual	6	10.02 (0.45)	10	10.34 (1.16)	22	9.98 (0.52)	24	9.70 (0.58)	28	9.70 (0.71)
P^2 buccolingual	8	10.37 (1.36)	11	10.00 (0.49)	22	9.82 (0.53)	23	9.77 (0.94)	28	9.54 (0.74)
M^1 buccolingual	8	12.21 (0.61)	13	12.01 (0.66)	22	12.03 (0.45)	24	11.56 (0.59)	25	11.51 (0.57)
M^2 buccolingual	6	12.48 (0.32)	14	12.11 (0.89)	22	12.04 (0.59)	22	11.52 (0.63)	27	11.44 (1.15)
M^3 buccolingual	6	13.43 (1.53)	11	11.85 (1.22)	14	11.21 (1.96)	7	11.43 (1.06)	18	10.38 (0.70)
I_1 mesiodistal	4	5.32 (0.52)	6	5.61 (0.39)	22	5.58 (0.30)	24	5.50 (0.34)	29	5.35 (0.40)
I_2 mesiodistal	5	6.01 (0.71)	6	5.95 (0.40)	22	6.19 (0.41)	24	6.13 (0.34)	29	5.98 (0.44)
C mesiodistal	7	6.93 (0.63)	13	6.94 (0.41)	22	7.04 (0.59)	24	7.18 (0.48)	29	7.00 (0.54)
P_1 mesiodistal	7	7.41 (0.40)	13	7.52 (0.52)	22	7.33 (0.48)	24	7.34 (0.49)	29	6.98 (0.71)
P_2 mesiodistal	7	7.61 (0.27)	8	7.80 (0.57)	22	7.45 (0.43)	23	7.26 (0.52)	29	6.87 (0.40)
M_1 mesiodistal	7	11.46 (0.95)	12	11.90 (0.69)	21	11.56 (0.48)	23	11.20 (0.59)	25	10.61 (0.59)
M_2 mesiodistal	9	11.56 (0.44)	12	11.24 (0.44)	20	10.63 (0.61)	20	9.80 (0.65)	19	9.38 (0.43)
M_3 mesiodistal	10	11.49 (0.71)	15	11.34 (0.86)	14	10.74 (0.93)	4	10.88 (0.90)	16	9.99 (0.79)

Table 6.2. (*cont.*)

Measurement	Hoabinhian		Neolithic/metal age		Semelai		Temiar Senoi		Semang	
	No.	Mean (mm (SD))	No.	Mean (mm (SD))	No.	Mean (mm (SD))	No.	Mean (mm (SD))	No.	Mean (mm (SD))
I_1 buccolingual	5	6.39 (0.20)	7	6.24 (0.54)	22	6.22 (0.54)	23	6.34 (0.76)	28	6.09 (0.41)
I_2 buccolingual	6	6.70 (0.55)	8	6.74 (0.45)	22	6.54 (0.36)	24	6.84 (0.46)	27	6.50 (0.48)
C buccolingual	7	8.37 (0.34)	13	8.31 (0.54)	22	7.71 (0.68)	24	7.84 (0.82)	29	7.88 (0.66)
P_1 buccolingual	7	9.06 (0.56)	13	8.75 (0.51)	22	8.39 (0.54)	24	8.24 (0.55)	29	8.25 (0.66)
P_2 buccolingual	7	9.11 (0.71)	9	9.04 (0.41)	22	8.77 (0.42)	23	8.64 (0.56)	29	8.43 (0.71)
M_1 buccolingual	7	11.51 (0.88)	14	11.04 (0.48)	21	11.08 (0.57)	23	11.12 (1.36)	25	10.89 (1.28)
M_2 buccolingual	9	11.35 (0.55)	13	10.79 (0.61)	20	10.76 (0.41)	20	10.30 (0.79)	19	10.38 (1.25)
M_3 buccolingual	10	11.31 (0.45)	16	10.08 (2.81)	14	10.52 (0.65)	4	10.48 (0.15)	16	10.52 (1.94)

I, incisor; C, canine; P, premolar; M, molar.

Table 6.3. *Statistically significant differences between tooth diameters in the Malayan samples*

Samples	No. cases	Specific diameters	
		Mesiodistal	Buccolingual
Hoabinhian versus neolithic/metal	1		M_2
Hoabinhian versus Semelai	9	M_2, M_3	I^2, $\underline{C}$, M^2, lower C, P_1, M_2, M_3
Hoabinhian versus Temiar	11	M^2, P_2, M_2	$\underline{C}$, M^1, M^2, M^3, lower C, P_1, M_2, M_3
Hoabinhian versus Semang	17	M^1, M^2, M^3, P_1, P_2, M_2, M_3	I^2, $\underline{C}$, M^1, M^2, M^3, I_1, lower C, P_1, P_2, M_2
Neolithic/metal versus Semelai	3	M^2	Lower C, P_1
Neolithic/metal versus Temiar	11	P^1, M^2, P_2, M_1, M_2	M^1, M^2, lower C, P_1, P_2, M_2
Neolithic/metal versus Semang	16	P^1, P^2, M^2, M^3, P_1, P_2, M_1, M_2, M_3	P^2, M^1, M^2, M^3, lower C, P_1, P_2
Semelai versus Temiar	8	M^2, M_1, M_2	I^1,* M^1, M^2, I_2,* M_2
Semelai versus Semang	16	I^1, P^1, P^2, M^1, M^2, M^3, I_1, P_1, P_2, M_1, M_2, M_3	M^1, M^2, M^3, P_2
Temiar versus Semang	9	I^1, P^2, M^1, P_1, P_2, M_1, M_2	I^1, I_2

I, incisor; C, canine; P, premolar; M, molar. Two-tailed *t*-tests employed using the *t*-test formula which does not assume equality of variance between the compared populations (Keller and Warrack 2003). In all the statistically significant differences, except for the two Semelai–Temiar comparisons indicated by *, the sample with larger teeth (according to the Penrose size statistic) displays the larger diameter than the sample with smaller teeth (see Table 6.4).

Table 6.4. *Square roots of Penrose size distances (male tooth diameters) in descending order*[a]

Sample	Penrose size distance (square root)
Murray Valley	1.408
Hoabinhian	1.111
Neolithic/metal phase	0.870
Khok Phanom Di	0.687
Semelai	0.605
Javanese	0.534
Senoi	0.443
North Chinese	0.189
Semang	0.0

[a] Size expressed as a value greater than Semang tooth size. The size distance between any two samples is the difference between their distances from the Semang.

Table 6.5. *Seriated square roots of Penrose shape distances: male tooth diameters*

	MV	HB	NM	Java	Sl	Sg	Sn	KPD	NCh
Murray Valley (MV)	–	0.626	0.734	0.735	0.760	0.880	1.008	0.934	0.977
Hoabinhian (HB)		–	0.569	0.691	0.714	0.841	0.848	0.881	0.953
Neolithic/metal (NM)			–	0.481	0.459	0.720	0.694	0.762	0.730
Javanese (Java)				–	0.314	0.549	0.503	0.398	0.647
Semelai (Sl)					–	0.512	0.448	0.517	0.559
Semang (Sg)						–	0.381	0.639	0.599
Senoi (Sn)							–	0.558	0.555
Khok Phanom Di (KPD)								–	0.414
North Chinese (NCh)									–

Comparative data sources are identified in Appendix 6.1.

tooth area. Hoabinhian tooth size approaches that of Murray Valley Aborigines, who are renowned for their massive teeth (Brace 1980); the pooled neolithic/metal phase teeth come next and are larger than those at the neolithic Thailand site of Khok Phanom Di. Semelai and Senoi teeth are about the same size as those of Batawi Javanese, while the Semang teeth are even smaller than those of North Chinese. At least in the case of the Malayan samples, the descending order of tooth size correlates with the descending order in stature, suggesting that general body size reduction may be a sufficient explanation for the differences in tooth size.

'Shape' differences resulting from variation between the compared populations in their tooth diameter proportions are also apparent and largely parallel the previously noted tooth size differences (Table 6.5). Murray Valley and Hoabinhians cluster together to make up the sister group to the main cluster, which itself has two branches. One of these includes all neolithic to recent Indo-Malaysian samples, and the other clusters Khok Phanom Di with North Chinese (Fig. 6.3a). Upon seriating the Penrose shape distances (see Bulbeck (1992) for discussion of seriated dendrograms), we can see that the Murray Valley and North Chinese samples stand at the Australoid and Mongoloid extremes in shape variation, and that tooth proportions in Malaya have become less Australoid between Hoabinhian and recent times (Table 6.5). Evidently, tooth size reduction in the Malay Peninsula specifically, and the Asia–Pacific region generally, has been more pronounced on certain tooth diameters than on others.

Bulbeck (1982) noted that Australian Aborigines stand out from other Oriental–Pacific populations by having second molars and, to a lesser degree, third molars that are both absolutely and relatively large. Bulbeck

(a)

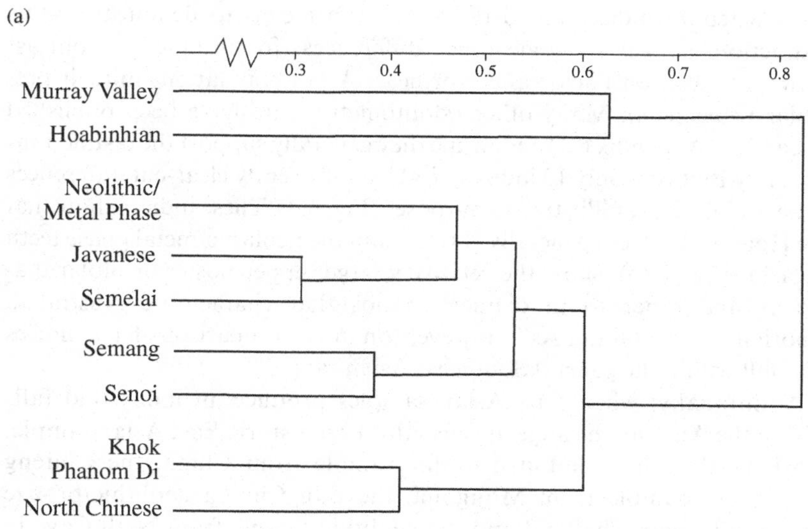

(b)

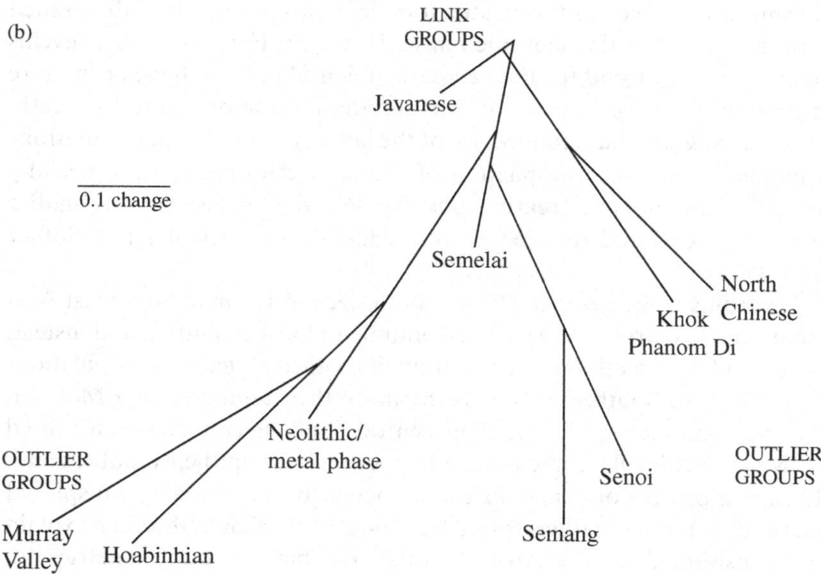

Figure 6.3. (a)Square roots of Penrose shape distances and male tooth diameters seriated average-linkage dendrogram (goodness of fit 92.3%). (b)Minimum spanning tree with branch lengths adjusted to best approximate intergroup distances and structure adjusted to retain seriated order (left to right).

constructed 75 indices based on population means to demonstrate this distinction, as well as some other differences (for instance, a contrast between southwest Pacific and Northeast Asian populations in their premolar robustness). Many other odontometric data have been published since 1982 (Appendix 6.1) and, while they generally support the distinctions noted by Bulbeck, only 15 indices produce sufficiently clear-cut differences to be useful here for illustrative purposes (Fig. 6.4). These indices show that the Hoabinhian teeth generally (11/15), and the neolithic/metal phase teeth occasionally (4/15), show the relatively large, upper posterior molar diameters and other tooth diameter ratios that characterise Australian Aborigines. The Orang Asli, however, on most comparisons have indices that fall within the general Southeast Asian range.

Additionally, when East Asian samples produce an index that falls within the Australian range, it is usually a prehistoric East Asian sample, most notably the Liu Lin neolithic sample from China, the Chifeng bronze age sample from Mongolia, the Ban Chiang neolithic/bronze age sample from Thailand and the neolithic sample from Niah Cave in Borneo (Fig. 6.4). Brace and colleagues (Brace 1976, Brace *et al.* 1984) showed that these regions generally witnessed tooth size reduction during the Holocene, which they (largely) attributed to the effects of the 'probable mutation effect' as the masticatory demands placed on teeth lessened with the advent of the neolithic phase. However, Fig. 6.4 further reveals an evolutionary trend for the decrease in dental mass to have been more pronounced on the posterior molars than on more anterior teeth. This may suggest that reduced use of the jaws by agricultural populations heightened the risk of impaction of the later erupting teeth, especially the posterior molars, creating positive selection pressures for smaller teeth in general and reduced posterior molars in particular (see Sofaer *et al.* 1971).

According to Brace *et al.* (1984), tooth size reduction in Southeast Asia occurred too sharply to be ascribed entirely to local evolution, and instead can be explained by the expansion of small-toothed agricultural populations from China to Southeast Asia. To rephrase their argument in a Malayan context, neolithic populations from central Thailand could have infiltrated the Malay Peninsula in the middle Holocene, built up their numbers and, through a process of demic diffusion, increasingly absorbed or displaced the resident hunter–gatherers (see Bellwood 1993). Could this then explain the transition from the Australian-like Hoabinhian odontometrics to, first, the intermediate neolithic/metal phase tooth diameters and finally the Orang Asli odontometrics, which in some ways resemble those at Khok Phanom Di? The answer is no.

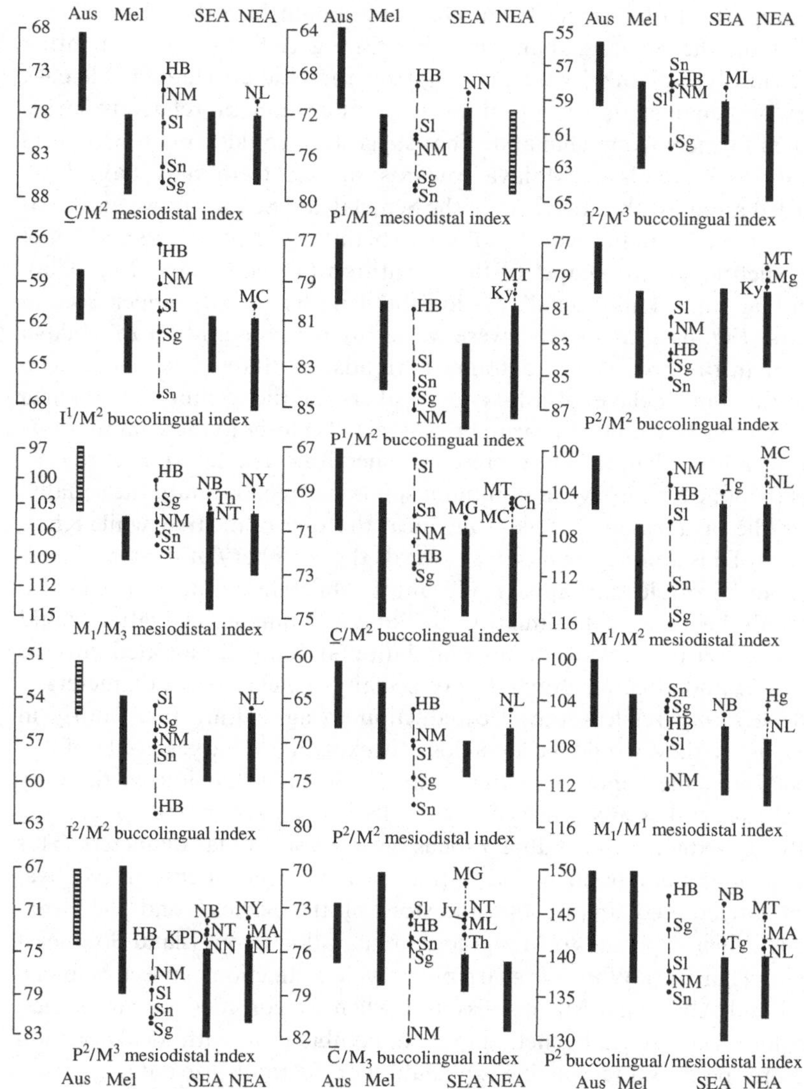

Figure 6.4. Main tooth shape indices based on samples' mean diameters. Aus, Australia; Mel, Melanesia; SEA, Southeast Asia; NEA, Northeast Asia. Malayan group: HB, Hoabinhian; NM, neolithic/metal phase; Sl, Semelai; Sg, Semang; Sn, Senoi. Southeast Asian groups within the Australian range: NN, neolithic Niah; ML, Leang Codong ('Leang Tjadang', metal phase), NB, neolithic/bronze age Ban Chiang; NT, neolithic/bronze age Non Nok Tha; MG, Gilimanuk (metal phase); KPD, Khok Phanom Di; Th, Thais; Tg, Tagalogs; Jv, Surabaya Javanese. Northeast Asian groups within the Australian range: NL, neolithic Liu Lin; MC, bronze age Chifeng Mongolians; MT, Tanegashima Yayoi (iron age); NY, Neolithic Yang Shao; MA, An Yang (bronze age); Ky, Kyoto Japanese; Mg, Urga Mongolians; Ch, Thailand Chinese. Teeth: I, incisor; C, canine; P, premolar; M, molar.

Immigration of farmers from Thailand should have had less genetic impact on the Semang than the other Orang Asli. So such an influx could hardly explain why the Semang have the smallest teeth (Table 6.4), unless we change the proposal to an immigration of relatively large-toothed farmers from Thailand. That suggestion could explain why local evolutionary trends in Malaya towards smaller teeth may have been retarded amongst the ancestors of the Semelai and Senoi. However, immigration from Thailand would still not explain why the Semang and Senoi have such a similar 'shape' to their dentition (Table 6.5 and Fig. 6.3a), including remarkably small mesiodistal lengths on the upper second molars (Fig. 6.4). Moreover, were we to invoke immigration to Malaya to explain the observed odontometric trends, the Batawi Javanese, who constitute an enclave of Malayic speakers in the vicinity of Jakarta (Adelaar 1992), would appear to be biologically closer to these immigrants than to Khok Phanom Di (representing neolithic Thailand). Figure 6.3b joins the nine samples via a minimum spanning tree; to retain the seriated order, the Javanese are a 'link group' near the apex of the tree, while Khok Phanom Di is among the outliers towards the periphery of the tree. Khok Phanom Di would thus appear relevant to Malayan odontometric evolution only because of its similarity to Batawi Javanese. The latter would appear closer to the roots of any population(s) that had migrated into the Peninsula and affected Orang Asli (especially Semelai) tooth diameters.

Models that relate tooth size reduction to agriculture fare poorly in Malaya, as they would be at a loss to explain the larger teeth of the Semelai and the Senoi compared with the Semang. Obviously, the only model that makes any sense of our results is body size reduction causing tooth size reduction, including a focus on posterior molar diameters. This local evolutionary trend would appear to have been intense in Malaya, albeit ameliorated among the ancestors of the Semelai and the Senoi through their exposure to Mongoloid immigration (from island Southeast Asia or Thailand). While the sharp phenotypic distinction between Semang, Senoi and Aboriginal Malays dissolves when we consider odontometrics, recognition of external genetic inputs in combination with local selection pressures permits a reasonable explanation of the tooth size data.

Craniology

Hanihara (1994) has employed six main cranial indices to illustrate East Asia's complex population history and to draw attention to the retention of various 'archaic' traits amongst Indo-Malaysian populations to the present day. Table 6.6 presents the available data for prehistoric Malayans and the Orang Asli on these indices, as well as the neurocranial module obtained by taking the cube root of cranial length, breadth and basion–bregma height. The data are presented as ranges of variation in view of the small available sample sizes, which are sometimes limited to one or two observations. Differences between the samples illustrate certain temporal trends.

The Orang Asli are characterised by small skulls (module usually less than 150 in males and 145 in females), which tend to be mesocranic (cranial index 75–80), orthocranic to hypsicranic (height/length index above 70), and metriocranic to acrocranic (height/breadth index above 92). Apart from a slight tendency for Senoi crania to be longer than those of Semang and Aboriginal Malays, this rather high-vaulted shape of intermediate breadth is remarkably similar across all three groups. Facial shape also barely differs, tending to be quite broad (facial index 45–50) with medium to tall orbits (orbital index above 76) and a broad nasal cavity (nasal index over 51). Applying the Penrose statistic to a slightly larger suite of measurements, Bulbeck (1996) similarly demonstrated the resemblance between Semang and Senoi crania in both size and shape. As in odontometrics, the Orang Asli display a characteristic cranial form that belies their differences in external phenotype.

The Hoabinhian crania (especially the Gua Peraling female) tended to be larger than Orang Asli crania, longer (all are dolichocranic) and higher vaulted (all are acrocranic, with height/breadth index above 98). Faces were broader (hypereurenic, facial index less than 45) and the nasal aperture was remarkably broad (nasal index above 67). The neolithic crania were highly variable and included the smallest neurocranial module (the Gua Tok Long female), the broadest cranial shape (Gua Cha As.33.5.2), the lowest height/length and height/breadth indices (Gua Tok Long and Gua Cha As.33.5.2, respectively) and the broadest face and broadest orbits (Gua Cha As.33.5.1) recorded in the Malay Peninsula. This variability might suggest expansion of the gene pool through the addition of an immigrant population, except that a suitable source population cannot be identified. For instance, Ban Chiang specimens (and related crania from Ban Kao and Khok Phanom Di) may have tended to be broad vaulted, but their vaults were also too large and tall, and their faces and orbits too moderate in shape, to explain any of the other Malayan neolithic extremes of variation (Hanihara 1994: Table 6.2, Pietrusewsky and Douglas 2002, p. 29; see also Sangvichien *et al.* 1969, Tayles 1999). Finally, the iron age skulls from Kuala Selinsing were in general similar to the Orang Asli skulls, although three of four neurocranial modules were slightly above

Table 6.6. *Prehistoric Malayan and Orang Asli ranges for cranial module and indices*[a]

Module/index[a]	Measurements[b] (sample size)					
	Hoabinhian	Neolithic	Kuala Selinsing	Semang	Senoi	Aboriginal Malays
Neurocranial module[c] Males	~149 to 151.6 (3)	~143.1 to ~151.2 (2)	153.2–153.5 (2)	140.3–152.3 (10)	140.4–149.4 (5)	142.0–147.0 (2)
Females	~ 138.5 to ~155 (2)	~ 133.3 (1)	144.3–149.9 (2)	138.8–148.7 (4)	134.6–145.0 (5)	138.4 (1)
Cranial index (100 B/L)	62.4–74.7 (7)	66.2 to ~85.3 (7)	69.8–79.2 (4)	71.2–85.0 (15)	69.4–79.4 (10)	77.2–80.3 (3)
Height/length index (100 H′/L)	~71.8 to ~83.0 (5)	~67.6 to 79.9 (3)	71.4–75.0 (4)	68.9–82.8 (14)	71.5–81.1 (10)	74.3–76.7 (3)
Height/breadth index (100 H′/B)	98.5 ~115.0 (6)	82.1–105.9 (3)	93.3–102.6 (4)	90.0–108.0 (14)	92.6–108.4 (10)	95.0–96.2 (3)
Facial index (100 G′H/J)	~43.3 to ~44.3 (2)	~41.9 to ~49.9 (2)	~50.4 to ~54.9 (2)	45.4–53.0 (12)	44.2–53.0 (7)	47.5–49.1 (3)
Orbital index (100 O_2/O'_1)	80.0 to ~88.9 (2)	~69.4 to ~85.2 (4)	82.6 (1)	80.0–87.2 (14)	77.8–95.0 (8)	84.2–94.1 (3)
Nasal index (100 NH′/NB)	67.0–70.5 (2)	56.2–60.4 (4)	58.7–66.0 (2)	51.0–59.1 (14)[d]	50.0–68.2 (8)	52.1–57.5 (3)

L, maximum cranial length; B, maximum cranial breadth; H′, basion–bregma cranial height; G′H, upper facial height; J, facial (bizygomatic) breadth, O'_1, orbital breadth to dacryon; O_2, orbital height; NH′, nasal height to nasiospinale; NB, nasal breadth.

[a]Measurements identified by their skull symbol in the Biometric School system (von Bonin 1931); exact definitions are unspecified for many Orang Asli measurements (taken from Martin 1905). Semang data include measurements from Schebesta and Lebzelter (1926) and Bulbeck's measurements on the Pangan Siong skull.

[b]Data presented as ranges of variation in view of the small sample sizes.

[c]Cube root of cranial length × cranial breadth × bason–bregma height.

[d]Range excludes an outlier at 44.8.

the Orang Asli range, and one specimen had a narrower face (facial index almost lepten, at 54.9) than has been found on any Orang Asli.

Overall, a progression from the Hoabinhian to the Orang Asli craniometric pattern could be observed in that cranial length and height (and thus neurocranial size) decreased over time, as did facial and especially nasal breadth. On the one hand, the extremes of neolithic variation may signal a combination of newcomers associated with the 'Ban Kao culture', *plus* the partitioning of Peninsular Malaysia into several distinct gene pools as a result of social disruptions caused by the immigrant newcomers (Bulbeck 2004). On the other hand, the relative homogeneity of the Orang Asli may reflect convergent adaptation to socio-economic differentiation, via the complementary processes of biological adjustments to restricted gene pool size *and* some role played by limited gene exchange.

Lauer (2002, pp. 111–118) used discriminant function analysis to compare the more complete skulls in terms of their main measurements (those in Table 6.6) with certain benchmark populations reported by Howells (1973a). The following results were obtained from classifying the specimens (DC) and plotting them on two main canonical discriminant functions (CDF). The Hoabinhian male Gua Cha As.33.6.11 came out closest to Tasmanians (DC, CDF) with a secondary relationship to the New Britain Tolai (CDF). The neolithic male Gua Cha As.33.5.1 was closest to Australian Aborigines (DC) but also approached Andaman Islanders and prehistoric Peruvians (CDF). All three prehistoric females were closest to the Zulu of Africa (DC, CDF) and then to the Senoi (CDF). This affinity was remote in the case of the Hoabinhian from Gua Peraling, but the two Kuala Selinsing females formed a 'bridge' between Gua Peraling and the Zulu/Senoi skulls. The Semang and Senoi skulls were usually classified correctly but occasionally associated with African (Teita), Melanesian (Tolai), and New World (Eskimo, Arikara) samples. On the canonical discriminant function plots, the Semang males fell close to the Hoabinhian male, whereas the Semang females lay close to the Senoi females. Overall, evidence emerged for a postneolithic transition from African or southwest Pacific craniometrics to the subtly different Orang Asli pattern.

Bulbeck (2005) identified the nearest neighbours for all Malayan skulls with three or more of the measurements using the Fordisc 2.0 computer program (Ousley and Jantz 1994) to compare individual skulls with the benchmark populations reported by Howells (1973a). Fordisc analysis allows a wider menu of measurements from Howells' comparative populations than were available for Lauer's analysis but has the drawbacks of using a different set of measurements in each comparison and not

Table 6.7. *Fordisc 2.0 nearest neighbour comparisons for Malayan skulls*[a]

	Hoabinhian	Neolithic	Metal phase	Orang Asli
Easter Island	2	0	1	1
Zulu	2	0	0	2
Other Africans	0	0	0	6
Tasmanians	1	3	0	0
Tolai	0	1	0	0
New World	0	0	2	1
Philippines/Atayal	0	0	1	6
Andamanese	0	0	0	7
Ainu	0	0	0	2
Hungarians	0	0	0	1
Total	5	4	4	26

[a] Summarised from Bulbeck (2005). Andamanese are found to be the nearest neighbour for a smaller proportion of the Semang skulls than either the Senoi or the Malay Aborigines.

supporting direct comparison between Orang Asli and prehistoric Malayan skulls. As summarised in Table 6.7, Fordisc comparisons would liken the Hoabinhian skulls to Easter Islanders and Zulu, the neolithic skulls exclusively to Tasmanians and Tolai, and the metal phase skulls to Mongoloids (in Easter Island, the Philippines and the New World). Orang Asli nearest neighbours are highly diverse but show a focus on Andamanese, African and Philippine/Atayal comparisons. The aberrant status of the neolithic skulls may reflect divergent evolution through genetic isolation, as noted above. Removing them, we may trace a continuity of Easter Island/African-like specimens through time, combined with New World/Philippine tendencies appearing in the metal phase and, finally, including Andamanese similarities with the present-day Orang Asli. Howells (1973b, p. 176) likened the Andamanese to 'little Africans' based on their infantile features and African-tending craniometrics. In this context, note that the Penrose shape analysis that clustered the Semang and Senoi together also placed the Andamanese in the same cluster (Bulbeck 1996).

Lauer (2002, in his Appendix C) summarised the morphological observations made by Bulbeck on prehistoric Malayan crania. The modal male expressions are depicted in Figs. 6.5–6.7 along with the craniometric trend towards a broader, lower vault and narrower face. They suggest the following Hoabinhian to iron age transitions. Robustness of the

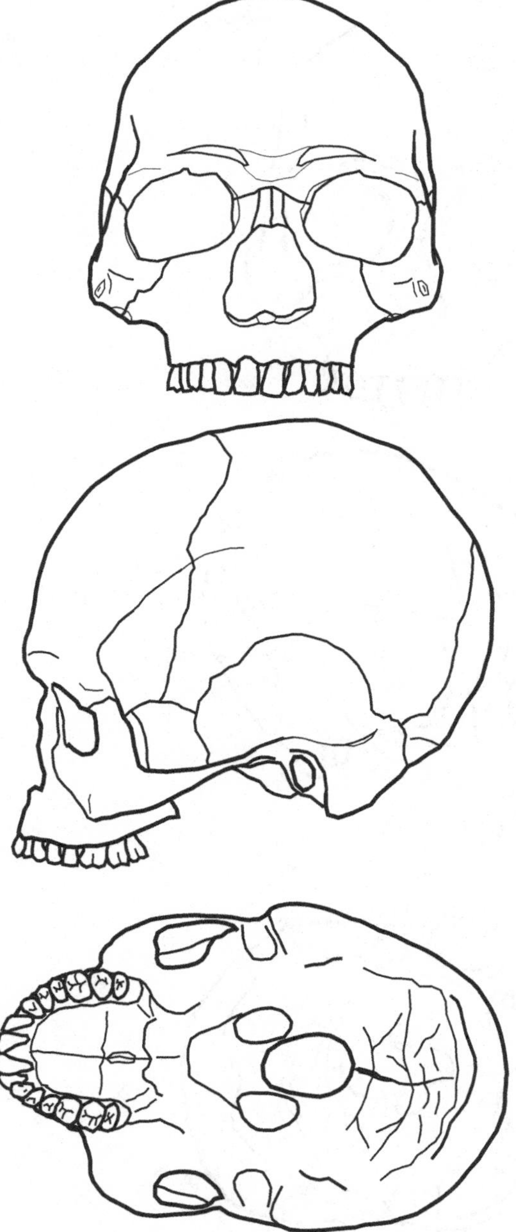

Figure 6.5. Modal Hoabinhian cranial morphology, showing masculine expressions for sexually dimorphic traits, based on Gua Cha As.33.6.11. (Adapted from data in Lauer (2002), Appendix 3.)

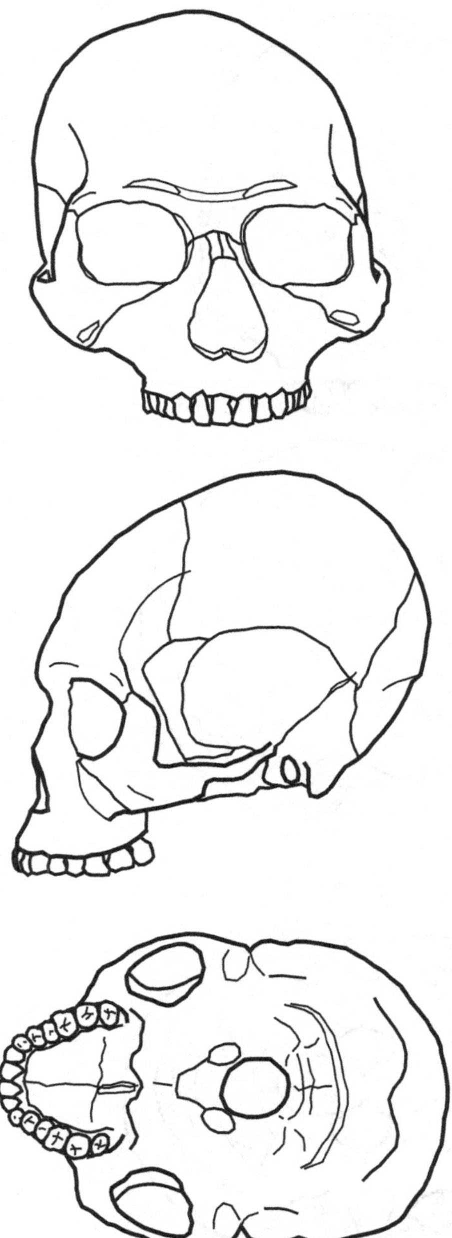

Figure 6.6. Modal neolithic cranial morphology, showing masculine expressions for sexually dimorphic traits, based on Gua Cha As.33.5.1. (Adapted from data in Lauer (2002), Appendix 3.)

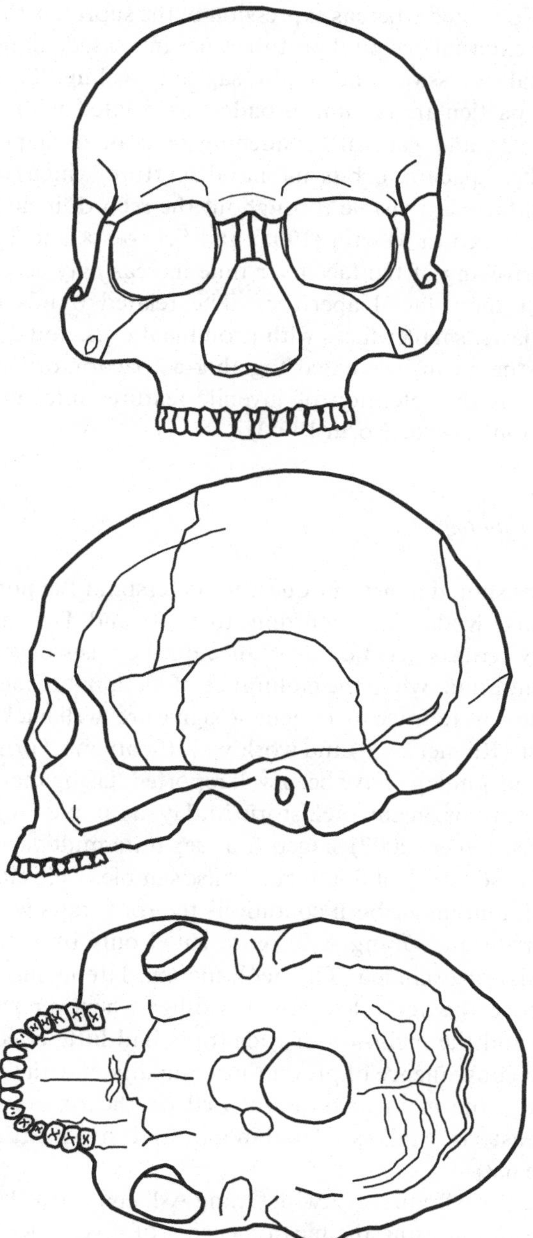

Figure 6.7. Modal iron age cranial morphology, showing masculine expressions for sexually dimorphic traits, based on Kuala Selinsing skull. (Adapted from data in Lauer (2002), Appendix 3.)

supraorbital region declined whereas expression of the supramastoid crest, parietal bosses and external occipital protuberance increased, along with a tendency to frontal recession and slight sagittal keeling. Palate size decreased and, in particular, became broader, associated with reduced subnasal prognathism and general broadening of cranial shape. Other features, notably the capacious orbits and nasal aperture, which dominate the relatively small face, are visible throughout the record, including the Orang Asli, as illustrated in Martin (1905) and Schebesta and Lebzelter (1926). Indeed, narrowing of the face over time increasingly accentuated the relative size of these facial apertures. The related trends towards reduced teeth and jaws, smaller faces with prominent eyes, and decreased supraorbital robustness can be treated together as paedomorphosis (i.e. the tendency towards the retention of juvenile features into adulthood (cf. Tobias 1957, Gould 1976, Storm 1995)).

Dental morphology

Dental morphology should assist our quest to understand the population history of Peninsular Malaysia. According to Scott and Turner (1997), dental morphology reflects genetic inheritance on the basis of long-term population differentiation, while the endurance of teeth in archaeological deposits should allow us to trace their 'genetic signature' well back in time. Further, Orang Asli (Rayner 2000) and worldwide (Scott and Turner 1997) patterns in crown morphology have been well reported. Using the criterion of at least six observations on any prehistoric Malayan sample for any trait included in analysis, Lauer (2002) staged four separate multivariate analyses. The neolithic and late neolithic/metal phase samples were either kept separate or pooled, and under both conditions the root traits were either included (thus barring the Orang Asli, represented only by their dental casts, from analysis) or excluded. The neolithic and late neolithic/metal phase teeth proved to be very close to each other whenever treated as separate samples, and the inclusion of root traits had little effect (apart from the unhelpful consequence of preventing comparison with the Orang Asli). Consequently, the main analysis focused on the six crown traits observed on at least six Malayan Hoabinhian and six neolithic/metal phase teeth (Table 6.8).

Lauer compared these frequencies with Orang Asli and extra-Peninsular frequencies of expression using the mean measure of divergence (MMD) statistic, along with the Tukey correction for small sample sizes (Table 6.9, bottom left half of matrix). Since the Hoabinhian data differed from the

Table 6.8. *Hoabinhian and neolithic/metal phase dental morphology frequencies used in the mean measure of divergence analysis*[a]

Trait	Frequencies (% (No. samples))	
	Hoabinhian	Neolithic/metal phase
Upper central incisor winging	0.0 (9)	0.0 (12)
Upper lateral incisor interruption grooves	57.1 (7)	50.0 (12)
Upper canine mesial ridge	0.0 (7)	0.0 (11)
Upper first molar Carabelli's cusp	0.0 (8)	20.0 (20)
Upper second molar hypocone absence	0.0 (7)	0.0 (16)
Lower second molar Y-groove pattern	0.0 (7)	6.7 (15)

[a] Presence/absence recognised in line with the break points recommended by Scott and Turner (1997).

neolithic/metal phase data more in terms of smaller sample size than trait frequencies (Table 6.8), yet the Hoabinhian MMD distances were consistently less than their neolithic/metal phase counterparts (Table 6.9), we may infer that the Tukey correction has deflated the Hoabinhian distances. To compensate for this and other sources of 'gross' differences between the samples in their MMD distances, Table 6.9 (top right half of the matrix) also expresses the distances in calibrated form (i.e. as a proportion of the 'expected' distance). The expected distance in any cell, defined here as the square root of the average MMD distance recorded for each of the two samples being compared in that cell, is placed in the denominator beneath the actual MMD distance put in the numerator. Therefore, if the distance between samples A and B is close to the average of A's and B's averaged MMD distances, the calibrated distance is close to 1. The value falls towards zero when A and B are particularly close to each other and rises above 1 (up to 2.54 in Table 6.9) when A and B are very unalike.

Inspection of the calibrated distances reveals they are less than 0.3 whenever we compare any two samples classified by Scott and Turner (1997) as sundadont: that is, Polynesia, early Southeast Asia, recent Southeast Asia and Micronesia. This same generalisation further extends to our three non-Negrito Orang Asli samples (the Temiar Senoi, Semelai and Temuan), supporting Rayner (2000) in characterising them as sundadont. Calibrated distances of 0.30 or less also applied whenever we compare any two of Scott and Turner's West Eurasian samples (North Africa, northern Europe and western Europe) and, with two exceptions, to all intersample comparisons involving Scott and Turner's sinodont

Table 6.9. *Mean measures of divergence distances (to two places) between prehistoric Malayans and comparative populations: original distances (bottom left section) and calibrated distances (top right section)*[a]

	NM	HB	WAf	Aus	Sn	Poly	Mel	Sl	ESEA	RSEA	Micr	SAf	Jom	Tmn	NGui	Jap	Khoi	SSib	Ch/M	Sg	NAf	Arc	NSib	NEur	Na-D	WEur	Amer
NM	–	**0.13**	0.45	0.51	0.92	1.11	1.14	1.50	1.71	1.71	1.36	1.31	1.49	2.02	1.34	1.72	1.30	2.01	1.72	1.88	1.92	1.68	1.85	1.79	1.70	2.03	1.89
HB	0.05	–	0.77	0.88	**0.13**	0.35	**0.22**	0.99	0.73	0.88	1.14	1.14	0.81	0.93	0.71	1.28	1.93	0.46	1.06	0.48	0.45	1.60	1.80	0.50	1.65	1.20	1.77
WAf	0.19	0.22	–	0.57	0.68	0.85	0.57	0.89	0.97	0.93	0.71	0.45	1.29	0.88	0.77	1.44	0.40	1.36	1.51	1.14	1.07	1.73	1.55	1.48	1.84	1.44	1.92
Aus	0.19	0.17	0.16	–	0.77	0.31	0.51	0.51	0.64	0.57	0.45	0.85	0.87	0.78	0.91	0.88	1.23	1.31	1.00	1.02	1.39	1.23	1.14	1.37	1.39	1.54	1.60
Sn	0.29	0.03	0.16	0.07	–	**0.06**	**0.17**	**0.06**	**0.06**	**0.13**	**0.14**	0.74	0.72	**0.26**	0.78	0.80	1.46	0.78	0.90	0.53	0.91	1.30	1.09	0.89	1.54	1.05	1.71
Poly	0.33	0.07	0.19	0.06	0.01	–	**0.18**	**0.06**	**0.13**	**0.20**	**0.20**	0.83	0.43	**0.14**	0.73	0.69	1.31	0.58	0.85	0.51	0.91	1.21	1.01	0.90	1.44	1.02	1.70
Mel	0.37	0.05	0.14	0.11	0.03	0.03	–	**0.17**	0.43	0.42	**0.14**	0.54	0.74	**0.13**	**0.29**	1.20	1.10	0.70	1.29	**0.28**	0.57	1.69	1.48	0.55	1.93	0.74	2.11
Sl	0.47	0.20	0.20	0.10	0.03	0.01	0.03	–	**0.07**	**0.20**	**0.15**	0.73	0.44	**0.14**	0.50	0.84	1.19	0.47	0.90	**0.25**	0.57	1.33	1.01	0.69	1.55	0.64	1.76
ESEA	0.49	0.14	0.21	0.12	0.01	0.02	0.07	0.01	–	**0.00**	**0.11**	0.76	0.51	**0.07**	1.00	0.38	1.32	**0.24**	0.42	0.58	0.85	0.87	0.57	1.07	1.00	1.07	1.18
RSEA	0.49	0.17	0.20	0.12	0.02	0.03	0.07	0.02	0.00	–	**0.05**	0.91	0.62	**0.00**	0.95	0.33	1.16	**0.30**	0.42	0.58	0.74	0.91	0.62	1.02	1.07	0.97	1.28
Micr	0.51	0.29	0.20	0.11	0.03	0.04	0.03	0.03	0.02	0.01	–	0.77	0.47	**0.06**	0.91	0.54	0.92	0.51	0.44	0.61	0.91	0.69	0.40	1.09	0.85	0.98	0.95
SAf	0.51	0.30	0.13	0.24	0.16	0.17	0.12	0.15	0.18	0.15	0.20	–	1.00	0.48	0.31	1.37	**0.29**	0.90	1.50	0.74	0.44	1.63	1.48	0.75	2.01	0.66	2.15
Jom	0.52	0.19	0.34	0.20	0.14	0.08	0.15	0.08	0.09	0.11	0.11	0.24	–	0.36	1.01	0.62	1.24	0.50	0.85	0.95	1.02	0.71	0.70	1.06	1.07	0.98	1.42
Tmn	0.55	0.17	0.18	0.14	0.04	0.02	0.02	-0.02	-0.01	0.00	0.01	0.09	0.06	–	0.50	0.63	1.01	**0.13**	0.77	**0.17**	0.31	1.14	0.81	0.61	1.46	0.51	1.65
NGui	0.55	0.20	0.24	0.25	0.18	0.16	0.07	0.11	0.21	0.20	0.25	0.09	0.26	0.10	–	0.98	0.72	0.88	1.84	**0.29**	**0.21**	2.13	2.02	**0.24**	2.39	0.34	2.54
Jap	0.62	0.31	0.39	0.21	0.16	0.13	0.25	0.16	0.07	0.06	0.07	0.34	0.14	0.11	0.47	–	1.50	0.58	**0.04**	1.26	1.39	**0.21**	**0.08**	1.63	**0.30**	1.62	0.47
Khoi	0.64	0.64	0.15	0.40	0.40	0.34	0.31	0.31	0.33	0.29	0.30	0.10	0.38	0.24	0.26	0.47	–	1.24	1.67	1.26	0.90	1.67	1.53	1.38	1.88	1.06	1.98
SSib	0.65	0.10	0.33	0.28	0.14	0.10	0.13	0.08	0.04	0.05	0.11	0.20	0.10	0.02	0.21	0.12	0.35	–	0.56	0.52	0.40	0.96	0.81	0.63	1.13	0.55	1.29
Ch/M	0.65	0.27	0.43	0.25	0.19	0.17	0.28	0.18	0.08	0.08	0.11	0.39	0.20	0.14	0.51	0.01	0.55	0.12	–	1.20	1.40	**0.30**	**0.18**	1.62	**0.26**	1.55	0.37
Smg	0.70	0.12	0.32	0.25	0.11	0.10	0.06	0.05	0.11	0.11	0.15	0.19	0.22	0.03	0.08	0.30	0.41	0.11	0.30	–	0.31	1.84	1.51	0.34	1.90	0.51	1.90
NAf	0.76	0.23	0.36	0.37	0.20	0.19	0.13	0.12	0.17	0.15	0.24	0.12	0.25	0.06	0.06	0.35	0.31	0.09	0.37	0.08	–	1.73	1.59	**0.13**	1.97	**0.06**	2.11
Arc	0.76	0.49	0.59	0.37	0.33	0.29	0.44	0.32	0.20	0.21	0.21	0.51	0.20	0.25	0.71	0.06	0.66	0.25	0.09	0.55	0.55	–	**0.03**	1.86	**0.10**	1.74	**0.26**
NSib	0.76	0.50	0.48	0.31	0.25	0.22	0.35	0.22	0.12	0.13	0.11	0.42	0.18	0.16	0.61	0.02	0.55	0.19	0.05	0.41	0.46	0.01	–	1.83	**0.13**	1.64	**0.24**
NEur	0.79	0.15	0.49	0.40	0.22	0.21	0.14	0.16	0.24	0.23	0.32	0.23	0.29	0.13	0.07	0.46	0.53	0.16	0.48	0.10	0.04	0.66	0.59	–	2.12	**0.09**	2.25
Na-D	0.87	0.57	0.71	0.47	0.44	0.39	0.57	0.42	0.26	0.28	0.29	0.71	0.34	0.36	0.90	0.10	0.84	0.33	0.09	0.64	0.71	0.04	0.05	0.85	–	2.03	**0.06**
WEur	0.90	0.36	0.48	0.45	0.26	0.24	0.19	0.15	0.24	0.22	0.29	0.20	0.27	0.11	0.11	0.46	0.41	0.14	0.46	0.15	0.02	0.62	0.53	0.03	0.82	–	2.20
Amer	1.07	0.68	0.82	0.60	0.54	0.51	0.69	0.53	0.34	0.37	0.36	0.84	0.50	0.45	1.06	0.17	0.98	0.42	0.14	0.71	0.84	0.12	0.10	1.00	0.03	0.98	–

NM, neolithic/metal phase; HB, Hoabinhian; WAf, West Africa; Aus, Australian Aborigines; Sn, Temiar Senoi; Poly, Polynesia; Mel, island Melanesia; Sl, Semelai; ESEA, early Southeast Asia; RSEA, recent Southeast Asia; Micr, Micronesia; SAf, South Africa; Jom, Jomon; Tmn, Temuan; NGui, New Guinea; Jap, Japanese; Khoi, Khoisans; SSib, South Siberia; Ch/M, China/Mongolia; Sg, Semang; NAf, North Africa; Arc, American Arctic; NSib, North Siberia; NEur, northern Europe; Na-D, Na-Dene; WEur, western Europe; Amer, Amerindians.

[a]Original distances from Lauer (2002: Table 4.4) employing comparative data in Scott and Turner (1997: Appendix A) and Rayner (2000). Calibrated values are given as a proportion of the 'expected' distance (see text). Bold indicates strong similarity (calibrated distance ≤0.3).

samples (Japanese, China/Mongolia, American Arctic, North Siberia, Na-Dene and Amerindians). Hence, despite the small number of traits employed, it would be reasonable to recognise provisional dental morphology complexes on the basis of all members having a calibrated MMD distance of 0.30 or less from each other in the current analysis. Consequently, we can accept a 'Melanesid' complex that includes Island Melanesia, New Guinea and the Semang (Table 6.9).

This affinity calls to mind Howells' (1973b) concept of an 'Old Melanesia', which used to incorporate the Indo-Malaysian archipelago as well as present-day Melanesia. On this basis, the Hoabinhian sample should exhibit a Melanesid dental morphology. However, only the island Melanesian sample resembled Hoabinhians; the New Guinea and, most critically, Semang samples were quite different. Indeed, Hoabinhian specimens had their closest affinities with the Temiar, who are a sundadont group, and the neolithic/metal phase teeth. The latter sample showed no other truly close association; it was quite close to West Africans and Australian Aborigines, and tolerably close to Temiar, but very distant from any other sample. This result is problematic because we would expect the neolithic/metal phase sample to be intermediate between the Hoabinhians and at least one of the Orang Asli. Instead, the neolithic/metal phase teeth clearly diverged away in the direction of West Africans and Australians. A multidimensional plot of the MMD distances (Figure 6.8) illustrates this and the other relationships discussed above.

The neolithic/metal phase sample, which was dominated by neolithic teeth (Lauer 2002), departed from the expected Hoabinhian to Orang Asli trajectory in the same way that the neolithic skulls seem to stand aside from a smooth craniometric transition (Table 6.7). In particular, no evidence emerges for a more Mongoloid affinity with the advent of the neolithic phase; if anything, a southwest Pacific affinity comes to the fore. However, we would misinterpret our results if we classified any prehistoric Malayan samples as southwest Pacific (or African), because the resemblances straddle both of these groups. Rather, it looks like prehistoric Malayans had their own range of variation that was not replicated anywhere else and, accordingly, produce sporadic similarities that are identified when individual specimens or small samples are compared with populations in other Old World tropical regions.

One salient result from dental morphology is that the Temiar, not the Semang, seem to be the group most closely related to our Hoabinhian sample, and whose ancestry can potentially be traced via the neolithic/metal phase sample. Support for this scenario is also found in the relatively large teeth and narrow cranial vault of the Senoi compared with the

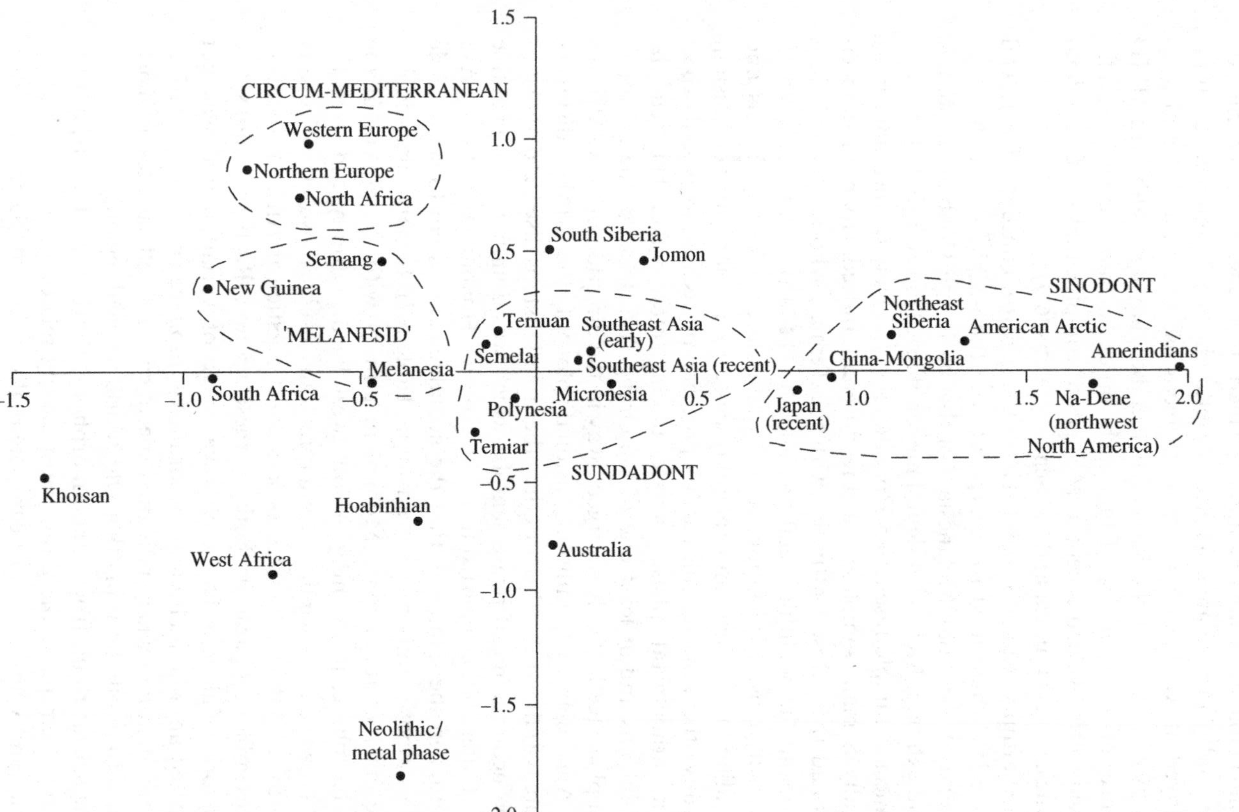

Figure 6.8. Two-dimensional scattergram of dental morphology mean measure of divergence distances shown in Table 6.5 (bottom left). (Adapted from Fig. 4.12 in Lauer (2000).)

Semang. However, the relevance here is less a question of ethnogenesis than the location of the prehistoric samples. The Temiar range includes Gua Cha, which has provided the majority of our Hoabinhian and neolithic specimens, as well as Gua Peraling. On its western margin, the Temiar range includes or abuts Gua Harimau, Gua Baik, Gua Gunung Runtuh and Gua Kerbau, four other sites with many burials between them (Fig. 6.1). As would be consistent with a local continuity scenario, we have some evidence that a larger proportion of the genes of the prehistoric inhabitants of northwestern Peninsular Malaysia are to be found amongst present-day Temiar than the Semang.

Discussion

In an earlier study of human evolution on the Malay Peninsula, Bulbeck (1996) faced the conundrum of explaining the osteological similarity between the Semang and Senoi when they are so different in external appearance, and when the neolithic skulls suggested minimal transition toward the Orang Asli condition. We now have at our disposal important resources unavailable in 1996 (useful data on Orang Asli teeth and mitochondrial DNA, and a far better coverage of prehistoric specimens) yet the same basic conundrum remains. This suggests that the problem is not some by-product of sample size or data failure but instead, if we can solve it, may hold the key to unlocking the 'black box' of late Holocene Malayan human evolution. Bulbeck's solution in 1996 was to propose convergent evolution: stature reduction and associated paedomorphosis for the ancestors of the Semang (albeit undocumented in the prehistoric skeletal record because the distributions of the relevant sites and the proto-Semang do not overlap) and the postneolithic, 'Mongoloid' genetic influence on the proto-Senoi who would have occupied the same region as the prehistoric burial sites. With the benefit of further data and analysis, we should now be able to improve on that solution.

Initially, the small number of available prehistoric specimens do not preclude clear evidence of osteological change in Malaya since the mid Holocene. The Orang Asli are represented by substantial samples of limb bones and crania, where the range of variation sometimes fails to overlap with the Hoabinhian range. Similarly, the teeth samples have smaller diameters that their prehistoric counterparts, differences that are often statistically significant. There is little reason to doubt that, since the middle Holocene, stature has reduced in Peninsular Malaysia along with tooth size (especially posterior molar size), and that vaults have broadened while faces and nasal apertures have narrowed, resulting in skulls that resemble Andamanese crania. Further, it seems unlikely that all three Orang Asli groups could be recent immigrants to Malaya. If they had arrived from different sources, why would they be osteologically so similar to each other, and why are they associated with Aslian languages that are spoken only in Malaya? If they immigrated as a bloc, which processes could account for their greater external phenotypic variability than found anywhere else in Southeast Asia? Even if it were argued that comparable phenotypic variability characterises the Philippines and Nicobar–Andaman Islands, there would still be a language mismatch as the indigenous languages of these island groups are neither Aslian nor Malayic (Adelaar 1992; Benjamin 2004).

However, if we accept that the small sample size for these prehistoric specimens does present a problem, could this account for the anomalies when we seek evolutionary trends? The available Hoabinhian sample presented a motley combination of wider comparisons: odontometrics were Australian-like, tooth morphology was Temiar-like, and craniometric comparisons aligned specimens with Easter Island, Zulu and southwest Pacific populations. The neolithic and metal phase sample showed an even wider span of comparisons: odontometrics were transitional; dental morphology was reminiscent of West African and Australian parameters, and craniometrics variably invoked southwest Pacific, African and Philippine comparisons. Consequently, population affinity is even more enigmatic for the neolithic/metal phase sample than the Hoabinhian sample, despite increased sample sizes in the format (Tables 6.2, 6.7 and 6.8), suggesting that sheer sample size limitations are not the cause of the anomalies. Further, consideration of individual neolithic crania reinforces the impression of expanded variation compared with the Hoabinhian sample, and the difficulties in linking this phenomenon to immigration from neolithic Thailand.

There are two good reasons why comparisons with prehistoric Malayans should produce unruly results and we predict that, as sample sizes augment in the future, they will exonerate our caveats. First, when we find prehistoric Malayan affinities in faraway places such as Africa or the southwest Pacific, this is not classificatory information but mainly evidence of differences from Southeast Asia's current inhabitants. Hence, we need not worry when slight differences in data input or analytical methods produce a switch (e.g. from Africa to the southwest Pacific) in the suggested affinity, as the conclusion to draw would be confirmation of atypicality by Southeast Asian standards. Second, inconsistency between comparisons can be expected where there is migration from outside locations. If we allow that immigrants did establish themselves in late Holocene Malaya, their arrival would have disrupted the original network of communications and mate exchange. One likely result is fragmentation of the original gene pool(s) into more isolated populations, and enhanced scope for runaway local evolution through genetic drift, especially among foraging groups with low population densities. As we would expect some gene exchange between immigrants and indigenes, even as local populations became genetically more isolated overall, we could expect some features on some samples to display evidence of convergence between immigrants and indigenes, and other features to show divergence. Continuity of this process into recent times could largely explain why the Orang Asli are osteologically similar yet diverse in their external phenotype, especially when, as Rayner and Bulbeck (2002) noted, mating preferences would have been affected by physical looks but not by features observable only on skeletons.

Segmentation of the indigenous hunter–gatherer population structure can also help to explain why the Semang, when compared with the Hoabinhians, appear to be the most derived Orang Asli group. Our reasoning adapts Storm's (1995) model, which attributed the Mongoloid features of reduced tooth size, rounded and gracile cranium, and narrower faces in late Holocene Java to body size reduction. Whereas Storm invoked Holocene climatic amelioration to explain these paedomorphic trends, adapting Brown's (1989) explanation for similar changes to Holocene Australian hunter–gatherers, we propose instead dietary change. Rock shelters in Malaya reveal a distinction between early to middle Holocene faunal assemblages, where pigs and other ungulates dominated, compared with late Holocene assemblages, which featured monkeys and other arboreal game. This change would appear to be associated with a transition from parties hunting along well-cleared trails to the Semang practice of one or several individuals foraging surreptitiously in the undergrowth, armed with blowpipes and subsisting mainly on a seasonally variable, vegetarian diet (Bulbeck 2003). The steep decrease in available high-quality protein, combined with foraging success of agile individuals for whom muscular strength was of minor importance, would have selected for smaller bodies and the associated paedomorphic trends (Bulbeck 1996). Thus, changed foraging strategies in the late Holocene would have exerted positive selection pressures towards a thriftier genotype, without which any hunter–gatherer group would have succumbed to fragmentation and outright loss of territory as non-foragers colonised the Peninsula (Bulbeck 2004). We draw an analogy with the Andamanese, who are also paedomorphic 'Negrito' foragers, even if their evolutionary challenge would have been adaptation to small islands rather than coexistence with non-foragers (Cooper 2002).

Evidence for immigration of non-forager groups comes from genetics, archaeology and history. Mitochondrial DNA data imply that the Temiar and Aboriginal Malays owe at least some of their ancestry to populations from mainland and island Southeast Asia, respectively. Changes in material culture in Malaya during the middle and late Holocene have clear parallels with Thailand and Sumatra, respectively (Bellwood 1993, 1997); even if the case for a demic diffusion of prehistoric farmers is yet to be detailed, we doubt that outsiders would donate technology and knowledge to local populations without expecting land, labour and sexual relations in return. Written statements and historical linguistics reflect a diversity of visitors to the Peninsula prior to 1,000 years BP, including South Asians, Persians, Mons, Khmers and early Malayic speakers (Bulbeck 2004). Some of the osteological diversity we see in the archaeological record may reflect the tangled genetic calling card left by this medley of visitors. Further, the commotion associated with Malaya's entry into the world of long-distance trade, including traffic in captive slaves, would have driven the proto-Senoi as well as proto-Semang into remote locations (Bulbeck 2004), increasing diversity through genetic isolation.

Bellwood (1993) stressed the importance of explaining the spread of Aslian languages across Malaya and cautions that population expansion should not be ignored. However, this would be no reason to reduce everything down to a question of agricultural populations in expansionary mode and hunter–gatherers in retreat. As we have shown, the Temiar are the Orang Asli with the strongest credentials for biological continuity with the people represented by our Hoabinhian and late Holocene burials. Although sampling may have robbed us of the chance to trace the ancestry peculiar to the Semang, it is also possible that the proto-Semang, far from retracting, could have expanded into Peninsular Malaysia from an immediate homeland north of the Thailand–Malaysia border, as their strong association with north Aslian languages suggests (Lauer 2002). Finally, the odontometric and (postneolithic) craniometric comparisons do reveal a consistent trend toward the Orang Asli condition, but body size reduction would be sufficient explanation for the changes involved. This is not to deny any genetic contribution by immigrants, only to suggest that these contributions were minor during the prehistoric period covered by our samples.

As for the comparative homogeneity of the Orang Asli in their skeletal traits, three complementary explanations may be offered. First, it may reflect late Holocene gene exchange between Orang Asli groups as the

Peninsula hinterland became increasingly crowded through the loss of land (and annexed populations) to politically dominant, coastally based polities. Mating preferences based on visible appearance could, however, have maintained or even increased the groups' differences in external phenotype. Second, to extend our argument on the Semang to the Senoi, these groups' osteological similarities may reflect parallel evolution towards a thrifty genotype in the face of increased social circumscription. Finally, the Aboriginal Malays adopted Malayic social organisation including large kin and domestic units (Benjamin 1985), buffering them against restrictions on group hunting (Hislop 1954) and other corporate economic activities (Benjamin 2002). However, a consequence of this would have been enhanced gene exchange with paedomorphic Mongoloid immigrants, notably Malayic speakers.

Conclusions

Fix (2002) outlined a model of *in situ*, cultural and genetic differentiation between the three Orang Asli groups, a model compatible both with limited gene flow between these groups and some level of migration into the Peninsula. Our osteological evidence adds time depth to Fix's model. This is less in the sense of revealing a deep ancestry to the tripartite division of the Orang Asli than in reconstructing a complex interplay of dialectical evolutionary processes that evidently operated well into the past. It allows us to explain, perhaps overexplain, both the similarities and the differences between the Orang Asli, which simple models based on migration (plus miscegenation) or local evolution fail to do. Study of prehistoric burials demonstrates body size reduction over the last five millennia, a suite of related craniodental paedomorphic trends and some anomalous findings that we attribute to disruptions in mating patterns created by late Holocene immigration. While our model can explain the anomalies, we also acknowledge the potential contribution of sampling distortions, given that our samples are small and amalgamated from an uncontrolled cross-section of sites. We look forward to when an increase in prehistoric skeletal samples will allow a more definitive test of the detail of our model.

Acknowledgements

Two small Australian Research Council grants to Bulbeck funded his fieldwork to record Malayan human remains in Malaysia and Europe

(Biological Anthropological Perspectives on Holocene Cultural Change in Peninsular Malaysia) and to collect dental casts in Malaysia (Study into Malay Peninsular Aborigines' Dental Differentiation). We thank Professor Rahimah Abdul Kadir, Dr Daw Mohammad Swessi, Dr Paula Nuti Pontes and Dr Zamri bin Radzi (University of Malaya) for taking the casts, and Dr Adi Haji Taha and Muhamad Mahfuz Nordin (Malaysia's Department of Museums and Antiquities) for organising the expedition. The following authorities graciously allowed Bulbeck to record prehistoric Malayan human remains under their care: Dr Adi Haji Taha, Professor Zuraina Majid (Universiti Sains Malaysia, Penang), Dr Robert Foley (Department of Human Biology, Cambridge University) and Dr John de Vos (Natural History Museum, Leiden). Professor Stephen Oppenheimer, Dr Martin Richards and three anonymous referees offered useful comments on the draft.

References

Adelaar K. A. 1992. *Proto-Malayic: A Reconstruction of its Phonology and Parts of its Morphology and Lexicon*. Canberra: Australian National University.

Ammerman A. J. and Cavalli-Sforza L. L. 1984. *The Neolithic Transition and the Genetics of Populations in Europe*. Princeton: Princeton University Press.

Bellwood P. 1985. *Prehistory of the Indo-Malaysian Archipelago*. Sydney: Academic Press.

1993. Cultural and biological differentiation in peninsular Malaysia: the last 10,000 years. *Asian Perspectives* **32**: 37–60.

1997. *Prehistory of the Indo-Malaysian Archipelago*, revised edn: Honolulu: University of Hawaii Press.

Benjamin G. 1985. In the long term: three themes in Malayan cultural ecology. In Hutterer K. L., Rambo A. T. and Lovelace G., eds., *Cultural Values and Human Ecology in Southeast Asia*. Ann Arbor MI: University of Michigan Papers on South and Southeast Asia, pp. 219–271.

1987. Ethnohistorical perspectives on Kelantan's prehistory. In Hassan Shuhaimi N. and Abdul Rahman N., eds., *Kelantan zaman awal: kajian arkeologi dan sejarah di Malaysia*. Kota Bahru: Perbadanan Muzium Negeri Kelantan Istana Jahar, pp. 108–153.

2002. On being tribal in the Malay world. In Benjamin G. and Chou C., eds., *Tribal Communities in the Malay world: Historical, Cultural and Social Perspectives*. Leiden: International Institute for Asian Studies/Singapore: Institute for Southeast Asian Studies, pp. 7–76.

2004. The Aslian languages: an assessment. In Steinhauer H. and Collins J. T., eds., *Endangered Languages and Literatures of Southeast Asia*. London: SOAS.

Brace C. L. 1976. Tooth reduction in the Orient. *Asian Perspectives* **19**: 203–219.

1980. Australian tooth-size clines and the death of a stereotype. Current *Anthropology* **21**: 141–164.

Brace C. L., Shao X. Q. and Zhang Z. B. 1984. Prehistoric and modern tooth size in China. In Smith F. H. and Spencer F., eds., *The Origin of Modern Humans*. New York: Liss, pp. 485–516.

Brown P. 1989. *Terra Australis 13: Coobool Creek*. Canberra: Australian National University.

Bulbeck F. D. 1982. Continuities in Southeast Asian evolution since the late Pleistocene. M.A. thesis. The Australian National University, Canberra, Australia.

1992. A tale of two cities: the historical archaeology of Gowa and Tallok, South Sulawesi, Indonesia. Appendix A: the South Sulawesi language group. Ph.D. thesis, Australian National University, Canberra.

1996. Holocene biological evolution of the Malay peninsula Aborigines (*Orang Asli*). *Perspectives in Human Biology* **2**: 37–61.

2003. Hunter–gatherer occupation of the Malay peninsula from the Ice Age to the Iron Age. In Mercader J., ed., *The Archaeology of Tropical Rain Forests*. New Brunswick, NJ: Rutgers University Press, pp. 119–160.

2004. Indigenous traditions and exogenous influences in the early history of Peninsular Malaysia. In Glover I. and Bellwood P., eds., *Southeast Asia: Origins to Civilization*. London: Routledge Curzon, pp. 314–336.

2005. The Gua Cha burials. In Zuraina M., ed., *The Perak Man and other Prehistoric Skeletons of Malaysia*. Penang: University of Science Malaysia Press, pp. 253–309.

Carey I. 1976. *The Orang Asli*. Kuala Lumpur: Oxford University Press.

Cooper Z. 2002. *Archaeology and History: Early Settlements in the Andaman Islands*. Delhi: Oxford University Press.

Fix A. 2002. Foragers, farmers, and traders in the Malayan Peninsula: origins of cultural and biological diversity. In Morrison K. D. and Junker L. L., eds., *Forager–Traders in South and Southeast Asia*. Cambridge, UK: Cambridge University Press, pp. 185–202.

Gould S. J. 1976. *Ontogeny and Phylogeny*. Cambridge, MA: Harvard University Press.

Hanihara T. 1994. Craniofacial continuity and discontinuity of Far Easterners in the late Pleistocene and Holocene. *Journal of Human Evolution* **27**: 417–441.

Hill C. A., Torroni C., Rengo W. *et al.* 2002. Mitochondrial DNA variation in the Orang Asli of the Malay Peninsula. In *The Fourth Meeting of the Human Origins and Disease Series*, October–November 2002. Cold Spring Harbor, NY: Cold Spring Harbor Laboratory Press.

Hislop J. A. 1954. Notes on the migration of bearded pig. *Federations Museum Journal* **1–2**: 134–137.

Howells W. W. 1973a. *Papers of the Peabody Museum of Archaeology and Ethnology*, Vol. 67: *Cranial Variation in Man*. Cambridge, MA: Harvard University Press.

1973b. *The Pacific Islanders*. London: Weidenfeld and Nicolson.

Keller G. and Warrack B. 2003. *Statistics for Management and Economics*, 6th edn. Melbourne: Brooks/Cole, pp. 397–398.

Lauer A. J. 2002. Craniometric measurements and tooth morphology of archaeologically derived skeletal remains from the Malay Peninsula. M.A. thesis, Australian National University, Canberra.

Martin R. 1905. *Die Inlandstämme der Malayischen Halbinsel: Wissenschaftliche Ergebnisse einer Reise durch die Vereinigten Malayischen Staaten.* Stuttgart: Gustav Fischer Verlag.

Martin R. and Saller K. 1959. *Lehrbuch der Anthropologie*, Vol. II. Stuttgart: Gustav Fischer Verlag.

Matsumura H. 1994. A microevolutionary history of the Japanese people from a dental characteristics perspective. *Anthropological Science* **102**: 93–118.

Ousley S. D. and Jantz R. L. 1994. *Fordisc 2.0 Personal Computer Forensic Discriminant Functions.* Knoxville, TN: University of Tennessee Department of Anthropology.

Penrose L. S. 1954. Distance, size and shape. *Annals of Eugenics* **18**: 337–343.

Pietrusewsky M. and Douglas M. T. 2002. *Ban Chiang, a Prehistoric Village Site in Northeast Thailand I: The Human Skeletal Remains.* Philadelphia: University of Pennsylvania Press.

Rambo A. T. 1988. Why are the Semang? Ecology and ethnogenesis of aboriginal groups in Peninsular Malaysia. In Rambo A. T, Gillogly K. and Hutterer K. L., eds., *Ethnic Diversity and the Control of Natural Resources in Southeast Asia.* Ann Arbor, MI: University of Michigan, pp. 19–36.

Rashid R. 1995. Introduction. In Rashid R., ed., *Indigenous Minorities of Peninsular Malaysia: Selected Issues and Ethnographies.* Kuala Lumpur: Intersocietal and Scientific, pp. 1–7.

Rayner D. R. T. 2000. The dental morphology of the indigenous peoples of the Malay Peninsula (*Orang Asli*). B.A. (Hons.) thesis, Australian National University, Canberra.

Rayner D. and Bulbeck D. 2002. Dental morphology of the 'Orang Asli' aborigines of the Malay peninsula. In Henneberg M., ed., *Causes and Effects of Human Variation.* Adelaide: Australasian Society for Human Biology, University of Adelaide, pp. 19–41.

Richerson P. J., Boyd R. and Bettinger R. L. 2001. Was agriculture impossible during the Pleistocene but mandatory during the Holocene? A climatic change hypothesis. *American Antiquity* **66**: 387–411.

Sangvichien S., Sirigaroon P. and Jørgenson J. B. 1969. *Archaeological Excavations in Thailand*, Vol. III: Part II: *The Prehistoric Thai Skeletons.* Copenhagen: Munksgaard.

Schebesta P. and Lebzelter V. 1926. Schädel und Skelettreste von drei Semang-Individuen. *Anthropos* **21**: 959–990.

Scott G. R. and Turner C. G. II. 1997. *The Anthropology of Modern Human Teeth.* Cambridge, UK: Cambridge University Press.

Sofaer J. L., Bailit H. L. and MacLean C. J. 1971. A developmental basis for differential tooth reduction during hominid evolution. *Evolution* **205**: 509–517.

Storm P. 1995. The evolutionary significance of the Wajak skulls. *Scripta Geologica* **110**: 1–247.

Tayles N. G. 1999. *Report of the Research Committee* LXI *The Excavation of Khok Phanom Di: A Prehistoric Site in Central Thailand,* Vol. V: *The People.* London: The Society of Antiquaries.

Tobias P. V. 1957. Bushmen of the Kalahari. *Man* **57**: 33–40.

Tuck Po L. 2000. Forest, Bateks, and degradation: environmental representations in a changing world. *Tonan Ajia Kenkyu* (*Southeast Asia Studies*) **38**: 165–184.

von Bonin G. 1931. Beitrag zur Kraniologie von Ost-Asien. *Biometrika* **23**: 52–113.

Appendix 6.1 Sources for comparative tooth metrics analysis

Samples used in summed tooth area and Penrose size analysis in bold.

Nasioi males, Bougainville Island, North Solomons: Bailit H. J., de Witt S. J. and Leigh R. A. 1968. The size and morphology of the Nasioi dentition. *American Journal of Physical Anthropology* **28**: 271–288.

An Yang (Bronze Age China), **Non Nok Tha** (neolithic/bronze age Thailand), Gilimanuk (metal phase Bali), recent Hong Kong Chinese, Thais and Yogyakarta Javanese (mixed-sex samples only): Brace C. L. 1976. Tooth reduction in the Orient. *Asian Perspectives* **19**: 203–219.

Kyoto Japanese, Koreans, metal phase Yayoi (males): Brace C. L. and Masafumi N. 1982. Japanese tooth size: past and present. *American Journal of Physical Anthropology* **59**: 399–411.

Neolithic and metal phase (recent) Borneo mixed-sex samples: Brace C. L. and Vitzthum V. 1984. Human tooth size at mesolithic, neolithic and modern levels at Niah Cave, Sarawak. *Sarawak Museum Journal* **33**: 75–82.

Yang Shao, Liu Lin, Xi Chang (neolithic China), Shanghai and Beijing (recent Chinese) (all males): Brace C. L., Shao X. Q. and Zhang Z. B. 1984. Prehistoric and modern tooth size in China. In Smith F. H. and Spencer F., eds., *The Origin of Modern Humans.* New York: Liss, pp. 485–516.

Euston Murray Valley, New Guinea Motupore (male data in both cases): Brown P. J. 1978. The ultrastructure of dental abrasion: its relationship to diet. B. A. (Hons.) thesis, Australian National University, Canberra.

Coobool Creek, Swanport, Broadbeach, Roonka Australian Aboriginal males (buccolingual breadths only): Brown P. 1989. *Terra Australis 13 Coobool Creek.* Canberra: Australian National University.

Central Australian Walpiri males, Sulawesi metal phase (Leang Buidane) (mixed-sex sample): Bulbeck F. D. 1982. Continuities in Southeast Asian evolution since the late Pleistocene. M. A. thesis, Australian National University, Canberra.

New Guinea Highland males: Doran G. A. and Freedman L. 1974. Metrical features of the dentition and arches of populations from Goroka and Lufa, Papua New Guinea. *Human Biology* **46**: 583–594.

Western Australian Aboriginal males: Freedman L. and Lofgren M. 1981. Odontometrics of Western Australian Aborigines. *Archaeology and Physical Anthropology in Oceania* **16**: 87–93.

Canton and Hong Kong Chinese males: Goose D. H. 1977. The dental conditions of Chinese living in Liverpool. In Dahlberg A. A. and Graber T., eds., *Orofacial Growth and Development*. Chicago: University of Chicago, pp. 183–194.

Sulawesi metal phase (Leang Tjadang, i.e. Leang Codong) (mixed-sex sample): Jacob T. 1967. *Some Problems Pertaining to the Racial History of the Indonesian Region*. Utrecht: Netherlands Bureau for Technical Assistance.

Metal phase Bali (mixed-sex sample; Gilimanuk): Jacob T. 1967. Racial identification of the bronze age human dentitions from Bali Indonesia. *Journal of Dental Research* **46**: 903–910.

New Britain males: Janzer O. 1927. Die Zähne der Neu-Pommern. *Vierteljahrsschrift für Zähnheilkunde* **43**: 289–319.

Bronze age Chifeng Mongolians, neolithic/metal phase Thailand (Ban Kao plus Ban Na Di), North Kyushu Yayoi, Tanegashima Yayoi, Kofun Japanese, Kamakura Japanese, Edo Japanese, modern Japanese, northern Chinese, Thai Chinese, recent Thais males: Matsumura H. 1994. A micro-evolutionary history of the Japanese people from a dental characteristics perspective. *Anthropological Science* **102**: 93–118.

Urga Mongolians, Indonesians, New Britain Melanesians (males): Matsumura H. 1995. Dental characteristics affinities of the prehistoric to modern Japanese with East Asians, American natives and Australo-Melanesians. *Anthropological Science* **103**: 235–261.

North Chinese males (incisor breadths taken from Coon C. S. 1962. *The Origin of Races*. New York: Alfred A. Knopf): Moorees C. A. F. 1957. *The Aleut Dentition*. Cambridge, MA: Harvard University Press.

Ban Chiang males (neolithic/bronze age Thailand): Pietrusewsky M. and Douglas M. T. 2002. *Ban Chiang, a Prehistoric Village Site in Northeast Thailand*. I: *The Human Skeletal Remains*. Philadelphia, PA: University of Pennsylvania Press.

Swanport, Broadbeach and Roonka Australian Aboriginal males (maxillary diameters only): Smith P. 1982. Dental reduction: selection or drift? In Kürten B., ed., *Tooth Form and Function*. New York: University of Columbia Press, pp. 366–379.

Broadbeach males: Smith P., Brown T. and Wood W. B. 1981. Tooth size and morphology in a recent Australian Aboriginal population from Broadbeach, Southeast Queensland. *American Journal of Physical Anthropology* **55**: 423–432.

Batawi Javanese (Batavian) males, Surabaya Javanese males: Snell C. A. R. D. 1938. *Menschelijke skeletresten uit de duin formationie van Java's zudkust nabij Poeger* (*Z. Banjoewangi*). Surabaya: G. Kolff.

Mixed-sex Thai sample: Srisopark S. S. 1973. A study on the size of permanent teeth, shovel-shaped incisors and paramolar tubercle in Thai skulls. *Journal of the Dental Association of Thailand* **22**: 199–205.

Khok Phanom Di males (neolithic Thailand): Tayles N. G. 1999. *Report of the Research Committee* LXI. *The Excavation of Khok Phanom Di, a Prehistoric Site in Central Thailand*, Vol. V: *The People*. London: Society of Antiquaries.

West Nakanai males: Turner C. G. II and Swindler D. R. 1978. The dentition of New Britain West Nakanai Melanesians. *American Journal of Physical Anthropology* **49**: 361–372.

Tagalog males, Luzon: Yap Potter R. H., Alcazaren A. B., Herbosa F. M. and Tomaneng J. 1981. Dimensional characteristics of the Filipino dentition. *American Journal of Physical Anthropology* **55**: 33–42.

7 *Dentition of the Batak people of Palawan Island, the Philippines: Southeast Asian Negrito origins*

CHRISTY G. TURNER II AND JAMES F. EDER
Arizona State University, Tempe, USA

Introduction

The Batak are an indigenous people of Palawan Island, the Philippines (Krieger 1942, Eder 1987; see Fig. 1.2, p. 6), who speak an Austronesian language related to that of other peoples in the Palawan region and classified with the Visayan group of Philippine languages. The Batak closely resemble other so-called 'Philippine Negritos', both in their mobile hunting and gathering lifestyle and in their physical attributes (short stature, dark skin colour and curly hair), which earned these distinctive-looking people their name. Approximately two dozen ethnolinguistically distinct groups of such peoples are found in the Philippines, including the Mamanwa of Mindanao and a series of groups known variously as Agta, Ayta, Aeta, Ata or Ati, which are scattered widely in northern, eastern and west central Luzon, the Bikol Peninsula and the islands of Panay and Negros. The total number of these peoples is difficult to estimate with any accuracy but is probably in the order of 15 000–20 000 persons (Eder 1987).

Philippine Negritos, in turn, resemble other small, dark, hunting and gathering folk of Southeast Asia, in particular the Andaman Islanders and the Semang of the Malay Peninsula. Collectively, Southeast Asian Negritos have long been presumed to represent the surviving remnants of what was once a more widespread and more racially and culturally homogeneous population (e.g. Cooper 1940, Fox 1952). Such claims remain speculative. A more recent, and probably sounder, view is that each group of Southeast Asian Negritos represents the outcome of long-term, local evolutionary development under similar ecological conditions (Solheim 1981, Rambo 1984; Ch. 6).

Bioarchaeology of Southeast Asia. Marc Oxenham and Nancy Tayles.
Published by Cambridge University Press.

The opportunity for long-term local evolution follows from the fact that human entry to Palawan, Andamans and other islands was repeatedly possible in the Pleistocene, if not entirely by dry land then by relatively short watercraft trips such as those postulated for the colonisation of Australia some 50,000 years BP (Macintosh 1949, Birdsell 1977, Mulvaney and Kamminga 1999). Palawan, a narrow island with a southwest–northeast trend, 435 km long, rises from near the now-submerged Sunda Shelf, which would have been a forested or open woodland plain joining or nearly joining Palawan, other Philippine islands, Borneo, Sumatra, Java and mainland Southeast Asia in each of the major sea-level-lowering Pleistocene glacial epochs.

It is easy to envision Palawan's Pleistocene climate and biota as similar, but not identical to, that found now (Bellwood 1985). Being near the equator, Pleistocene Palawan would have had a hot humid tropical climate but with perhaps less than today's annual rainfall of over 200 cm. However, given the great extent of the Pleistocene Sunda Shelf, some local conditions might have been more tropical and rainforest like. Although inspection of Southeast Asian maps often gives the impression of environmental homogeneity, substantial regional variation occurs with distance from sea coasts, elevation, soil types and other factors (Philip 2000). Hence, there are good environmental reasons to argue for Batak local evolution even though identifying the selective pressure for reduced body size usually relies on explanations of limited availability of nutrients favouring smaller body size and a shorter period of growth. Nevertheless, even such a reasonable possibility is far from proven, and there is no body of evidence that overwhelmingly refutes the old idea of an early entry of small-statured people, as mentioned above.

Birdsell (1977, p. 132), famous for his controversial theory of a three-wave colonisation of Sahulland, wrote: 'the early peopling of the Philippines may give some evidence as to what kinds of people first moved out from the Sunda Shelf toward the Sahul . . .'. Earlier, Birdsell (1949) envisioned these first people as being oceanic Negritos whose dispersal ultimately reached Tasmania. Moreover, according to Ruhlen (1976, p. 104): 'Around 7,000 BC . . . there was not yet a single Austronesian speaker anywhere in Oceania. Rather the people who were to become the Austronesians were still living in what is today south central China, and they were speaking a language which we may call Austro-Tai.' Ruhlen (1994, pp. 143–144) maintained this view in his more recent writing. By comparison, the grand Pacific scholar Howells (1973, p. 104), at about the same time and discussing the same Austronesian origin issue, warned: 'On present appearances Formosa [Taiwan] could well be the

Austronesian "homeland" for the Pacific, as many think. We should, however, not forget the possibility – I would say the likelihood – that some other area, for example the Philippines, once contained a similar diversity, since blotted out by spreads of junior forms of Austronesian. In that case Formosa would simply represent a relic of a former state of affairs.' As we shall see, our Batak dental sample could be used to support either view.

Montillo-Burton (2000) has recently reviewed the issue of the peopling of the Philippines, concluding that there were people in the Philippines before the gradual entry of Austronesian speakers and that the archaeological and other evidence points to a dual origin, one from Taiwan in the north and a southerly source that started in mainland Southeast Asia, crossing the Sunda plain to Borneo, and then to Palawan and the other Philippine islands.

There is abundant stone artefact evidence in the Philippines for a pre-Austronesian population that may well have included groups of middle Pleistocene *Homo erectus* as Shutler has long suggested (Shutler and Shutler 1975). There are numerous later Pleistocene sites in the Philippines (Bellwood 1985), and at least one late Pleistocene Palawan site with human remains. Excavated by Fox (1970), the earliest Tabon Cave human remains were found in association with upper Palaeolithic unifacial flake tools (Flake Assemblage II). The human remains had associated charcoal that was carbon-14 dated to 24,000–22,000 years BP UCLA C-14 288, 699, 283 (Fox 1970;). The senior author examined these remains on 2 May 1984, at the National Museum in Manila, and found that one of the two mandible fragments (66-V-121) possessed a three-rooted first permanent molar socket, while the other (P-XIII-T-431) had the more common two-rooted polymorphism. The specimen 66-V-121 still stands as the oldest example we know of a three-rooted lower first molar in *Homo sapiens*. None is known to occur in any fossil hominid from Africa or Europe before 15,000 years BP. There is one example of a *H. erectus* three-rooted lower first permanent molar found in the Chinese Lower Cave of Zhoukoudian (C. G. Turner II, D. Swindler and W. Liu, unpublished data).

Hence, the concern of our study of Palawan dentition revolves around the hoary and controversial issue of the anatomically modern human colonisation of, and micro-evolution in, Sundaland, and around understanding the part played by small-statured indigenous Palawan people in this question. Our standardised description of modern Batak dental morphology provides a basis for future research when archaeologists discover additional human skeletal and dental remains in Palawan and elsewhere in the Philippines.

Materials and methods

The descriptions of 18 upper and lower crown traits herein are based on dental impressions of 29 individuals (12 males, mean age ~33 years (range, 19 to ~47)); 17 females, mean age ~25 years (range, 15 to ~49) obtained by Eder in 1981 using Jeltrate and Castone materials. Sample size was limited mainly to the amount of casting materials taken to the field. Most of the casts were obtained from young adults. Where repeated sampling of a region has been done, for example the American southwest, Europe and elsewhere, a sample of 30 individuals has almost always proven to be highly representative (Scott and Turner 1997). Descriptions follow the Arizona State University Dental Anthropology System (ASU DAS) (Turner *et al.* 1991). While casts cannot add to the knowledge of Batak root variation, the sample is unique (we know of no other Batak dental study) and provides a modern baseline from which crown morphology can be used to assess affinity with past and contemporary Southeast Asian and Oceanic populations (such as those postulated in the north and south entry models for the peopling of the Philippines). The 18 traits were selected because they are easily and reliably scored (Nichol and Turner 1986); they have low to non-significant inter-trait correlations, sexual dimorphism or age changes (Nichol 1990, Scott and Turner 1997); and they stand up to harsh aboriginal tooth use and wear for several years after eruption.

All the dental observations for the Batak and comparative samples were made by Turner in numerous institutions identified and acknowledged in previous publications (Turner 1985, 1990, 1992a,b, Scott and Turner 1997).

Univariate and multivariate comparisons are provided. There has long been concern about the interpretations of multivariate statistical studies on measures of intergroup distance or divergence. Criticisms have ranged from not knowing if the distances of the early 'racial likeness' formula were statistically significant, to questions about which statistic is the best. Penrose (1954) showed that his relatively simple D2 statistic corresponded well with the complicated statistic of Rao. More recently, Scott and Turner (1997) reached a similar conclusion about other distance or divergence statistics that provide approximately the same results, although Mizoguchi (1998) has urged further consideration. We have decided to stay with C. A. B. Smith's mean measure of divergence (MMD) statistic for five practical reasons. First, it produces affinity assessments usually in line with independent lines of evidence from history, archaeology, ethnography, geographic distance, genetics, linguistics and other multivariate distance studies. This is most notable in the worldwide bootstrap genetic affinity studies by Cavalli-Sforza *et al.* (1988, 1994) and their very satisfactory concordance with dental affinity evaluations based on MMD (Scott and Turner 1997). Second, MMD accommodates the problem of missing data, which is always associated with archaeologically derived samples of teeth and bones. Third, MMD is simple and, therefore, eloquent in a mathematical sense (see remarks by Penrose 1954, p. 342) and, fourth, there have been many studies using MMD, suggesting that it has met the approval of many researchers. Finally, for obvious reasons of continuity and comparability, the worldwide dental anthropology study carried out over a 35-year period by Turner and colleagues needed to be based on standardised procedures. Our earliest study employing MMD was on Yap dentition (Harris *et al.* 1975). Our findings have stood the test of time as have most of our other studies employing Smith's MMD statistic. We found that Yap teeth more closely resembled those of Polynesians than Melanesians but not as greatly as Japanese and Papago Indians used in the MMD matrix. Cavalli-Sforza *et al.* (1994) found the same general relationship. Hence, the development of ASU DAS, used worldwide, and the continued employment of the simple and reliable MMD statistic, the results of which are checked in each new study with as many lines of independent evidence as possible; this provides standardised analyses throughout the world and over many years of study.

Dental morphology and intergroup affinity

Table 7.1. provides the frequencies for the 18 crown traits used for the MMD divergence comparisons. Table 7.2 provides the frequencies for 9 of the 18 traits that provided the best intergroup discrimination between the Batak and the major comparative populations used to assess epigenetic affinity. Such highly discriminating traits are those that have the largest intergroup frequency differences. For example, upper incisor 1 shovelling has a relatively high frequency in the Batak: at least twice as frequent as in Australia–Tasmania, Melanesia, Nubia and West Africa. Just the opposite occurs for the upper canine mesial ridge (termed Bushman canine by Morris (1975)).

Univariately, the Batak dentition fitted within the Southeast Asian sundadont dental pattern, which is a generalised, somewhat simplified and retained condition compared with the more specialised and intensified sinodont pattern of Northeast Asia and the Americas. Sinodonty had to have evolved out of sundadonty, and must have done so in Northeast Asia. Sundadonty, in turn, resembles teeth from Australia–Tasmania and Melanesia more than those from Africa, as expected on the basis of

Table 7.1. *Batak dental morphology*

Trait	Break point number	No. individuals	Breakpoint number frequencies[a]							
			1	2	3	4	5	6	7	8
Winging UI1	1	29	10.3	0.0	89.7					
Shovelling UI1	3	26	0.0	15.4	46.2	23.1	15.4			
Double shovelling UI1	2	12	83.3	0.0	16.7					
Interruption grooves UI2	1	25	76.0	24.0						
Tuberculum dentale U12	1	26	88.5	0.0	0.0	3.8	0.0	0.0	0.0	7.7
Mesial ridge UC	1	25	92.0	4.0	4.0					
Distal accessory ridge UC	2	10	20.0	0.0	10.0	40.0	30.0			
Hypocone UM2	2	27	29.6	0.0	3.7	22.2	44.4			
Cusp 5 UM1	1	15	93.3	0.0	0.0	6.7				
Carabelli's UM1	2	24	45.8	29.2	8.3	4.2	0.0	8.3	0.0	4.2
Lingual cusp No. LP2	2	27	0.0	44.4	55.6					
Y groove pattern LM2	2	24	29.2	58.3	12.5					
Cusp No. LM1	2	13	7.7	53.8	38.5					
Cusp No. LM2	1	26	26.9	42.3	3.8	26.9				
Deflecting wrinkle LM1	3	14	57.1	0.0	0.0	42.9				
Trigonid crest LM1	1	17	94.1	5.9						
Protostylid LM1	1	26	96.2	3.8						
Cusp 7 LM1	1	26	100.0							

[a] Arizona State University Dental Anthropology System dichotomised (breakpoint) scores: winging = 1; shovelling = 4–7; double-shovelling = 3–7; interruption groove = 2; tuberculum dentale = 2–8; canine mesial ridge = 2–4; distal accessory ridge UC = 3–6; hypocone = 3–7; M1 cusp 5 = 2–7; Carabelli = 3–8; lingual cusp number P2 = 3–4; groove pattern M2 = 2; cusp number M1 = 3; cusp number M2 = 2–4; deflecting wrinkle = 4; distal trigonid crest = 2; protostylid = 2–7; M1 cusp 7 = 2–6. U, upper; L, lower; I, incisor; C, canine; P, premolar; M, molar.

geographic distance. The same can be said for sundadonty and the European dental pattern (Scott and Turner 1997).

Table 7.3 provides the MMD matrix for all of the 13 comparative groups. In the analysis, the order of computer comparison was based on random selection, not geographic, archaeological, linguistic or ethnographic ordering. From Table 7.3, the dendrogram of Fig. 7.1 was prepared using the unweighted pair group arithmetic averages method.

Table 7.2. *Key trait frequencies that characterise and distinguish the Batak and illustrate the relationship between frequencies and mean measure of divergence (MMD) values*[a]

	Batak	Taiwan	Philippine	Early Malay	Australian Tasmanian	Melanesian	Nubian	West African
MMD		0.0353	0.0599	0.0661	0.1917	0.2731	0.3092	0.3851
Trait frequency								
Shovelling UI1	38.4	59.1	46.4	29.6	15.9	9.3	4.6	10.5
Mesial ridge UC	8.0	0.0	0.0	9.7	5.0	3.2	22.3	32.4
Hypocone UM2	70.4	85.2	86.7	89.1	98.1	93.5	96.3	96.4
Cusp 5 UM1	6.7	22.2	30.8	24.4	68.2	62.4	50.0	32.3
Carabelli UM1	25.0	33.3	39.4	23.0	42.8	41.9	46.1	58.3
Lingual cusp LP2 > 1	55.6	79.2	69.9	85.7	81.0	88.1	93.3	82.1
Four-cusped LM2	26.9	19.0	28.1	24.6	7.2	53.0	6.1	16.7
Deflecting wrinkle LM1	42.9	44.4	13.3	10.6	17.1	17.6	10.0	20.0
Cusp 7 LM1	0.0	6.1	7.8	4.6	8.2	11.7	21.4	67.4
Sample size range	14–27	9–33	28–122	66–156	35–156	118–253	14–33	19–56

[a]Dichotomizing breakpoints are the same as in Table 7.1. Information on the comparative samples can be obtained from C. G. Turner. U, upper; L, Lower; I, incisor; C, canine; P, premolar; M, molar.

Ward's clustering method produced a nearly identical dendrogram except that Sri Lanka linked remotely with Melanesia, which, as here, was part of the Oceanic–African deep branch. Both methods produce good dental clustering as assessed from independent evidence whenever available (linguistic, genetic, archaeological, geographic, ethnographic; see Turner (1985) for references on methods). The Batak clearly associated using both clustering methods with the sundadont Southeast Asians, and not the Oceanic Australo-Melanesians or Africans.

The Batak sample had only one pairwise MMD value that was statistically non-significant, namely with prehistoric Taiwan. Taiwan had five non-significant MMD values and Borneo had the most, with eight. This

Table 7.3. *Mean measures of divergence matrix*

	WAFR	BATK	SRIL	TAIW	THAI	NUBI	BANC	PHIL	MELA	BORN	SEAS	MALA	AUST
WAFR													
BATK	0.385			ns									
SRIL	0.167	0.095								ns			
TAIW	0.370	0.035	0.188		ns		ns	ns		ns			
THAI	0.254	0.104	0.068	0.006				ns		ns			
NUBI	0.063	0.309	0.153	0.280	0.213					ns			ns
BANC	0.291	0.085	0.144	0.070	0.048	0.201		ns		ns	ns		
PHIL	0.263	0.060	0.063	0.021	0.005	0.204	0.042			ns	ns	ns	
MELA	0.176	0.273	0.090	0.207	0.088	0.121	0.111	0.124					
BORN	0.141	0.084	0.028	0.059	0.032	0.056	0.029	0.005	0.106		ns		ns
SEAS	0.200	0.110	0.065	0.060	0.034	0.116	0.013	0.025	0.074	0.005			
MALA	0.256	0.066	0.091	0.080	0.047	0.166	0.090	0.017	0.124	0.029	0.062		
AUST	0.125	0.192	0.132	0.125	0.078	0.042	0.117	0.077	0.081	0.044	0.075	0.084	

WAFR, West Africa; BATK, Batak; SRIL, Sri Lanka; TAIW, Taiwan; THAI, Thailand; NUBI, Nubia; BANC, Ban Chiang; PHIL, Philippines; MELA, Melanesia; BORN, Borneo; SEAS, Southeast Asia; MALA, Malay Peninsula; AUST, Australia–Tasmania; ns, not statistically significant ($p > 0.02$)

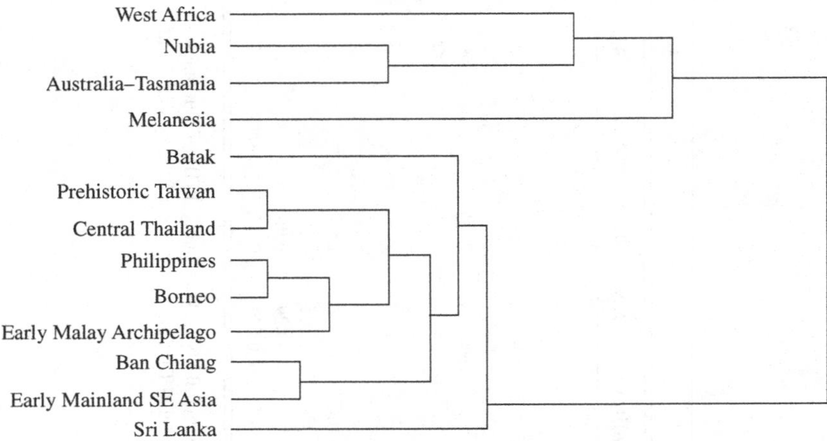

Figure 7.1. Dendrogram of mean measure of divergence values for dental morphology clustered with the unweighted pair group, arithmetic averages method. There are two major clusters, an upper one containing the Australo-Melanesians and Africans, and a lower one containing Southeast Asians. Within the Southeast Asian cluster, the Batak and Sri Lanka values are the most divergent. Like the groups in the upper branch, Sri Lanka is understandably divergent on grounds of simple geographic distance. The less-understandable relatively large Batak divergence is discussed in the text (computer file: Live Batak 2).

does not seem to reflect inadequate sample size; rather it appears to indicate the very intermediate nature of Borneo dental morphology, which is related to their central waterway location in island Southeast Asia.

Table 7.4 provides the ranking of the MMD values between the Batak and comparative samples. The smallest MMD (0.0353) was with prehistoric Taiwan, but this value was not significant: the small value is not unexpected given the introductory remarks about origins. Next most similar with the Batak is the Philippine sample, which is expected on geographic grounds. In fact, expectation on the basis of geographic distance is well founded in the remaining MMD values. Thus, the Batak–African MMD values are the most divergent.

What is most interesting from an origins perspective is the next most similar MMD value, between the Batak and the early Malay archipelago samples (0.0661). This relationship is not maintained in the dendrogram (Fig. 7.1); although this clustered all of the Southeast Asian samples together in one major branch, it did position the Batak as relatively remote in the Southeast Asian cluster. It is interesting that Sri Lanka clustered with Southeast Asian sundadonts, a result found also by Hawkey (1998) using a much older Sri Lanka sample. The MMD values do not support

Table 7.4. *Ranking of Batak mean measures of divergence*

Region	Mean measure of divergence
Prehistoric Taiwan	0.0353
Philippines	0.0599
Early Malay archipelago	0.0661
Borneo	0.0842
Ban Chiang	0.0853
Sri Lanka	0.0953
Central Thailand	0.1037
Early mainland Southeast Asia	0.1097
Australia–Tasmania	0.1917
Melanesia	0.2731
Nubia 117 and 67/80	0.3092
West Africa	0.3851

the view that the Batak are simply short relatives of Australians and Tasmanians, as suggested by Birdsell (1949, 1977). The Australo-Melanesians clustered (as they do in a variety of dental, cranial and genetic studies) with the Africans (for a brief discussion of this curious relationship see Turner (1992b, pp. 148–149)).

Congenital absence

Congenital absence (CA) can be considered a morphological trait inasmuch as a tooth can be normal sized, reduced in size or totally absent; similarly a morphological trait such as Carabelli's trait can be big, variously intermediate or absent. However, identifying CA from casts is not as dependable as it is with skeletal remains. For this reason, we have separated CA from the other crown traits discussed here. Still, with caution, reasonable estimates can be obtained. In our sample, both male and female Batak had one or more CA teeth (33.3%; 9/19 individuals). All CA teeth were upper or lower third molars except in one unusual adult male with seven CA teeth, a peg-shaped upper molar 3 and an odd-shaped tooth between his upper central incisors. There was no significant difference between males with CA (46.1%; 6/13 individuals) and females with CA (21.4%; 3/14 individuals). Third molars erupt at and after 15 years of age. Occurrence of CA, which is one of the defining features of sinodonty and sundadonty (Turner 1990), was not used in the multivariate comparisons discussed above because of the scoring reliability concern. Nevertheless, the occurrence of one or both CA–peg-reduced upper third molars was 25.9% (7/27 individuals). This value is similar to that found in Hokkaido Ainu and Borneo samples, which are at the upper limit for sundadonts (Turner 1990).

Discussion

Frequencies of dental crown morphological traits place our Batak sample within the Southeast Asian sundadont dental pattern, as expected on geographic and linguistic grounds. The cluster dendrogram sets the Batak sample apart from the other Southeast Asians, indicating that the latter have smaller MMD values amongst themselves than they do with the Batak. Even the Philippine sample is more similar to six other Southeast Asian samples (Borneo, central Thailand, early Malay archipelago, prehistoric Taiwan, early mainland Southeast Asia and Ban Chiang) than it is with the Batak sample.

Setting aside the possibility of sampling error, the micro-evolutionary explanation that most quickly comes to mind is that the Batak have undergone some genetic drift as a result of their isolation (geographic as well as cultural) and small local population size, resulting from their hunting and gathering lifestyle. However, given there are suggestions in the older anthropological literature for a once larger and more connected population of small-statured Southeast Asians, referred to as Australoids as well as Negritos, we should consider the possibility that the Batak dental morphology does reflect deeper prehistory: that is, morphogenetic prehistory near the branch point when ancestral Australo-Melanesians began colonising their part of Oceania (sometime before 50,000 years BP). This is probably before the time when the sundadont dental pattern began to evolve out of a hypothetical pattern called proto-sundadonty, which Turner proposed at a 1990 conference in Tokyo (Turner 1992a,b, Hanihara 1992).

Omoto's (1985) genetic studies of the Philippine Negritos found that the Batak most closely resembled the Aeta Negritos of Luzon. Using the Aeta for further comparisons, he found them to be more like Micronesians, Tagalog (Manila) and Japanese, and least like Australians and New Guineans. Omoto summarised his study with a model that portrayed a major evolutionary lineage in Sundaland beginning with 'proto-Australoids', who evolved into Negritos. They, in turn, received genetic input from 'proto-Mongoloids' in Asia 20,000 years BP, followed by Austro-Melanesian influences starting about 6,000 years BP. This model identifies with local evolution for the Southeast Asian Negritos (he explicitly rejected an African origin), coupled with later Austro-Melanesian linguistic and gene flow.

Omoto later (1995) added that he favoured at least four largely independent centres for Mongoloid evolution after a separation from ancestral Caucasoids. The Negritos were still viewed as a local evolutionary product in Southeast Asia. Another genetic study, by Matsumoto (1988), also found the Philippine Negritos very similar to one another and to other groups in the immediate vicinity, providing further support for local evolution.

As for Aeta Negrito dental studies, Hanihara (1990, p. 24) suggested that: 'the lineage leading to Negritos, which may be nearer to the hypothetical ancestral type of the Proto-Mongoloid population, might have evolved into some phyletic lines, one of which was the Neolithic Jomon . . .'. He added that his dental studies showed no direct relationship between Negritos and Australian Aborigines, as had been proposed by Birdsell (1977 and earlier). As our dendrogram shows, we would agree entirely with Hanihara's suggestion. In both works, however, diachronic evidence is most desirable in order to assess the degree of post-Pleistocene admixture that may have shifted the Batak dental pattern away from that of Australo-Melanesians. Although Pietrusewsky (1996) lacked a Batak cranial series, it is noteworthy that his Philippine sample clustered with a well-defined group, all various Southeast Asian populations.

Bulbeck (1981, 1993, 2000, Rayner and Bulbeck 2001; see also Ch. 6) has probably given more attention to the bioarchaeological question of the formation of southern Mongoloids and representative cultures than most other workers. In his earlier work, he accepted the idea of a local Southeast Asian evolution for the southern Mongoloids, but in more recent years he has expanded his database and has incorporated datasets from other workers, creating a type of mega-analysis of Southeast Asian cranial and dental variation. While the major result of this enterprise was to expand affinity inferences far beyond Southeast Asia, Bulbeck (1993) still maintained that the Negritos are a consequence of local evolution that was completely independent of the Australian lineage.

From a global perspective, the Batak fit in well with Scott and Turner's (1997) Sunda–Pacific macro-cluster. In addition, the Batak dental pattern may well approximate the hypothetical 50,000-year-old proto-sundadont dental pattern: the stem pattern out of which evolved Scott and Turner's (1997) Sahul–Pacific and Sunda–Pacific geogenetic branches. Turner (1992b) proposed that proto-sundadonty might have looked like the average dental trait frequencies for several Australian Aborigine dental samples, based on the assumption that modern differences are not a result of post-Pleistocene admixture or directionally driven shifts, but instead derive from random genetic drift. Given the evolutionarily conservative nature of dental morphology, it is not surprising that the substantially isolated Batak have crown trait frequencies similar to, but not identical with, those of the hypothetical proto-sundadonts (see also Turner and Swindler (1978) for similar views on Melanesians).

In summary, our study of Batak crown morphology agrees with workers who feel that a local evolution model best explains their morphological features and those of other Southeast Asian Negritos. In addition, we offer that the genetic isolation of at least the Batak Negritos has favoured the partial retention of a dental pattern with an antiquity that may be in the order of 50,000 years. There is nothing extraordinary in this suggestion inasmuch as the modern European dental pattern was established no more recently that 25,000 years ago (Haeussler and Turner 2000), and Northeast Asian sinodonty may be as old as 30,000 years (Turner *et al.* 2000). Finally, as with the genetic studies mentioned above, there is no dental evidence to support a close relationship between the Batak and African populations.

Acknowledgements

The dental impressions were collected with the assistance of Rafaelita Fernandez and Raul Fernandez, from the Batak inhabiting the Langogan River valley of central Palawan Island. Computer data entry and statistical analyses were carried out by Linda Nuss with aid from the National Science Foundation (BNS-8303786). Sources of the comparative dental information have been credited and acknowledged previously. We thank the editors and the two anonymous reviewers for their helpful comments.

References

Bellwood P. 1985. *Prehistory of the Indo-Malaysian Archipelago*. Sydney: Academic Press.

Birdsell J. H. 1949. The racial origin of the extinct Tasmanians. *Records of the Queen Victoria Museum, Launceston* **2**: 105–122.

1977. The recalibration of a paradigm for the first peopling of Greater Australia. In Allen J., Golson J. and Jones R., eds., *Sunda and Sahul: Prehistoric Studies in Southeast Asia, Melanesia and Australia*. London: Academic Press, pp. 113–167.

Bulbeck F. D. 1981. Continuities in Southeast Asian evolution since the late Pleistocene: some new material described and some old questions reviewed. M.A. thesis, Australian National University, Canberra.

1993. Enigmas for Southeast Asian human evolution posed by the Negritos. In *Annual Conference of the Australasian Society of Human Biology*, Adelaide.

2000. Dental morphology at Gua Cha, West Malaysia, and the implications for 'Sundadonty'. *Indo-Pacific Prehistory Association Bulletin* **19**(Melaka Papers 3): 17–41.

Cavalli-Sforza L. L., Piazza A., Menozzi P. and Mountain J. 1988. Reconstruction of human evolution: bringing together genetic, archeological and linguistic data. *Proceedings of the National Academy of Sciences of the USA* **85**: 6002–6006.

Cavalli-Sforza L. L., Menozzi P. and Piazza A. 1994. *The History and Geography of Human Genes*. Princeton, NJ: Princeton University Press.

Cooper J. M. 1940. Andamanese–Semang–Eta cultural relations. *Primitive Man* **13**: 29–47.

Eder J. F. 1987. *On the Road to Extinction: Depopulation, Deculturation, and Aadaptive Well-being among the Batak of the Philippines*. Berkeley, CA: University of California Press.

Fox R. B. 1952. The Pinatubo Negritos: their useful plants and material culture. *Philippine Journal of Science* **81**: 173–414.

1970. *National Museum Monograph 1: The Tabon Caves*. Manila: National Museum of the Philippines.

Haeussler A. M. and Turner C. G. II. 2000. The upper Paleolithic children of Russia: comparative dental anthropological analysis of the permanent teeth. *American Journal of Physical Anthropology Supplement* **30**: 169–170.

Hanihara T. 1990. Affinities of the Philippine Negritos with Japanese and Pacific populations based on dental measurements: the basic populations in East Asia, I. *Journal of the Anthropological Society of Nippon* **98**: 13–27.

1992. Negritos, Australian Aborigines, and the proto-sundadont dental pattern: the basic populations in East Asia, V. *American Journal of Physical Anthropology* **88**: 183–196.

Harris E. F., Turner C. G. II and Underwood J. H. 1975. Dental morphology of living Yap islanders, Micronesia. *Archaeology and Physical Anthropology in Oceania* **10**: 218–234.

Hawkey D. E. 1998. Out of Asia: dental evidence for microevolution and affinities of early populations from India/Sri Lanka. Ph.D. thesis, Department of Anthropology, Arizona State University at Tempe.

Howells W. 1973. *The Pacific Islanders*. New York: Charles Scribner.

Krieger H. W. 1942. *Smithsonian Institution War Background Studies*, No. 4: *Peoples of the Philippines*. Washington, DC: Smithsonian Institution Press.

Macintosh N. W. G. 1949. A survey of possible sea routes available to the Tasmanian Aborigines. *Records of the Queen Victoria Museum* **2**: 123–144.

Matsumoto H. 1988. Characteristics of Mongoloid and neighboring populations based on the genetic markers of human immunoglobulins. *Human Genetics* **80**: 207–218.

Mizoguchi Y. 1998. Book review *The Anthropology of Modern Human Teeth: Dental Morphology and its Variation in Recent Human Populations*, by Scott G. R. and Turner C. G. II. Cambridge: Cambridge University Press. *Anthropological Science* **105:** 247–250.

Montillo-Burton L. 2000. The peopling of the Philippines in the light of archaeological and linguistic evidence. *Italia-Filippine Maynilad* **3**: 4–9.

Morris D. H. 1975. Bushman maxillary canine polymorphism. *American Journal of Physical Anthropology* **54**: 431–433.

Mulvaney J. and Kamminga J. 1999. *Prehistory of Australia*. Washington, DC: Smithsonian Institution Press.

Nichol C. R. 1990. Dental genetics and biological relationships of the Pima Indians of Arizona. Ph.D. thesis, Department of Anthropology, Arizona State University at Tempe.

Nichol C. R. and Turner C. G. II. 1986. Intra- and inter-observer concordance in classifying dental morphology. *American Journal of Physical Anthropology* **69**: 299–315.

Omoto K. 1985. The Negritos: genetic origins and microevolution. In Kirk R. and Szathmary E., eds., *Out of Asia: Peopling the Americas and the Pacific*. Canberra: Journal of Pacific History, pp. 123–131.

1995. Genetic diversity and the origins of the 'Mongoloids'. In Brenner S. and Hanihara K., eds., *The Origin and Past of Modern Humans as Viewed from DNA*. Singapore: World Scientific, pp. 92–109.

Penrose L. S. 1954. Distance, size and shape. *Annals of Eugenics* **18**: 337–343.

Philip G Limited. 2000. *Oxford Atlas of the World*, 8th edn. New York: Oxford University Press.

Pietrusewsky M. 1996. Multivariate craniometric investigations of Japanese, Asians, and Pacific islanders. In Omoto K., ed., *International Symposiums on Interdisciplinary Perspectives on the Origins of the Japanese*. Kyoto: International Research Center for Japanese Studies, pp. 65–104.

Rambo A. T. 1984. Why are the Semang? Ecology and ethnogenesis in peninsular Malaysia. In Rambo A. T., Gillogly K. and Hutterer K. L., eds., *Ethnic Diversity and Control of Natural Resources in Southeast Asia*. Ann Arbor, MI: Center for South and Southeast Asian Studies, University of Michigan, pp. 19–31.

Rayner D. and Bulbeck D. 2001. Dental morphology of the 'Orang Asli' aborigines of the Malay Peninsula. In Henneberg M., ed., *Causes and Effects of Human Variation*. Adelaide: Australian Society of Human Biology, pp. 19–41.

Ruhlen M. 1976. *A Guide to the Languages of the World*. Stanford, CA: Stanford University Press.

1994. *The Origin of Language: Tracing the Evolution of Mother Tongue*. New York: Wiley.

Scott G. R. and Turner C. G. II. 1997. *The Anthropology of Modern Human Teeth: Dental Morphology and its Variation in Recent Human Populations*. Cambridge, UK: Cambridge University Press.

Shutler R. S. Jr and Shutler M. E. 1975. *Oceanic Prehistory*. Menlo Park, CA: Cummings.

Solheim W. G. II. 1981. Philippine prehistory. In Casal G., Casino E., Ellis G., Rose R. and Solheim W. G., eds., *The People and Art of the Philippines*. Los Angeles, CA: Museum of Cultural History, University of California, pp. 17–83.

Turner C. G. II. 1985. The dental search for native American origins. In Kirk R. and Szathmary E., eds., *Out of Asia: Peopling the Americas and the Pacific*. Canberra: Journal of Pacific History, pp. 31–78.

1990. Major features of sundadonty and sinodonty, including suggestions about East Asian microevolution, population history, and late Pleistocene relationships with Australian aboriginals. *American Journal of Physical Anthropology* **82**: 295–317.

1992a. Microevolution of East Asian and European populations: a dental perspective. In Akazawa T., Aoki K. and Kimura T., eds., *The Evolution and Dispersal of Modern Humans in Asia*. Hokusensha: Tokyo, pp. 415–438.

1992b. The dental bridge between Australia and Asia: following Macintosh into the East Asian hearth of humanity. *Perspectives in Human Biology* **2**: 143–152.

Turner C. G. II. and Swindler D. R. 1978. The dentition of New Britain West Nakanai Melanesians. VIII. Peopling of the Pacific. *American Journal of Physical Anthropology* **49**: 361–372.

Turner C. G. II, Nichol C. R. and Scott G. R. 1991. Scoring procedures for key morphological traits of the permanent dentition: the Arizona State University dental anthropology system. In Kelley M. A. and Larsen C. S., eds., *Advances in Dental Anthropology*. New York: Wiley-Liss, pp. 13–31.

Turner C. G. II, Manabe Y. and Hawkey D. E. 2000. The Zhoukoudian Upper Cave dentition. *Acta Anthropologica Sinica* **19**: 253–268.

Part II *Health, disease and quality of life*

8 *Subsistence change and dental health in the people of Non Nok Tha, northeast Thailand*

MICHELE TOOMAY DOUGLAS
University of Hawaii at Manoa, Hawaii, USA

Introduction

Dentition is an information-rich mainstay of research in skeletal biology, forensic science and bioarchaeology, contributing information on sex, age at death, growth and development, and diet, among others. The steadily accumulating global database of dental pathological conditions has enabled comparative research across space, time, technology and subsistence base (e.g. summary in Larsen (1997) and Pechenkina *et al.* (2000)). Lukacs (1989) proposed standardisation of recording and reporting dental pathological conditions and the use of a dental pathology profile (DPP) for comparative research.

The DPP was conceived to improve the comparison of dental assemblages and to suggest modes of subsistence in situations where archaeological evidence was lacking (Lukacs 1989, p. 273). Among the variables in the DPP are caries, enamel hypoplasia, calculus, alveolar resorption, severe attrition, antemortem tooth loss, periapical cavities (indicative of tooth infection) and variables of jaw robusticity and size. The transitions from hunting and gathering to agriculture and then to intensified agriculture resulted in shifts in the components of the DPP that can be chronicled and compared (Lukacs 1989, p. 276). For example, hunter–gatherers typically have low frequencies of caries, calculus, malocclusion and alveolar resorption, a high frequency of severe attrition, and large jaw size. Agricultural populations typically have the opposite profile, low rates of severe attrition (except in cases where food contains abrasives) and high rates of caries, calculus, resorption, dental crowding and malocclusion.

This approach, applied to many of the extant prehistoric skeletal series from Thailand (e.g. Ban Chiang, Ban Na Di, Ban Lum Khao, Khok Phanom Di, Noen U-Loke, Nong Nor and Non Nok Tha), has shown a

Bioarchaeology of Southeast Asia. Marc Oxenham and Nancy Tayles.
Published by Cambridge University Press.

DPP consistent with broadly based subsistence systems even in the presence of sedentism and intensifying agriculture, and inconsistent with other populations of the world during the same transitions (Douglas 1996, Nelsen 1999, Tayles 1999, Domett 2001, Pietrusewsky and Douglas 2001, 2002). The causes of this inconsistency are problems for further research but one relevant fact is that rice, the staple starch of the region, is not as cariogenic as the grasses and grains used by other agricultural populations (Tayles *et al.* 2000).

As the database of skeletal biology in Thailand has expanded and the basic questions about the prehistoric inhabitants (e.g. How tall where they? What were their proportions and skull shape?) are answered, more detail can be added to the research foundation and more specific questions can be addressed. Among these are the documentation and assessment of sex differences in the occurrence of dental pathological conditions, cultural modifications to the dentition, physiological stress indicators, and so on (e.g. Grauer and Stuart-Macadam 1998). Differences in skeletal morphology between the sexes are commonly used for skeletal sex estimation; these differences have a biological basis and are nearly universal. For example, the female pelvic girdle morphology has been shaped by evolution to accommodate childbearing, with ensuing changes to the female os coxae that are distinguishable from the male os coxae. Male skeletons are generally larger and more robust than female skeletons. What about differences between the sexes in the prevalence of dental pathological conditions?

Sex differences in dental pathological conditions are known from both clinical and archaeological studies. Dental caries, the most studied of the dental pathological conditions, is more common in females than males in the majority of these studies (Hillson 2000, p. 263). While suggestions have been made that greater caries rates in females have a biological basis – the result of earlier tooth eruption, differences in oral flora and so on – no evidence for such differences has been forthcoming (Larsen 1997 (p. 75), 1998 (p. 175)). Consequently, sex differences in caries rates require explanations of differences in diet, activity, behaviour or other gender-based causes. Lukacs (1996) proposed that sex differences in dental caries frequencies in a South Asian skeletal sample were the result of differences in diet and, by extension, a division of labour. The greater caries frequency in females occurred because they were more closely associated with agricultural products and consequently ate more carbohydrates than males. Males, by comparison, were more closely associated with wild and domesticated protein sources and thus ate more protein than females.

The goal of this chapter is to present selected components of the DPP of the little known northeastern Thailand skeletal series from Non Nok Tha.

The DPP for the combined series is presented, along with an examination of sex differences in the DPP and how these sex differences changed over time.

Non Nok Tha ('Partridge Mound')

The Non Nok Tha site, the first large-scale joint project by Thai and US researchers in Thailand, was excavated in the late 1960s (Bayard 1971). The site is located on the northern part of the Khorat plateau, a region bounded on the north and east by the Mekong river and on the south by two ranges of mountains, with marked seasonality in climate (Fig. 1.1, p. 5). The 'plateau' is actually two basins: the Khorat and Sakon Nakhon, which are drained by the Mun and Chi rivers, and the Songkhram river, respectively.

Non Nok Tha is located in the Phu Wiang region of the western plateau, near low terrace ridges, ranging from 180 to 210 m in elevation, along a tributary of the Chi river. The mountains, which include triple-layer rainforest, are 2–3 km from the site. Water is unpredictable in this region, because of the shadowing effect of the Phetchabun mountains (1375 mm mean annual rainfall), but there are three permanent watercourses nearby. There are typically three seasons: the cool season from November to February (17–31 °C), hot season of March and April (22–36 °C) and rainy season from May to October (24–33 °C). These environs provided a variety of resources to a population with beginning 'wet-rice' technology: the mountains supported limited hunting and gathering, the rolling lowlands allowed inundation rice cultivation, and swampy areas around the perennial streams provided wet-rice capabilities (Bayard 1971, Bayard and Solheim 1992). Sourcing studies of the excavated stone tools and the presence of numerous mammal and fish bones suggest that the early inhabitants of the site availed themselves of the proximate resources. Even today, the modern inhabitants of the nearby village of Ban Na Di are 'recreational hunter–gatherers', collecting edible plants and animals in the course of tending to the rice paddies, gardening and travelling. In addition to local resources, burial goods included exotic shell beads and adzes of non-local stone, providing evidence for established trade networks during the initial mortuary phases of Non Nok Tha (Bayard 1971, Bayard and Solheim 1992).

Evidence of rice, pottery and domesticated animals (i.e. pig, dog, cattle) is present from the earliest levels (Higham 1975, 1989). The presence of small fish vertebrae in the bottoms of many of the funerary vessels suggests that fermented fish paste, a staple of the modern diet, was a part of the diet during the early phases of the site. Wild animal remains, including deer, frog, hare, mongoose, crocodile, fish, turtle and shellfish are present but because of the lack of screening, no specific statements can be made as to the relative abundance of these animals over time.

Archaeological investigations at Non Nok Tha began in the early 1960s as salvage explorations prior to dam construction and development of the Mekong river valley for water management. Excavations at NP-7, a mound covering 1 ha and rising 80–150 cm from the rice paddy fields near the modern village of Ban Na Di, began in 1965–1966 under the direction of Wilhelm Solheim and Donn Bayard of the University of Hawaii, and Hamilton Parker of New Zealand. During the course of the excavation, it was discovered that the local villagers had a name for the mound: 'Non Nok Tha' or Partridge Mound, so this name came to be the designation commonly used for the site. In the first season, a total area of 150 m^2 was excavated close to the centre of the mound, and 85 burial features were identified (though not all contained skeletal material). In 1968, Bayard returned for a second excavation, north of the first, of 189.5 m^2 from which cultural deposits were removed, with a total of 125 burial feature numbers assigned (again, not all contained skeletal material). Although the final report of the Non Nok Tha excavations has not been published, it is available as a manuscript (Bayard and Solheim 1992) and is used to augment the published record as cited. Pietrusewsky (1974a,b) previously described the 1966 skeletal series, and Brooks and Brooks (1983, 1987) previously described the 1968 skeletal series, but this chapter presents results of reanalysis of both collections by the same researcher (Douglas 1996).

Excavations during both seasons at Non Nok Tha were conducted within the excavation zone only: that is, if the feet of a burial extended into the designated area, the feet were excavated and nothing more. Although primary interment was the usual burial practice, there was significant grave intercutting and subsequent commingling of remains. During the Early Period (EP) and initial phases of the Middle Period (MP), graves were sometimes topped with substantial mounds containing pots, animal bone and additional human bone. Many of the skeletal elements had heavy mineral concretions, and in some cases preservative was applied to skeletal remains en bloc. Still, many of the skeletons were substantially complete and preservation overall is fair to good.

Mortuary patterns and chronology

The excavations at Non Nok Tha provide evidence of the 'neolithic', the beginnings of the bronze age, as well as the introduction of iron on a timeline extending from approximately 5000/4500 to 1800 BP (Table 8.1). Even though cultural and natural processes (e.g. roots) disturbed the deposits and soil distinctions were subtle, the excavators recognised a fairly comprehensive soil stratigraphy. Within the sequence, Bayard (1971) identified three major periods: the EP, with three phases (EP1, EP2, EP3), the MP, with eight phases, and the Late Period (LP), with six phases. Although the chronology of the site has been disputed since the initial publications (e.g. Higham 1983, 1989, Bayard 1996–97, Bayard and Solheim 1992), for the purposes of this discussion, the exact dates of the various periods are not as important as ordering the burials through time, and associating them with identified cultural traditions.

Non Nok Tha functioned as a cemetery, with intermittent use for industry and very little evidence of occupation. Intermittent use of the site, such as would have been practised by swidden agriculturists is suggested by the relatively small size of the site; the high density of burials; the absence of middens, fire pits, hearths and evidence of structures; the shallow depth of cultural deposits; and the low sherd density (Bayard 1971, 1984, 1996–97, p. 906). There is evidence for continuity in the mortuary features over time, with three episodes of cultural and/or technological change, but not population replacement.

The period EP1–EP2 represents the pre-metal phases at Non Nok Tha. Stone tools were abundant, and no evidence for metal has been found. A single copper tool and small fragments of metal were recovered from EP3 sections. Early Period mortuary ritual included primary, supine and extended interments, often with mounding over the grave. Grave goods associated with children suggest ascribed status, and the range in number and quality of grave goods may reflect simple status differences among the burials (Bayard 1984, p. 98).

The first episode of culture change occured at the EP3/MP1–MP2 transition with a change in burial ritual (male burials are disarticulated); bronze was relatively abundant, with evidence for casting; clay spindle whorls suggest the advent or increase in cotton weaving, and clay pellets for use with the 'pellet bow' were found. Groups of distinctive vessels were placed with the burial, on the surface of the filled grave, as well as within large mounds built up over the larger graves. More types of pottery were introduced, including the complementary distribution of type 'C' and 'L' vessels, suggesting the presence of at least two 'affiliative' groups, each with a range of 'wealth', burying their dead at Non Nok Tha (Bayard 1984, p. 99). Bayard grouped MP1 and MP2 as a cluster of shared mortuary and cultural features (Bayard and Solheim 1992, pp. 8–9).

The MP3 period has also been interpreted as a transition zone, as the mounding ritual disappeared. Bronze casting continued and the first bronze

Table 8.1. *Mortuary features of the Non Nok Tha excavations*

Timeline[a]	Phases	Mortuary style	Cultural remains and comments
1000 BP to present	LP1–6	Cremation, burial of ashes in pottery vessels	Abandonment of site between MP8 and LP1; raiding of MP graves for pots, evidence for houses, iron common, celadon; decline in wild animals, increase in fowl and pig; charred water buffalo bones
*Late group (*n = *39)*			
3000–2000 BP	MP8[b] MP7[b] MP6[c]	Primary, supine, extended burials in shallower graves; pot nests in 1968 excavation; no animal bones	Decline in number of grave goods; single water buffalo bone from non-burial context
3500–3000 BP	MP5 MP4	Primary, supine, extended burials in graves of medium depth with groups of pots in fill and on top of fill; head to the southwest in MP4, to the northwest in MP5 and to the southwest in MP6; some skulls missing and replaced with clamshells, portions of domesticated animals included	Bronze sporadic. 'C' and 'L' pots continue; in later periods, an increase in domestic wares is seen as opposed to funerary wares
*Early group (*n = *37)*			
4200–3500 BP	MP3[b]	Primary, supine, extended burials in narrow, trench-like graves; No mounding; Head to west or southwest	First bronze ornaments
	MP2 MP1	Primary, supine, extended burials; smaller mounds of grave goods including portions of domesticated animals; secondary burials and commemorative deposits of grave goods but no skeleton	Spatial distinction between burials with vessel types 'C' and 'L' suggests affiliated groups; shell bead strings, bronze-casting evidence, pottery midden, spindle whorls, fish bone in vessels, stone tools
5000–4200 BP	EP3	Disarticulation in males (decomposed prior to burial?); mounds over graves, including human bone and portions of domesticated animals; fish bones in bottom of vessels	Beads and stone adzes absent; first bronze artefact from this phase, novel vessels, clay pellets
	EP2 EP1	Primary, supine, extended in deep graves with pots at head and foot; mounds over graves, including human bone and portions of domesticated animals; child burials segregated	Vessels of food, pots, stone adzes, strings of shell beads, rice chaff temper, evidence for social ranking; single water buffalo bone

[a]Extrapolated from Bayard (1996–97, p. 895 and his Fig. 2) and Bayard and Solheim (1992, p. 29).
[b]Found only in the 1966 excavations.
[c]Found only in the 1968 excavations.
Sources: Bayard and Solheim (1992), Higham (1975).

ornament was recovered from this context. As the MP progressed (MP4–MP8), there was a decrease in the number and quality of grave goods and a decrease in the depth and size of graves, suggesting a decline in differences in status and/or rank (Bayard 1984, p. 115). In addition, evidence of a decline in the number of large wild animal bones (e.g. deer) suggests a greater reliance on domesticated resources and/or a declining availability of wild resources. The third transition occurred between the MP and LP. The parahistoric transition, begun by a period of 'abandonment', followed with the appearance of cremation burials, iron, celadon and water buffalo bones, leading to the presumption of the presence of Buddhist religion and intensified paddy rice agriculture. This last technological and cultural transition cannot be examined for human health changes because of the practice of cremation and thus the lack of useable skeletal remains.

Material and methods

The two Non Nok Tha excavation samples were treated as a single series, with recognition of the presence of all of the biases of archaeological skeletal collections. For the temporal analysis, the whole skeletal sample was divided between MP3 and MP4. This point, although not ideal, was chosen in order to avoid splitting the MP1–MP2 grouping made by Bayard, because the major changes of the MP progressed after MP3, and in the interest of evening the sample sizes. The EP1–EP3 and MP1–MP3 burials were designated the 'early group', while the MP4–MP8 burials were designated the 'late group'. Thus, the early group represented hunter–gatherer–cultivator using wild and domesticated resources and living, perhaps transiently, in small groups near Non Nok Tha. The late group represented hunter–gatherer–cultivators with a greater, though not exclusive, emphasis on domesticated flora and fauna than on wild resources, and living, perhaps more sedentary, in larger hamlets near Non Nok Tha.

The Non Nok Tha skeletal series numbered 180 burials including a fair representation of subadults less than 15 years of age (27%). Douglas (1996) presented details of methods, palaeodemography and skeletal indicators of stress, including the dental conditions discussed here, in both the deciduous and permanent teeth in the combined series. Pietrusewsky and Douglas (2002) presented summary data on other aspects of the Non Nok Tha skeletal series. Of concern here is the sample of adult individuals (>15 years of age) with sex estimates and preservation of some portion of the dentition (Tables 8.2–8.4). Consistent with other researchers interested in Southeast Asia (e.g. Tayles 1999, Oxenham 2000), the individual was

Table 8.2. *Age, sex and group distribution of the Non Nok Tha adults*

Age (years)	Males (No. (%))	Females (No. (%))	Total (No. (%))	Early group (EP1–EP3, MP1–MP3)			Late group (MP4–MP8)		
				Males	Females	Total	Males	Females	Total
15–19.9	1 (2.5)	3 (7.1)	4 (4.9)		3	3	1		1
20–24.9	2 (5.0)	4 (9.5)	6 (7.3)	1	3	4		1	1
25–29.9	2 (5.0)	5 (11.9)	7 (8.5)		2	2	2	3	5
30–34.9	3 (7.5)	2 (4.8)	5 (6.1)	2	2	4	1		1
35–39.9	10 (25.0)	5 (11.9)	15 (18.3)	4	1	5	6	4	10
40–44.9	4 (10.0)	10 (23.8)	14 (17.1)	2	7	9	2	3	5
45–49.9	5 (12.5)	1 (2.4)	6 (7.3)	1		1	4	1	5
≥50	8 (20.0)	6 (14.3)	14 (17.1)	3	3	6	5	2	7
Middle-aged	3 (7.5)	4 (9.5)	7 (8.5)	2		2		4	4
Adult	2 (5.0)	2 (4.8)	4 (4.9)	1	1	2	1	1	2
Total	40	42	82	16	22	38	22	19	41

Three burials (1966–12A, 1966–42I, 1966–62) with unknown provenance were omitted from the temporal samples. Kolmogorov–Smirnov (KS) two-sample test of the male–female age distribution was not statistically significant ($D = 0.190$; $D_{\alpha=0.05} = 0.323$). Chi-square test of the samples of males and females aged <35 years and those aged >35 years was not statistically significant ($\chi^2 = 2.133$; d.f. = 1; $\alpha = 0.05$). KS two-sample test of the early group male–female age distribution was not statistically significant ($D = 0.304$; $D_{\alpha=0.05} = 0.480$). KS two-sample test of the late group male–female age distribution was not statistically significant ($D = 0.215$, $D_{\alpha=0.05} = 0.480$). Chi-square tests of the samples of males and females aged <35 years and those aged >35 years were not statistically significant in the early group ($\chi^2 = 2.048$; d.f. = 1; $\alpha = 0.05$) or the late group ($\chi^2 = 0.432$; d.f. = 1; $\alpha = 0.05$).

Table 8.3. *Non Nok Tha adult tooth sample by sex*

Tooth	Male			Female			Total		
	Max.	Mand.	Total	Max.	Mand.	Total	Max.	Mand.	Total
I1	31	29	60	36	31	67	67	60	127
I2	36	39	75	37	39	76	73	78	151
C	40	51	91	41	47	88	81	98	179
P3	50	55	105	53	47	100	103	102	205
P4	49	48	97	54	49	103	103	97	200
M1	46	44	90	54	50	104	100	94	194
M2	46	48	94	46	50	96	92	98	190
M3	25	40	65	27	41	68	52	81	133
Sum	323	354	677	348	354	702	671	708	1379

I, incisor, C, canine, P, premolar, M, molar, Max., maxilla, Mand., mandible.

Table 8.4. *Non Nok Tha adult tooth sample by phase*[a]

Tooth	EP1	EP2	EP3	MP1	MP2	MP3	Early group	MP4	MP5	MP6	MP7	MP8	Late group	Sum	Unk.	Total
I1	8	2	13	25	12	6	66	16	21	11	4	7	59	125	2	127
I2	12	3	17	30	12	8	82	24	19	14	4	6	67	149	2	151
C	13	4	20	40	15	10	102	30	20	14	4	6	74	176	3	179
P3	12	6	21	43	16	12	110	34	24	19	3	11	91	201	4	205
P4	14	5	21	43	16	11	110	29	24	18	4	11	86	196	4	200
M1	14	8	21	43	13	12	111	24	21	20	4	10	79	190	4	194
M2	12	8	19	42	15	12	108	26	21	21	4	6	78	186	4	190
M3	7	4	13	29	12	11	76	16	16	14	1	7	54	130	3	133
Sum	92	40	145	295	111	82	765	199	166	131	28	64	588	1353	26	1379
Indv.	4	2	6	23		3	38	15	10	11	1	4	41	79	3	82

I, incisor; C, canine; P, premolar; M, molar; EP, Early Period; MP, Middle Period; Unk., unknown provenance (burials 1966–12A, 1966–42I and 1966–62); Indv., number of individuals with at least one tooth or alveolus.

[a] Sums and totals are for observed teeth only and do not include observed alveoli.

included in the sample if at least one tooth or alveolus in either the maxilla or mandible was present.

Age and sex estimations were based on Douglas (1996). Each tooth or alveolus was assessed for the presence of antemortem tooth loss (AMTL), carious lesions, periapical cavities, calculus, alveolar resorption and attrition. One characteristic of the Non Nok Tha dental sample was the high rate of broken teeth. It is impossible to determine if the broken teeth are evidence of attempted tooth ablation (Sangvichien *et al.* 1969), the use of the teeth as tools or damage to the teeth during pre-excavation disturbances, excavation, cleaning and/or handling. No effort has been made to distinguish broken teeth from intact teeth in these analyses.

There are a variety of reasons for AMTL, including caries, advanced wear with pulp exposure, accident and deliberate removal. Distinguishing between dental agenesis and AMTL at a given tooth position can be difficult. Criteria such as evidence for remodelling of the alveolar bone and the presence of dental disease in other teeth suggest a tooth has been lost, while the complete absence of a tooth, unremodelled alveolar bone and the presence of the antimere suggest dental agenesis. In the Non Nok Tha sample, agenesis was noted only in third molars and these were counted as absent in this analysis.

Each tooth received a single score for carious lesions: absent or present with notation of the tooth surface location (e.g. occlusal, interproximal, buccal, lingual, etc.). All locations were combined for this discussion. This method for observed caries does not record multiple lesions on a single tooth. Each alveolus was evaluated for periapical cavities and scored as present or absent. Calculus was scored on a four-point scale (none, slight, moderate, marked) by tooth (Brothwell 1981, p. 155). Alveolar resorption, identified by porosity and resorption of the alveolar margin (Clarke 1990, 1993, Clarke and Hirsch 1991, Hildebolt and Molnar 1991) was also scored on a four-point scale at each alveolus. Each tooth was evaluated for attrition and scored following Brothwell (1981, p. 72): none, enamel wear, dentin exposure, pulp exposure and wear to the roots.

Teeth lost before death, regardless of the cause, were not included in an observed caries rate but they may have been lost because of caries. Lukacs (1992, 1995) advocated the use of a 'corrected caries rate' for an improved estimate of the frequency of caries in a dental sample. The calculation incorporates an estimate of the number of teeth that *may* have been lost antemortem because of caries, and it requires an evaluation of the cause of pulp exposure within each tooth. The methods of data collection used in the Non Nok Tha analysis did not distinguish causes for pulp exposure. Careful examination of the original recording forms shows that in the majority of cases of advanced dental attrition (pulp exposure or wear to the roots) no carious lesion was present, suggesting the cause of pulp exposure is attrition rather than caries. However, attrition was typically not scored in the presence of huge carious lesions of the crown where the pulp is obviously exposed, so these teeth would not be included in the calculation. Under these circumstances, a modification to the 'corrected caries rates' calculation was proposed that used the number of teeth with huge caries to estimate the proportion of teeth with pulp exposure as a result of caries (Tayles *et al.* 2000 (p. 73), Domett 2001 (p. 123)). Corrected caries rates reported here are based on this amendment to Lukacs' (1995, p. 153) procedure.

Rates for these dental pathological conditions in the Non Nok Tha dental sample have been presented in two ways: individual counts and tooth counts. The advantages and disadvantages of these two methods are obvious to anyone who has reported dental data. The tooth count method, which improves sample sizes and allows statistical analyses, disassociates the teeth from the individual, resulting in loss of epidemiological data. The individual count method gives equal weight to an individual with one tooth and an individual with a complete dentition, but results in smaller sample sizes. The discipline has made a call for reporting by both methods (Lukacs 1989, Hillson 2000).

Inferential statistics used in this analysis include the Kolmogorov–Smirnov test, the chi-square statistic, and Fisher's exact test (Thomas 1986). For statistical testing of frequencies 'by individual', the level of significance was set at 0.10, rather than the more common 0.05 level, a choice made to permit a wider allowance for possible differences in the face of small sample sizes, although, of course there is an increased risk of error. For statistical testing of frequencies 'by tooth', the level of significance was set at 0.05 since sample sizes are much greater.

Results

Since pathological conditions of the dentition are age progressive, it is important that the age distributions of the samples compared are similar. Preliminary examination of the sex and age distributions of individuals (Table 8.2) suggests that the female sample may be younger than the male sample, but statistical testing shows no significant differences in the male–female age distributions in either the whole sample or the temporal subsamples. There is a fairly even representation of teeth from both jaws and all classes in males and females in the whole sample (Table 8.3), as well as good representation of tooth classes in the temporal subsamples (Table 8.4). Results for the entire Non Nok Tha adult sample, including presentation of the DPP by age and by temporal subsample, will be presented first. The male and female DPP results, including the DPP by age and by temporal subsample, follow.

Dental pathology profile of the Non Nok Tha people

The per tooth DPP in Non Nok Tha adults (Table 8.5) shows the low rates generally consistent with a pre-agricultural economy and other northeastern Thailand skeletal series (Pietrusewsky and Douglas 2002). In the Non Nok Tha dental sample, AMTL, scored by alveolus, was uncommon in observed alveoli (7.4%) but more common in individuals (40.2%). No AMTL was noted in individuals less than 15 years of age (0/70 permanent teeth). Most adult individuals with AMTL had lost more than one tooth. Of the 82 individuals, 33 had AMTL and 11(33%) had lost a single tooth, 16 (48.5%) had lost from two to five teeth and the remaining six individuals (18.2%) had lost in excess of five teeth. The average number of teeth lost per person was 1.5 teeth (125/82).

Carious teeth were uncommon in the Non Nok Tha permanent teeth. No carious teeth were noted in the 13 individuals less than 15 years of age (0/52 teeth). Carious teeth were more common in the mandible than the maxilla and more common in the molars than any other tooth class. Carious lesions commonly were interproximal in origin (13/34; 38.2%) with massive (9/34; 26.5%) and occlusal (8/34; 23.5%) categories following. No obvious sex differences in lesion origin were noted (Douglas 1996, p. 387). Occlusal surface lesions occurred more often in younger individuals while interproximal lesions were more common in the older age groups. Caries affected 24.1% of individuals. The mean number of caries per person serves to reconcile the discrepancy between the frequency of affected individuals and the frequency of affected teeth (Lukacs 1989, p. 280). The mean number of carious teeth per person in the Non Nok Tha sample was 0.43 (corrected 0.60), extremely low values that are consistent with a non-agricultural economy (Lukacs 1989, p. 280).

Periapical cavities were uncommon in the Non Nok Tha dental sample, occurring in 1.9% of alveoli and in 15.4% of individuals. The mean number of periapical cavities per person was also quite low (0.35). Periapical cavities were more common in the mandible than the maxilla and more common in the molar and premolar alveoli than in the anterior alveoli (Douglas 1996, p. 397).

Table 8.5. *Dental pathology profile of Non Nok Tha adults*[a]

Disease	Adults		Individuals		Teeth	
	Individual A/O (%)	Teeth A/O (%)	Early group A/O (%)	Late group A/O (%)	Early group A/O (%)	Late group A/O (%)
Antemortem tooth loss	33/82 (40.2)	125/1698 (7.4)	15/38 (39.5)	18/41 (44.0)	45/900 (5.0)*	80/766 (10.4)*
Observed caries	19/79 (24.1)	34/1233 (2.8)	6/38 (15.8)	10/38 (26.3)	11/666 (1.7)*	22/536 (4.1)*
Corrected caries[b]	30/79 (38.0)	47.3/1358 (3.5)	11/38 (28.9)	16/38 (42.1)	13.3/711 (1.9)*	32.4/619 (5.2)*
Periapical cavity	12/78 (15.4)	27/1398 (1.9)	7/36 (19.4)	5/38 (13.2)	11/767 (1.4)	16/599 (2.7)
Calculus[c]	37/66 (56.1)	288/945 (30.5)	16/33 (48.5)	20/30 (66.7)	146/497 (29.4)	138/426 (32.4)
Resorption[c]	22/75 (29.3)	93/1022 (9.1)	8/34 (23.5)	13/37 (35.1)	39/540 (7.2)*	53/454 (11.7)*
Attrition[d]	17/80 (21.3)	76/1199 (6.3)	8/36 (22.2)	9/40 (22.5)	35/642 (5.5)	41/536 (7.6)

A/O, affected/observed teeth and/or alveoli.

[a]Any burial with at least one tooth or alveolus was counted as an individual. Three burials (1966–12 A, 1966–42I, 1966–62) with unknown provenance were omitted from the temporal samples.

[b]Corrected caries rate after Lukacs (1995, p. 153) modified by Domett (2001, p. 123), see text.

[c]Advanced (moderate and marked).

[d]Advanced (pulp exposure or wear to roots).

*Statistically significant differences ($\alpha = 0.05$). None of the individual rate differences was statistically significant ($\alpha = 0.10$).

Two categories of periodontal disease were included in the DPP: alveolar resorption and calculus. Four-point scale qualitative scoring of these conditions necessitates reduction of the data, so moderate and marked degrees of disease were combined to create an 'advanced' category. Advanced calculus was common when assessed by adult individuals (56.1%) as well as by teeth (30.5%). Although calculus is an age-related disorder, it is noteworthy that burial 1968–116, a possibly female individual (12–14 years of age), had slight calculus on seven teeth (7/28; 25.0%), otherwise no calculus was noted in the subadult sample (Douglas 1996, p. 411). Advanced calculus is more common in the mandibular than the maxillary teeth, but all tooth classes were affected about equally in this sample. No alveolar resorption was noted in the subadult sample of alveoli (0/27).

Advanced alveolar absorption was found in 29.3% of adult individuals and in 9.1% of teeth. Advanced resorption was more common in the maxilla than in the mandible and was more common in the posterior alveoli than in the anterior alveoli (Douglas 1996, p. 404).

General observations of dental wear were also scored in the permanent teeth from Non Nok Tha. Exposure of the dentin was noted in five teeth (5/53; 9.4%) in the subadult population: in the mandibular incisors and first molar of burial 1968–116, a possibly female aged 12–14 years (Douglas 1996, p. 419). In the adult sample, advanced attrition (pulp exposure and roots only) was found in 21.3% of individuals and 6.3% of teeth. Advanced attrition was more common in the mandibular teeth (7.3%) than in the maxillary teeth (5.3%) and was more common in the canines (9.0%), incisors (8.2%) and premolars (8.0%) than in the posterior teeth (Douglas 1996, p. 420). This advanced attrition of the anterior teeth, more common in male teeth (8.1%) than female teeth (4.6%), suggests that task-related use of the teeth may be contributing to the general wear pattern.

Examination of the DPP in individuals by age interval (Table 8.6) allows some interesting observations, although the sample sizes are uneven. AMTL, carious lesions and advanced calculus all appear in the youngest age intervals. Advanced attrition and advanced alveolar resorption (with the exception of an enigmatic 20–25-year-old female) were noted in individuals over 35 years of age. Periapical cavities appeared throughout the age distribution. The older age at occurrence of advanced attrition suggests that caries, rather than severe attrition, was the more likely cause of AMTL in the Non Nok Tha series, at least in the younger age intervals.

Based on the DPP model (Lukacs 1989, p. 276), as subsistence changes from hunting and gathering to agriculture, rates of caries, calculus and resorption are expected to increase, while rates of advanced attrition are expected to decline. Periapical cavities and AMTL are less predictable variables. The direction of AMTL rates might be predictable if the proximate cause of AMTL, either caries or severe attrition, can be inferred from the dental sample.

Therefore, the majority of the DPP rates should be lower in the early group from Non Nok Tha since they are presumed to have had a broader diet that was more reliant upon wild resources, and to have been more mobile than the late group. That is the case in both teeth and individuals (Table 8.5), with several exceptions. The rates of advanced attrition were remarkably consistent over time, rather than declining as the model predicts. The tooth rates of periapical cavities and calculus increased as expected, as did the individual rate of calculus, but the individual rate of

Table 8.6. *Dental pathology profile by age of Non Nok Tha adults*[a]

Disease	Age category (years) for observed pathology No. (%) affected								All ages observed pathology No. (%) affected
	15–19.9	20–24.9	25–29.9	30–34.9	35–39.9	40–44.9	45–49.9	50+	
Antemortem tooth loss	4 (25.0)	6 (33.3)	7 (0.0)	5 (20.0)	15 (33.3)	14 (57.1)	6 (50.0)	14 (50.0)	71 (40.8)
Caries	4 (25.0)	6 (16.7)	7 (14.3)	4 (0.0)	15 (26.7)	14 (28.6)	6 (16.7)	13 (23.1)	69 (21.7)
Periapical cavity	4 (0.0)	6 (0.0)	7 (14.3)	5 (20.0)	14 (21.4)	13 (23.1)	6 (0.0)	13 (23.1)	68 (16.2)
Calculus[b]	4 (25.0)	6 (33.3)	6 (66.7)	4 (25.0)	13 (53.8)	12 (66.7)	5 (80.0)	8 (87.5)	58 (58.6)
Resorption[b]	4 (0.0)	6 (16.7)	7 (0.0)	5 (0.0)	14 (57.1)	11 (18.2)	6 (16.7)	13 (69.2)	66 (31.8)
Attrition[c]	4 (0.0)	6 (0.0)	7 (0.0)	5 (0.0)	15 (13.3)	13 (23.1)	6 (33.3)	12 (58.3)	69 (20.3)

[a]Individuals with broad age estimates (i.e. 'adult', 'middle-aged') were not included. Individuals with unknown provenance were included.
[b]Advanced (moderate and marked).
[c]Advanced (pulp exposure or wear to roots).

periapical cavities decreased (none of these changes is statistically significant). The mean number of periapical cavities per individual also declined (3.27 to 2.38). Explaining these exceptions is difficult and they may ultimately be related to sampling error and the problems of incomplete individual dentitions, but they may also reflect an increase in inter-individual variation in diet. The AMTL rate increased over time, along with the caries rate, while rates of advanced attrition remained stable, supporting the conclusion that tooth loss in the series as a whole is related to caries rather than to severe attrition.

Dental pathology profile of Non Nok Tha males

The majority of the observations of advanced calculus and resorption in the Non Nok Tha dental series can be attributed to the males (Table 8.7); more than 70% of male individuals had advanced calculus and 40% or more were affected by AMTL, caries (corrected) and advanced resorption. The mean number of carious teeth in males was 0.38 (corrected 0.51).

Although sample sizes are very small in the younger age intervals (Table 8.8), there was little AMTL, caries, advanced resorption and advanced attrition in males prior to the age of 35 years. In contrast, advanced calculus was seen in more than half of male individuals after the age of 20 years. By old age, the majority of males have advanced alveolar resorption, calculus, attrition and AMTL.

The one statistically significant temporal change in the male DPP was an increase in the rate of advanced attrition (Table 8.7). The male DPP tooth rates overall were consistent with model expectations, except for a decline in calculus and an increase in attrition. As in the whole sample, there were inconsistencies in the direction of change between the individual rates and the tooth rates, which may be explained by small sample sizes. The tooth rate of AMTL increased in the late group, though the individual rates remain nearly equal. This inequality was reflected in an increase in the number of affected teeth per individual (0.88 to 1.62 teeth) and a change in the involved tooth classes from the incisors in the early group to the premolars and canines in the late group (Douglas 1996, p. 423). In contrast to the whole Non Nok Tha sample where AMTL seemed to be related to caries, the AMTL rate in males seemed to vary with the rate of advanced attrition while the caries rate stayed even over time, suggesting attrition as a cause of AMTL in males in the late group.

Dental pathology profile of Non Nok Tha females

The Non Nok Tha female tooth count data (Table 8.9) were consistent with a pre-agricultural economy though the relatively low rate of advanced

Table 8.7. *Dental pathology profile of Non Nok Tha adult males*[a]

Disease	Males		Individuals		Teeth	
	Individual count A/O (%)	Tooth count A/O (%)	Early group A/O (%)	Late group A/O (%)	Early group A/O (%)	Late group A/O (%)
AMTL	16/40 (40.0)	79/864 (9.1)	7/16 (43.8)	9/22 (40.9)	29/391 (7.4)*	50/449 (11.1)*
Observed caries	10/37 (27.0)	14/597 (2.3)	4/16 (25.0)	4/19 (21.1)	6/284 (2.1)	8/292 (2.7)
Corrected caries[b]	16/37 (43.2)	18.7/676 (2.8)	6/16 (37.5)	8/19 (42.1)	8/313 (2.6)	11/342 (3.2)
Periapical cavity	8/37 (21.6)	19/721 (2.6)	5/15 (33.3)	3/20 (15.0)	8/334 (2.4)	11/363 (3.0)
Calculus[c]	25/35 (71.4)	223/526 (42.4)	11/16 (68.8)	13/17 (76.5)	117/248 (47.2)*	102/262 (38.9)*
Resorption[c]	15/37 (40.5)	72/566 (12.7)	5/15 (33.3)	10/20 (50.0)	32/239 (13.4)	40/306 (13.1)
Attrition[d]	12/39 (30.8)	48/591 (8.1)	5/16 (31.3)	7/21 (33.3)	14/278 (5.0)**	34/297 (11.4)

A/O, affected/observed; AMTL, antemortem tooth loss.
[a]Any burial with at least one tooth or tooth socket was counted as an individual. Two burials (1966–12 A, 1966–421) with unknown provenance were omitted from the temporal samples.
[b]Corrected caries rate after Lukacs (1995, p. 153) modified by Domett (2001, p. 123) see text.
[c]Advanced (moderate and marked).
[d]Advanced (pulp exposure or wear to roots).
*Statistically significant differences: *$\alpha = 0.10$;
**$\alpha = 0.05$. None of the individual rate changes was statistically significant ($\alpha = 0.10$).

Table 8.8. *Dental pathology profile by age of adult males*[a]

Disease	Age category (years) for observed pathology No. (%) affected								All ages observed pathology No. (%) affected
	15–19.9	20–24.9	25–29.9	30–34.9	35–39.9	40–44.9	45–49.9	50+	
Antemortem tooth loss	1 (0.0)	2 (0.0)	2 (0.0)	3 (0.0)	10 (30.0)	4 (50.0)	5 (60.0)	8 (62.5)	35 (37.1)
Caries	1 (0.0)	2 (0.0)	2 (0.0)	2 (0.0)	10 (30.0)	4 (50.0)	5 (20.0)	7 (28.6)	33 (24.2)
Periapical cavity	1 (0.0)	2 (0.0)	2 (0.0)	3 (33.3)	9 (33.3)	4 (25.0)	5 (0.0)	7 (28.6)	33 (21.1)
Calculus[b]	1 (0.0)	2 (100.0)	2 (100.0)	2 (50.0)	9 (66.7)	4 (50.0)	5 (80.0)	6 (100.0)	31 (74.2)
Resorption[b]	1 (0.0)	2 (0.0)	2 (0.0)	3 (0.0)	9 (66.7)	4 (25.0)	5 (20.0)	7 (71.4)	33 (39.4)
Attrition[c]	1 (0.0)	2 (0.0)	2 (0.0)	3 (0.0)	10 (20.0)	4 (25.0)	5 (40.0)	7 (71.4)	34 (29.4)

[a]Individuals with broad age estimates (i.e. 'adult', 'middle-aged') were not included. Individuals with unknown provenance were included.
[b]Advanced (moderate and marked).
[c]Advanced (pulp exposure or wear to roots).

Table 8.9. *Dental pathology profile of Non Nok Tha adult females*[a]

Disease	Females		Individuals		Teeth	
	Individual count A/O (%)	Tooth count A/O (%)	Early group A/O (%)	Late group A/O (%)	Early group A/O (%)	Late group A/O (%)
AMTL	17/42 (40.5)	46/834 (5.5)	8/22 (36.4)	9/19 (47.4)	16/509 (3.1)**	30/317 (9.5)**
Observed caries	9/42 (21.4)	20/636 (3.1)	2/22 (9.1)*	6/19 (31.6)*	5/382 (1.3)**	14/244 (5.7)**
Corrected caries[b]	14/42 (33.3)	28.3/682 (4.1)	5/22 (22.7)	8/19 (42.1)	5.8/398 (1.5)**	24.8/274 (9.1)**
Periapical cavity	4/40 (10.0)	8/677 (1.2)	2/21 (9.5)	2/18 (11.1)	3/433 (0.7)	5/236 (2.1)
Calculus[c]	12/31 (38.7)	65/419 (15.5)	5/17 (29.4)	7/13 (53.8)	29/249 (11.6)**	36/164 (22.0)**
Resorption[c]	7/37 (18.9)	21/456 (4.6)	3/19 (15.8)	3/17 (17.6)	7/301 (2.3)**	13/148 (8.8)**
Attrition[d]	5/40 (12.5)	28/608 (4.6)	3/20 (15.0)	2/19 (10.5)	21/364 (5.8)**	7/239 (2.9)**

A/O, affected/observed; AMTL, antemortem tooth loss.

[a]Any burial with at least one tooth or tooth socket was counted as an individual. One burial (1966–62) with unknown provenance was omitted from the temporal samples.

[b]Corrected caries rate after Lukacs (1995:153) modified by Domett (2001:123) see text.

[c]Advanced (moderate and marked).

[d]Advanced (pulp exposure or wear to roots).

*Statistically significant differences: $*\alpha = 0.10$; $**\alpha = 0.05$.

calculus (15.5% teeth) is noteworthy. More than one-quarter of the individual females were affected with AMTL, advanced calculus and caries (corrected). The low number of individuals with advanced attrition relative to the number with caries suggests that AMTL in females was related to caries rather than attrition. The mean number of carious teeth in females was 0.45 (corrected 0.67).

Examination of the female DPP by age (Table 8.10) shows that the young age at occurrence of AMTL, carious lesions and calculus evident in the entire sample may be the result of disease in females. Since dental pathological conditions are age progressive and interdependent, tooth ablation should be suspected in young individuals with AMTL tooth loss. The youngest female with AMTL is burial 1968–74, aged 17–22 years. The right maxillary incisors, canine and third premolar were missing, though two tiny root fragments remained *in situ* at the canine and lateral incisor positions. This suggests deliberate tooth ablation or trauma. Tooth ablation has been documented in other skeletal series from Thailand (e.g. Sangvichien *et al.* 1969, Tayles 1996) by identifying patterns of loss, typically of the incisors, that are independent of dental pathology. Additional young females with AMTL (see footnotes to Table 8.10) were found in this series, but no obvious patterns of missing teeth were identifiable. Three of the seven females with caries were under the age of 35 years. Alveolar resorption, severe attrition and periapical cavities (with the exception of one individual) did not become common in females until middle age.

Female dental health declined over time at Non Nok Tha (Table 8.9), exactly as predicted by the model. The temporal changes in the female DPP were much greater and more dramatic than in the male DPP. Statistically significant increases were noted in the tooth rates of AMTL, caries, calculus and resorption. In contrast, there was a statistically significant decrease in the tooth rate of advanced attrition. This decline in advanced attrition was reflected primarily in the incisors (Douglas 1996, p. 423) and included a decline in the mean number of affected teeth per female (1.07 to 0.37). The increase in AMTL in the females of the late group occurred primarily in the mandibular molars and was associated with an increase in carious lesions of the molars (Douglas 1996, pp. 382, 393). The individual rates of dental disease also followed model predictions.

Comparison of male and female profiles

To streamline this discussion of sex differences in DPP, the focus will be on the tooth rates of the pathological conditions rather than on the individual rates. Non Nok Tha male and female DPPs showed statistically significant

Table 8.10. *Dental pathology profile by age of adult females*[a]

Disease	Age category (years) for observed pathology No. (%) affected								All ages observed pathology No. (%) affected
	15–19.9	20–24.9	25–29.9	30–34.9	35–39.9	40–44.9	45–49.9	50+	
Antemortem tooth loss	3 (33.3)[b]	4 (50.0)[c]	5 (0.0)	2 (50.0)[d]	5 (40.0)	10 (60.0)	1 (0.0)	6 (66.7)	36 (44.4)
Caries	3 (33.3)	4 (25.0)	5 (20.0)	2 (0.0)	5 (20.0)	10 (20.0)	1 (0.0)	6 (16.7)	36 (19.4)
Periapical cavity	3 (0.0)	4 (0.0)	5 (20.0)[e]	2 (0.0)	5 (0.0)	9 (22.2)	1 (0.0)	6 (16.7)	35 (11.4)
Calculus[f]	3 (33.3)	4 (0.0)	4 (50.0)	2 (0.0))	4 (25.0)	8 (75.0)		2 (50.0)	27 (40.7)
Resorption[f]	3 (0.0)	4 (25.0)	5 (0.0)	2(0.0)	5 (40.0)	7 (14.3)	1 (0.0)	6 (66.7)	33 (24.4)
Attrition[g]	3 (0.0)	4 (0.0)	5 (0.0)	2 (0.0)	5 (0.0)	10 (20.0)	1 (0.0)	5 (40.0)	35 (11.4)

[a]Individuals with broad age estimates (i.e. 'adult', 'middle-aged') were not included. Individuals with unknown provenance were included.
[b]A single individual with three lost teeth that may be deliberate ablation.
[c]Two individuals missing one tooth each (fourth premolar and second molar).
[d]One individual missing a third molar.
[e]One individual with abscessing of the left mandibular first molar socket.
[f]Advanced (moderate and marked).
[g]Advanced (pulp exposure or wear to roots).

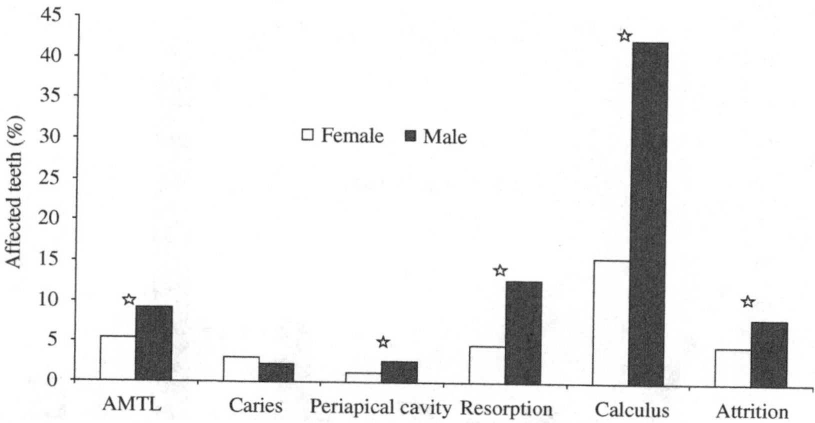

Figure 8.1. Dental pathology profile for sex differences. (Asterisks indicate statistically significant sex differences; $\alpha = 0.05$). AMTL, antemortem tooth loss.

differences in all but one of the conditions examined here (Fig. 8.1). Some of these differences may result from varying age distributions (the female sample was younger than the male sample), although statistical testing of the distributions was negative and, in general, the differences held up when examined in collapsed age categories (Douglas 1996). Certainly, the sex differences were consistent; males had higher rates of advanced calculus and alveolar resorption, AMTL and advanced attrition: the constellation of conditions that interrelate and are expected in an older population.

The high rate of advanced calculus in males is noteworthy. Plaque is a layer of protein produced by bacteria in the mouth, which may become mineralised to form calculus (Hillson 1986, 1996). Contributors to calculus formation include genetic predisposition, tooth crowding, diet and environmental factors such as water hardness, while good oral hygiene will limit calculus deposits. It appears that the Non Nok Tha males had a dietary or behavioural practice that contributed to an alkaline oral environment, or perhaps had poorer oral hygiene than the Non Nok Tha females.

In the early group at Non Nok Tha (Fig. 8.2), males had significantly higher rates of all of the dental conditions except caries and attrition. Again, the calculus rate in males is noteworthy, suggesting some behaviour distinct from females. The difference may not be dietary since the relatively equal rates of carious teeth and advanced attrition suggest equal access to carbohydrates and starches and shared grittiness, fibrousness or preparation methods.

As noted above, the male DPP remained fairly constant over time, though there was a significant increase in advanced attrition. In contrast,

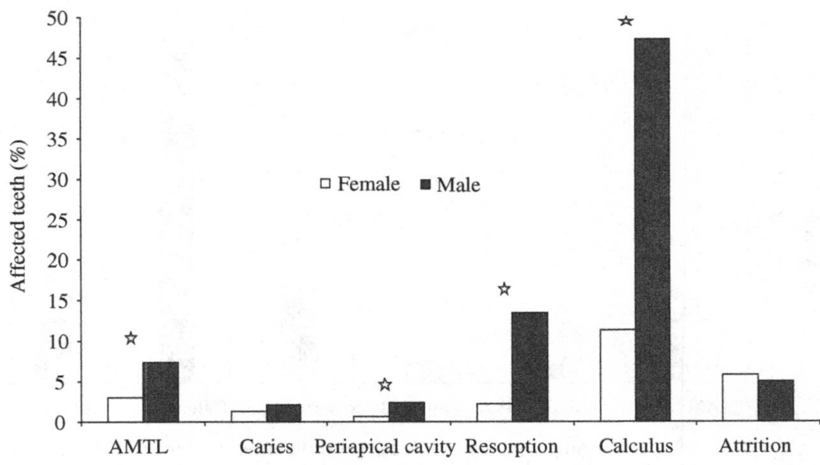

Figure 8.2. Dental pathology profile for the early group. (Asterisks indicate statistical significance differences; $\alpha = 0.05$). AMTL, antemortem tooth loss.

the female DPP changed dramatically over time, with significant increases in AMTL, caries, calculus and resorption, and a decline in advanced attrition. Although there were still statistically significant sex differences in the late group, the females appeared to 'catch up' to the males (Fig. 8.3). For caries, the female rate surpassed the male rate.

Discussion

Overall, the tooth count rates of dental pathological conditions in the Non Nok Tha adult (>15 years of age) sample were low and were consistent with a broadly based subsistence economy reliant upon hunting–gathering–cultivating activities in a wide-ranging environment. The rates of all of the pathological conditions fall at the lower end of the range of other skeletal samples from Thailand (Pietrusewsky and Douglas 2002, pp. 245–247), a finding consistent with the early pre-metal component of the site and the less-nucleated settlement pattern suggested by the archaeology (Bayard 1971, 1984, Wilen 1989). Indications of good dental health in the Non Nok Tha skeletal series as a whole are consistent with low levels of childhood stress suggested by the prevalence of linear enamel hypoplasia and by low levels of other stress indicators (e.g. adult stature, cribra orbitalia, infectious disease, trauma) found in additional analyses (Douglas 1996).

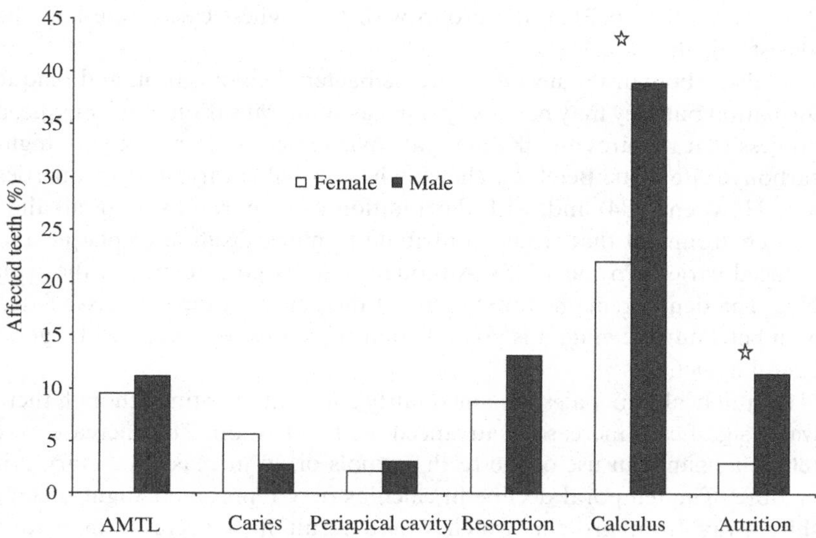

Figure 8.3. Dental pathology profile for the late group. (Asterisks indicate statistically significant sex differences; $\alpha = 0.05$). AMTL, antemortem tooth loss.

The individual count data are more complicated to interpret but provide more information on the epidemiology of these dental diseases. Periodontal disease appears to have been a significant health factor for many individuals at Non Nok Tha, with advanced calculus in more than half of the individuals and advanced alveolar resorption in nearly one-third of the individuals. Both of these disorders are likely contributors to AMTL, found in 39% of the individuals sampled. Still, the individual rates of dental disease at Non Nok Tha were in the low-to-moderate range of the DPP based on literature survey (Lukacs 1989, p. 276), again consistent with the suggested subsistence system.

Non Nok Tha males have significantly poorer dental health than Non Nok Tha females, a difference that seems to revolve around the rates of advanced calculus. Lieverse (1999) suggested that both high-carbohydrate diets, through increased bacterial colonisation, and high-protein diets, through increased alkalinity of the saliva, could increase the amount of calculus. Calculus formation may also be enhanced by behaviour, such as chewing betel nut, and other components of the diet, such as water hardness or silicon content (present in beer and rice), but these have not been proven (Lieverse 1999, p. 225). Although both high-carbohydrate and high-protein diets are thought to increase calculus formation, it is revealing that, in an analysis of Southeast Asian archaeological samples

(Oxenham 2000, p. 212), the group with the highest caries rate had the lowest calculus rate.

High-carbohydrate diets do increase bacterial colonisation and plaque formation but they may not always increase mineralisation of the plaque, a process that requires an alkaline oral environment not present with high-carbohydrate diets. Betel nut chewing has probable cariostatic properties (e.g. Howden 1984) and, with the addition of lime, results in an alkaline oral environment that would contribute to mineralisation of plaque and reduced caries (Powell 1985). Although there is no evidence in the Non Nok Tha dental remains for staining of the dentition typically associated with betel nut chewing, it is possible that the stains were avoided by good dental hygiene.

Dental health in males remained fairly constant over time, though there was a significant increase in advanced dental attrition. This increase may reflect a change in use of the teeth as tools or an increase in dietary grit or fibre. The temporal decline in calculus may represent a slight dietary shift away from protein, which would result in a decrease in plaque mineralisation.

Females seem to have experienced more of a temporal decline in dental health than males. This observation is consistent with a sexual division of labour, where females adopt more activities associated with agriculture than males, making agricultural foodstuffs more accessible to females than males. The decline in female dental attrition over time is also consistent with a shift toward softer, more processed agricultural foodstuffs. The frequency of advanced osteoarthritis of the cervical and upper thoracic spine increased in both males and females over time, suggesting changes in physical activity that might follow a change in subsistence regimen (Douglas 1996, pp. 469–470). The transition toward intensified agriculture was not all bad for Non Nok Tha females because they also experienced a temporal increase in stature, suggesting an increase in total caloric intake (Douglas 1996, p. 317).

Conclusions

The DPP for the whole Non Nok Tha series provides valuable and interesting information for the reconstruction of prehistoric health and disease and observations of a general decline in dental health over time. Sex differences noted in the sample provide evidence for possible sexual division of labour, differential access to food groups and/or differential oral behaviours. However, the full picture of dental health and disease in this series is only revealed by examination of temporal changes in the DPP by sex. While male DPP remained fairly stable, a dramatic temporal decline in female dental health is evident. Fuller comparisons with other skeletal series from northeastern Thailand and the Southeast Asian region, and palaeodietary research on these series, are required to elucidate further the possible causes for these sex differences.

Acknowledgements

A National Science Foundation Dissertation Improvement Grant (1993–1994) supported this research. Thanks are due to Dr Sanjai Sangvichien, Professor of Anatomy, Mahidol University Medical School, Bangkok, Thailand for research assistance with the 1966 Non Nok Tha skeletal series. Ms Vadhana Subhavan and Mr Somsak Pramankij were invaluable resources in Bangkok. Dr Sheilagh Brooks, Mr Richard Brooks, and Dr Bernardo Arriaza granted permission and provided research space for work on the 1968 Non Nok Tha collection at the University of Nevada-Las Vegas. The late Dr Donn Bayard was gracious and accommodating in all of our long-distance correspondence. My appreciation to Michael Pietrusewsky, Christine Sherman, Ann Lucy Stodder and Joyce White who made valuable comments and suggestions on previous drafts of this chapter. Finally, thanks to the editors, Nancy Tayles and Marc Oxenham, as well as two anonymous reviewers, for their time, attention and additional worthwhile suggestions.

References

Bayard D. T. 1971. *University of Otago Studies in Prehistoric Anthropology*, Vol. 4. *Non Nok Tha. The 1968 Excavation: Procedure, Stratigraphy, and Summary of the Evidence*. Dunedin: University of Otago Press.

1984. Rank and wealth at Non Nok Tha: the mortuary evidence. In Bayard D., ed., *Otago University Studies in Prehistoric Anthropology*, Vol. 17: *Southeast Asian Archaeology at the XV Pacific Science Congress*. Dunedin: University of Otago Press, pp. 87–122.

1996–97. Bones of contention: the Non Nok Tha burials and the chronology and context of early Southeast Asian bronze. In Bulbeck F. D., ed., *Ancient Chinese and Southeast Asian Bronze Age Cultures*, Vol. 2. Taipei: SMC, pp. 889–940.

Bayard D. T. and Solheim II W. G. 1992. Archaeological excavations at Non Nok Tha, Northeast Thailand, 1966–1968. Manuscript on file, Department of Anthropology, University of Otago.

Brooks S. T. and Brooks R. H. 1983. The incidence of paleopathological occurrences in the skeletal series from Non Nok Tha, Thailand. In *The 15th Pacific Science Congress*. Dunedin: University of Otago Press.

1987. An analysis of the 1968 Non Nok Tha skeletal series, Thailand, a summary of anthropometric and anthroposcopic results. Manuscript on file, Physical Anthropology Laboratory, University of Hawaii-Manoa, Hawaii.

Brothwell D. R. 1981. *Digging up Bones: The Excavation, Treatment, and Study of Human Skeletal Remains*, 3rd edn. Ithica, NY: Cornell University Press.

Clarke N. G. 1990. Periodontal defects of pulpal origin: evidence in early man. *American Journal of Physical Anthropology* **82**: 371–376.

1993. Periodontitis in dry skulls. *Dental Anthropology Newsletter* **7**: 1–4.

Clarke N. G. and Hirsch R. S. 1991. Physiological, pulpal, and periodontal factors influencing alveolar bone. In Kelly M. A. and Larsen C. S., eds., *Advances in Dental Anthropology*. New York: Wiley-Liss, pp. 241–266.

Domett K. M. 2001. *British Archaeological Reports International Series*, No. 946: *Health in Late Prehistoric Thailand.* Oxford: Archaeopress.

Douglas M. T. 1996. Paleopathology in human skeletal remains from the pre-metal, bronze, and iron ages, Northeastern Thailand. Ph.D. thesis University of Hawaii-Manoa. [Ann Arbor, MI: University Microfilms.]

Grauer A. L. and Stuart-Macadam P. 1998. *Sex and Gender in a Paleopathological Perspective*. Cambridge, UK: Cambridge University Press.

Higham C. F. W. 1975. *Otago University Studies in Prehistoric Anthropology*, Vol. 7: *Non Nok Tha: The Faunal Remains from the 1966 and 1968 Excavations at Non Nok Tha, Northeastern Thailand.* Dunedin: University of Otago Press.

1983. The Ban Chiang culture in wider perspective. *Proceedings of the British Academy of London* **69**: 229–261.

1989. *The archaeology of Mainland Southeast Asia: from 10,000 BC to the Fall of Angkor*. Cambridge, UK: Cambridge University Press.

Hildebolt D. R. and Molnar S. 1991. Measurement and description of periodontal disease in anthropological studies. In Kelley M. A. and Larsen C. S., eds., *Advances in Dental Anthropology*. New York: Wiley-Liss, pp. 225–240.

Hillson S. 1986. *Teeth*. Cambridge, UK: Cambridge University Press.

1996. *Dental Anthropology*. Cambridge, UK: Cambridge University Press.

2000. Dental pathology. In Katzenberg M. A. and Saunders S. R., eds., *Biological Anthropology of the Human Skeleton*. New York: Wiley-Liss, pp. 249–286.

Howden G. F. 1984. The cariostatic effect of betel nut chewing. *Papua New Guinea Medical Journal* **27**: 123–131.

Larsen C. S. 1997. *Bioarchaeology: Interpreting Behavior from the Human Skeleton*. Cambridge, UK: Cambridge University Press.

1998. Gender, health, and activity in foragers and farmers in the American southeast: implications for social organization. In Grauer A. L. and Stuart-Macadam P., eds., *Sex and Gender in Paleopathological Perspective*. Cambridge, UK: Cambridge University Press, pp. 165–187.

Lieverse A. R. 1999. Diet and the aetiology of dental calculus. *International Journal of Osteoarchaeology* **9**: 219–232.

Lukacs J. R. 1989. Dental paleopathology: methods for reconstructing dietary patterns. In Iscan M. Y. and Kennedy K. A. R., eds., *Reconstruction of Life from the Skeleton*. New York: Liss, pp. 261–286.

1992. Dental paleopathology and agricultural intensification in South Asia: new evidence from bronze age Harappa. *American Journal of Physical Anthropology* **87**: 133–150.

1995. The 'caries correction factor': a new method of calibrating dental caries rates to compensate for antemortem loss of teeth. *International Journal of Osteoarchaeology* **5**: 151–156.

1996. Sex differences in dental caries rates with the origin of agriculture in South Asia. *Current Anthropology* **37**: 147–153.

Nelsen K. M. 1999. The dental health of the people from Noen U-Loke, a prehistoric iron age site in Northeast Thailand. M.A. thesis, University of Otago, Dunedin.

Oxenham M. F. 2000. Health and behavior during the mid Holocene and metal period of Northern Viet Nam. Ph.D. thesis, Northern Territory University, Australia.

Pechenkina E. A., Benfer R. A. and Zhijun W. 2000. Diet and health changes at the end of the Chinese neolithic: the Yangshao/Longshan transition in Shaanxi Province. *American Journal of Physical Anthropology* **117**: 15–36.

Pietrusewsky M. 1974a. *Otago University Studies in Prehistoric Anthropology*, Vol. 6. *Non Nok Tha: The Human Skeletal Remains from the 1966 Excavations at Non Nok Tha, Northeastern Thailand.* Dunedin: University of Otago Press.

1974b. The paleodemography of a prehistoric Thai population: Non Nok Tha. *Asian Perspectives* **17**: 25–40.

Pietrusewsky M. and Douglas M. T. 2001. Intensification of agriculture at Ban Chiang: is there evidence from the skeletons? *Asian Perspectives* **40**: 157–178.

2002. *University Museum Monograph 111: Ban Chiang, A Prehistoric Village Site in Northeast Thailand I: The Human Skeletal Remains.* Philadelphia: University of Pennsylvania Museum of Archaeology and Anthropology.

Powell M. L. 1985. The analysis of dental wear and caries for dietary reconstruction. In Gilbert R. I. and Mielke J. H., eds., *The Analysis of Prehistoric Diets.* Orlando, FL: Academic Press, pp. 301–338.

Sangvichien S., Sirigaroon P. and Jørgensen J. B. 1969. *Archaeological excavations in Thailand*, Vol. III. *Ban-Kao: Neolithic Cemeteries in the Kanchanaburi Province.* Part Two: *The Prehistoric Thai Skeletons.* Munksgaard: Copenhagen.

Tayles N. 1996. Tooth ablation in prehistoric Southeast Asia. *International Journal of Osteoarchaeology* **6**: 333–345.

1999. *Report of the Research Committee* LXI. *The Excavation of Khok Phanom Di, A Prehistoric Site in Central Thailand*, Vol. V: *The People.* London: Society of Antiquaries.

Tayles N., Domett K. and Nelsen K. 2000. Agriculture and dental caries? The case of rice in prehistoric Southeast Asia. *World Archaeology* **32**: 68–83.

Thomas D. H. 1986 *Refiguring Anthropology: First Principles of Probability and Statistics.* Prospect Heights, IL: Waveland Press.

Wilen R. N. 1989. *British Archaeological Reports International Series*, No. 517: *Excavations at Non Pa Kluay, Northeast Thailand.* Oxford: Archaeopress.

9 *Human biology from the bronze age to the iron age in the Mun River valley of northeast Thailand*

KATE DOMETT
James Cook University, Townsville, Australia

NANCY TAYLES
University of Otago, Dunedin, New Zealand

Introduction

Prehistory

Recent excavations of two sites in the Mun River valley in Northeast Thailand (Fig. 1.1, p. 5) as part of The Origins of Angkor Project have uncovered two large cemeteries. The site of Ban Lum Khao includes a bronze age cemetery in use between approximately 3400 and 2500 BP (Higham 2002). A 10 m × 14 m square was excavated to a maximum depth of 1.7 m on the edge of this mounded site as other areas of the site had been subjected to looting (Higham 2002). The cemetery site of Noen U-Loke, only a few kilometres to the west, is dated to the iron age, approximately 2300–2200 BP to 1700–1600 BP (Higham 2002). This iron age site was considerably more extensive than earlier sites and covered up to 12 ha (Higham 2002). A total of 220 m^2 were excavated to a depth of 5 m (Higham 2002). Given the size of these sites, neither was excavated to their full extent. These two sites present an opportunity to compare health changes through time with the significant advantage that the populations lived in similar natural environments.

The aim of the Origins of Angkor Project is to investigate the social, cultural and technological developments in the Mun River valley that led to autonomous communities in this area undergoing the transition to more centralised and hierarchical societies (Higham and Thosarat 1998). The descendants of these communities built large-scale monuments and temples, some of which are still standing in the northeast today. These represent the northeastern extension of the Angkorian state (Higham 2002). It is

Bioarchaeology of Southeast Asia. Marc Oxenham and Nancy Tayles.
Published by Cambridge University Press.

likely that these pre-Angkorian populations underwent extensive changes over time, for example in the intensification of rice agriculture, the development of technology, in particular metalworking, and an increase in social complexity and exchange with other communities. The impact these factors can have on human health is the focus of this chapter. The environmental situation of a prehistoric site is also an important factor to be considered in an assessment of human health, but with the similarity in natural environments of these samples this variable is largely removed. There is some evidence of an environmental change during the iron age within this region (Boyd and McGrath, 2001a). The significance of this will be discussed.

After 2500 BP on the Khorat plateau of northeast Thailand, the first use of copper alloy artefacts began during the bronze age. Although no bronzes were recovered from Ban Lum Khao, the trading of ingots and casting of implements was evident from other archaeological finds. Subsistence was varied but included an increasing reliance on rice agriculture (Higham, 2002). The initial occupation layer (late neolithic) at Ban Lum Khao included a series of pits containing rich organic remains, including shellfish, fish, turtles and mammalian bones such as water buffalo, deer, pig and domesticated dogs (Higham and Thosarat, 1998). With the exception of water buffalo, these faunal species continued into the bronze age occupation layers (Higham and Thosarat 1998). The surrounding environment would have been suitable for intensive irrigated wet rice agriculture (Welch 1985). The availability of rice is confirmed by pottery temper analysis, which has indicated rice chaff was commonly used in the later phases at Ban Lum Khao (Voelker 2002).

Archaeological evidence shows that social structure became increasingly complex during the iron age, beginning around 500 BC (Higham, 2002). This period saw the establishment of hierarchical and centralised societies with a widening exchange network, including contact with Indian traders. Iron-working techniques were adopted and assisted with a further intensification of rice agriculture (Higham 1996). Noen U-Loke is one of numerous sites dating from the iron age in the area (Higham 2002). These appear to have been substantial villages sited on relatively elevated areas on the banks of rivers. Mortuary ritual at Noen U-Loke included, in some cases, burying the dead in beds of rice. Although this had a detrimental effect on the bone (Tayles *et al.* 1998), it is consistent with other evidence of the abundance of rice at this site (Higham 2002). Geoarchaeological studies support the presence of rice throughout the iron age in this region (Boyd and McGrath 2001b). This is all strongly suggestive of a successful rice-based economy. Animal bones with evidence of butchery found at the site include pig, cattle and water buffalo (McCaw 2000). Together with rice, some fish bones (Thosarat 2000) and other plant foods (Boyd and McGrath 2001b), this implies the availability of a nutritious and balanced diet.

The geoarchaeological record is suggestive of an environmental change, possibly as a result of the increase in human settlement and change in hydrological conditions experienced during the iron age in the Mun River valley (Boyd and McGrath 2001a). Human impact on the environment is evident through the management of rice and arboreal resources (Boyd and McGrath 2001a). Despite the physical proximity of the two sites considered in this study, the temporal differences saw an increase in the manipulation of the physical environment, which may be a factor in the consideration of human health. This would be expected with cultural and technological developments over time.

Health and disease with cultural change

Changes in health are expected over time. In order for complex societies to develop into such large-scale civilisations as that of Angkor, the preceding populations must have been able to feed and support themselves, at least to the extent that allowed the development of an elite class. This suggests that the people should have potentially been healthy. The societies must also have had an appropriate social structure and sufficient resources to support specialists who did not contribute directly to the survival of the group. It has been suggested that this state was reached during the iron age from the mid first millennium BC in Thailand, enabling communities to become more centralised politically and to support and organise larger numbers of people, leading to the development of state-level organisation (Higham 1996). Given that there is evidence of considerable cultural development in this region, it is fair to hypothesise that the preceding communities were healthy enough to instigate such development.

Through time, the health of the prehistoric people may be expected to have improved as their ability to exploit their environment developed. However, such cultural changes as the development and intensification of agriculture, the domestication of animals, sedentism, increasing social complexity and the improvement of metal working and tool production, while making subsistence activities more efficient, may also have detrimental effects on population health, through the increased population size and density they permit. The detrimental effects have been postulated for some regions of the world, especially the Americas (Cohen and Armelagos 1984, Larsen 1997). These analyses documenting the effects of intensifying agriculture on population health indicate a decline in health occurred (Larsen 1997, Steckel and Rose 2002). However, current evidence from Southeast Asia does not consistently fit this pattern (Tayles *et al.* 2000, Domett 2001, Pietrusewsky and Douglas 2002a). The nature of the cultural changes in the environment of the Southeast Asia region may have resulted in a different pattern of health changes. For example, the nutritional value of rice as a dietary staple is recognised to be considerably higher than that of maize (Juliano 1993).

In order to interpret the skeletal evidence and the effects of the events taking place during the bronze and iron ages in Thailand, a biocultural approach is taken, with archaeological evidence considered along with the assessment of the state of health from the skeletal remains. The human skeleton can provide information regarding health and disease and a skeletal collection, as a sample of individuals from a community, can provide information that suggests their general quality of life. This will be a reflection of both the environment they lived in and their ability to exploit it. Skeletal and dental parameters that may help to indicate any differences in the patterns of health between the two samples were selected on their ability first to provide a representation of health and second to be statistically comparable between samples.

Genetic heterogeneity between the samples is a potential source of variation in individual and population health. While it is an underlying assumption that the skeletal samples are relatively genetically homogenous, both between the samples and within each sample, it is acknowledged that genetic heterogeneity may be a possible explanation for some health differences. Future research on bone chemistry may clarify the relationships between and within the populations.

Materials and methods

The Ban Lum Khao skeletal remains were excavated in 1996 and represent 110 individuals. The material is generally well preserved, but not all individuals are represented by a complete skeleton; approximately 45% of skeletons are complete or near complete. A surface concretion of calcium carbonate on the bone obscures the subperiosteal surface on much of the material and could be only partially removed from some bones. A detailed investigation of this material has previously been undertaken and reported in Domett (2001). The data in this present study has been derived from this research, although some minor amendments have been made as a consequence of a recent reinvestigation of the skeletal material.

The material is currently stored in the Archaeological Laboratory at the Ninth Office of the Fine Arts Department, Phimai, Thailand.

The Noen U-Loke material was excavated over two seasons in 1997–1998. It represents 120 individuals and is also stored at the Phimai Fine Arts Department. This material is not as well preserved as that of Ban Lum Khao. Much of the bone is fragmented and many individuals are represented by incomplete skeletons. The skeletons from the deeper layers of the site tend to be better preserved. In the middle layers, a significant number of the interments were in graves filled with unthreshed rice. This became silicified over time and, as yet unidentified, chemical reaction between the grain and the bone resulted in the demineralisation of the bone tissue. The outlines of bones were visible but no detail identifiable. The organic fraction of the tissue that survived was so lacking in structure that in the worst cases it was impossible to recover anything for further analysis. The upper-layer burials tended to be fragmented. Overall, only 39 (32.5%) of the skeletons could be classified as more-or-less complete. Previous reported work on this material includes an investigation of the dental health (Nelsen 1999). As with the Ban Lum Khao material, a recent reassessment of this data has led to some minor differences between this analysis of the dentition and that of Nelsen (1999).

Parameters permitting a representation of the state of health of these two communities were selected based on previous studies by the authors (Tayles 1999, Domett 2001) and following the biocultural model of Goodman and Armelagos (1989). Included are demographic indicators, stature, linear enamel hypoplasia and dental pathologies. Skeletal pathology was not included in this comparison of the two sites and will be presented in a future publication. In general, the parameters chosen can all be statistically compared between the samples. Methods used for assessing these variables on both samples are as detailed in Domett (2001). Statistical tests used for non-metric variables are the chi-square test and Fisher's exact test (FET); Student *t*-tests were used for metric variables. The critical *p* value was set at 0.05. In order to maximise the statistical validity of testing for dental variables, teeth rather than individuals were used as units of analysis.

Results

Palaeodemography

Demographic summaries are provided in Table 9.1 for each sample. A number of individuals in both samples were not assigned an age estimate

Table 9.1. *The age and sex distribution of Ban Lum Khao and Noen U-Loke samples*

Age (years)	Ban Lum Khao (No. (%))			Noen U-Loke (No. (%))			
	Females	Males	Total	Females	Males	Unknown sex	Total
Under 15							
0–0.9			21 (19.1)				37 (30.8)[a]
1–4			14 (12.7)				9 (7.5)
5–9			11 (10.0)				3 (2.5)
10–14			5 (4.5)				3 (2.5)
Unknown age			0 (0.0)				1 (0.8)
Subtotal			51 (46.4)				53 (44.2)
Over 15							
15–19	2 (1.8)	3 (2.7)	5 (4.5)	0 (0.0)	2 (1.7)	2 (1.7)	4 (3.3)
20–29	17 (15.5)	7 (6.4)	24 (21.8)	8 (6.7)	13 (10.8)	4 (3.3)	25 (20.8)
30–39	7 (6.4)	9 (8.2)	16 (14.5)	7 (5.8)	5 (4.2)	3 (2.5)	15 (12.5)
40–49 +	4 (3.6)	7 (6.4)	11 (10.0)	6 (5.0)	4 (3.3)	1 (0.8)	11 (9.2)
Unknown age	2 (1.8)	1 (0.9)	3 (2.7)	0 (0.0)	3 (2.5)	9 (7.5)	12 (10.0)
Subtotal	32 (29.1)	27 (24.5)	59 (53.6)	21 (17.5)	27 (22.5)	19 (15.8)	67 (55.8)
Overall total			110 (100.0)				120 (100.0)

[a]Including two late fetal skeletons.

because of incompleteness and/or poor preservation. For the Noen U-Loke sample, this accounted for over 10% of the total sample, compared with just 3% for the Ban Lum Khao sample.

The overall level of infant and child mortality was very similar in the Ban Lum Khao and Noen U-Loke samples, at 46% and 44%, respectively (Table 9.1 and Fig. 9.1). These levels are at the upper range for prehistoric Southeast Asian sites (Douglas 1996, Domett 2001) and are suggestive of a good recovery rate for these small skeletons.

Despite a similar overall level of subadult mortality between the samples, there were some significant differences between subadult subgroups (Table 9.1 and Fig. 9.1). It is apparent that the Ban Lum Khao sample had more children dying at an older age rather than in infancy compared with Noen U-Loke. The proportion of infants (those aged 0–0.9 years) was significantly higher in the Noen U-Loke sample (30.8%) than in the Ban Lum Khao sample (19.1%) (FET $p = 0.048$). In contrast, the Ban Lum Khao sample had a significantly higher proportion of children in the 5–9-year age category (10%) compared with Noen U-Loke (2.5%) (FET

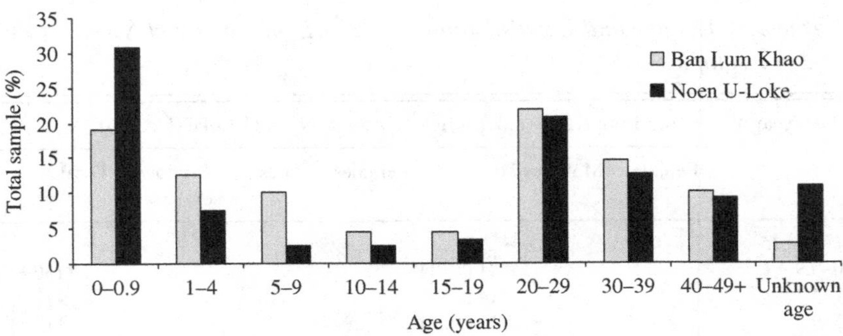

Figure 9.1. The age distribution of the Ban Lum Khao and Noen U-Loke samples.

$p = 0.025$). Ban Lum Khao mortality was also higher in all other childhood subgroups, but not to a statistically significant level.

Estimates of sex are available for all adults from Ban Lum Khao compared with only 70.6% of adults at Noen U-Loke. Despite the lack of data for some individuals, the female-to-male ratios were near equal within both samples, implying a good representation of both sexes. The sex ratio was 1.2:1 in the Ban Lum Khao sample and 0.8:1 in the Noen U-Loke sample. This is not significantly different between the samples (FET $p = 0.33$).

Overall, the adult age structures were similar in both samples (Table 9.1 and Fig. 9.1): both samples showed few late adolescents dying and the highest proportion of adults dying in the young adult age range (20–29 years), with a decreasing proportion through to 'old age' (40–49 years). However, there was a significantly higher proportion of young adult females in the Ban Lum Khao sample (15.5%) compared with the Noen U-Loke sample (6.7%) (FET $p = 0.035$) (Table 9.1). This difference was carried through to the total proportion of females in the sample so that there are more females represented overall at Ban Lum Khao (29.1%) than at Noen U-Loke (17.5%) (FET $p = 0.042$).

The juvenile–adult ratio (JA) and the mean childhood mortality (MCM) are used as indicators of fertility and the level of population growth (negative, positive or stationary) (Jackes 1992). They circumvent the problems of potential under-representation of infants and the difficulties of age estimation in adults (Jackes 1992). The JA ratio is measured as the ratio of individuals aged between 5 and 15 years divided by the number of individuals over the age of 20 years. The MCM is calculated by averaging the probability measures from 5 to 15 years (5q5, 5q10, 5q15). The results for the two samples are provided in Table 9.2. The JA ratio at Ban Lum Khao was 0.30, which is high compared with Noen U-Loke at 0.11. This

Table 9.2. *Juvenile–adult ratio and mean childhood mortality data for Ban Lum Khao and Noen U-Loke*

	D(5–14) / D(20+)[a]	JA	5q5, 5q10, 5q15[b]	MCM
Ban Lum Khao	16/54	0.30	0.15, 0.08, 0.08	0.10
Noen U-Loke	7/63	0.11	0.04, 0.06, 0.06	0.05

JA, juvenile–adult ratio, MCM, mean childhood mortality.
[a]Individuals aged 5–14 years/individuals aged over 20 years.
[b]Probability measures from 5 to 15 years.

reflects a decline in fertility, with a reduced proportion of juveniles to adults at Noen U-Loke. The MCM rate was also higher at Ban Lum Khao (0.10317) compared with that at Noen U-Loke (0.05219); this also signifies that a decline in fertility had occurred from the bronze age to the iron age. These statistics independently suggest higher fertility at Ban Lum Khao than at Noen U-Loke. Plotting the two statistics (JA and MCM) for the two sites, as recommended by Jackes (1992, 1994), suggests that the Ban Lum Khao population was growing at a rapid rate, whereas the Noen U-Loke population was increasing at a comparatively slower rate.

Growth and growth disturbances

Stature

Stature could be estimated for nearly three-quarters (73%) of adults in the Ban Lum Khao sample. In contrast, the poor preservation of the Noen U-Loke material meant that only 19% of adults had complete or reconstructable long bones that were measurable. With such contrasting proportions between the samples, the following results are interpreted with caution.

The summary results comparing male and female statures between Ban Lum Khao and Noen U-Loke are presented in Table 9.3. The female means were almost identical, with no significant difference. The male mean statures showed a significant difference between the samples, with the Noen U-Loke males being on average 4.6 cm taller (169.3 cm) than Ban Lum Khao males (164.7 cm) (t-test $p = 0.049$). The Ban Lum Khao male sample showed a wider range of statures, with individuals both taller and shorter (over 12 cm shorter in one case) than the Noen U-Loke sample. The smaller sample size from Noen U-Loke is likely to have influenced the limited range identified in their statures.

Table 9.3. *Adult stature summary statistics for Ban Lum Khao and Noen U-Loke adults*

	No.	Mean (cm)	SD (cm)	Range (cm)[a]	Sexual dimorphism (% of male stature)
Female[b]					
Ban Lum Khao	25	154.7	3.8	147.9–162.2 (14.3)	6.1
Noen U-Loke	4	154.6	4.7	151.5–161.6 (10.1)	8.7
Male[c]					
Ban Lum Khao	18	164.7	6.2	152.4–174.9 (22.5)	
Noen U-Loke	9	169.3	3.1	165.3–173.7 (8.4)	

SD, standard deviation.
[a]Number in parentheses is the size of the range (e.g. 162.2 – 147.9 = 14.3).
[b]Female comparison: $t = 0.039$; $p = 0.97$ (not significant).
[c]Male comparison: $t = -2.073$; $p = 0.049$ (significant at 5% level).

The level of sexual dimorphism in stature in both samples was within the normal range of 5–10% (Molnar, 1992). The Noen U-Loke sample had a slightly higher level of sexual dimorphism (8.7%) than Ban Lum Khao (6.1%) (Table 9.3).

Enamel hypoplasia

It is accepted that canines and incisors may form hypoplasias at higher frequencies than other teeth (Goodman *et al.* 1980). Therefore, to ensure that one sample does not have a significantly higher proportion of these teeth, the percentage of anterior teeth with linear enamel hypoplasia (LEH) was compared between the samples. Both samples had very similar percentages of canines and incisors, both mandibular and maxillary: between 35 and 38%. Therefore, we can assume that the data presented below and in Table 9.4 are not biased.

Subadults No deciduous teeth in the Ban Lum Khao sample had LEH. In the permanent teeth of the Ban Lum Khao subadults (<15 years), 28.3% (30/106) of teeth had LEH (Table 9.4). No deciduous teeth in the Noen U-Loke sample had evidence of LEH, although one subadult has evidence of non-linear enamel hypoplasia. In the permanent teeth of the Noen U-Loke subadults, 7.0% (4/57) of teeth had evidence of LEH. These results indicate a four-fold larger percentage of LEH in the permanent subadult teeth in the Ban Lum Khao sample. This difference is statistically significant (χ^2 $p < 0.01$).

Table 9.4. *The proportion of teeth with linear enamel hypoplasia defects at Ban Lum Khao and Noen U-Loke*

	Affected/total observed teeth (%)		Site difference
	Ban Lum Khao	Noen U-Loke	p value[a]
Subadults[b]	30/106 (28.3)	4/57 (7.0)	<0.01
Adults[c]			
Female	54/429 (12.6)	39/258 (15.1)	0.35
Male	35/308 (11.4)	20/304 (6.6)	0.04
Uncertain sex		13/114 (11.4)	
Total	89/737 (12.1)	72/676 (10.7)	0.40

[a]Chi-square test.
[b]Permanent teeth only.
[c]Teeth with advanced attrition have been removed from the calculation for adult teeth.

Adults The Noen U-Loke sample had a significantly higher level of advanced attrition (see Table 9.7, below), which may have eliminated some evidence of LEH in this sample. The proportion of LEH in adults is, therefore, presented after the teeth with advanced attrition are removed from the calculation for both sites (Table 9.4). Comparing the total samples, both Ban Lum Khao and Noen U-Loke showed no significant differences in the presence of LEH, with 12.1% and 10.7% of teeth affected, respectively. The two female groups showed no significant differences in the proportion of LEH, with Ban Lum Khao females exhibiting slightly less (12.6%) LEH than Noen U-Loke (15.1%) (χ^2 $p=0.35$). However, Ban Lum Khao males had almost twice the proportion of LEH (11.4%) compared with Noen U-Loke (6.6%) and this was significant (χ^2 $p=0.04$).

Enamel hypoplasia and stature

It is possible that an individual with enamel hypoplasia, as an indication of stress in childhood, may be affected long term. Depending on the individual's ability to catch-up in growth after the stress period (Stini 1985), those with enamel hypoplasia may not reach their genetic potential for stature. The correlation between low stature and high levels of enamel hypoplasia has been demonstrated elsewhere (Pechenkina *et al.* 2002, Stodder and Douglas 2003).

Here we compare the mean stature of those individuals with LEH with those without LEH. Within both samples, individuals with LEH had lower mean statures than those without LEH (Table 9.5); however, no significant differences between males and females within each sample were found.

Table 9.5. *Mean stature for Ban Lum Khao and Noen U-Loke adult individuals with and without linear enamel hypoplasia*

	Mean stature (cm) with LEH				Mean stature (cm) without LEH			
	Males	No.	Females	No.	Males	No.	Females	No.
Ban Lum Khao	162.8	5	153.9	7	165.5	9	154.3	12
Noen U-Loke	169.4	4	151.7	2	169.9	4	157.5	2
t-test *p* value	<0.01		0.07		0.222		0.585	

LEH, linear enamel hypoplasia.

Comparison of the two samples with LEH showed that Noen U-Loke males had a significantly taller stature (169.4 cm; $n=4$) compared with Ban Lum Khao males (162.8 cm; $n=5$) although sample sizes were small (t-test $p<0.01$). This follows the pattern seen in the complete samples (Table 9.4). Females with LEH showed no significant differences in their mean statures between the samples.

Dental health

Many conditions associated with the dentition are age related, including advanced attrition and antemortem tooth loss. Therefore, the age structures of the subsamples with teeth were compared to determine whether age differences influenced the final results. In the comparison of the age structure of the individuals contributing teeth or tooth positions observable for each of the conditions assessed, significant differences were identified when the female and male age structures were compared separately. However, no significant age-structure differences were apparent when the total samples were compared. Given these findings, only the total percentages of each condition were compared between the two samples. Although this runs the risk of losing sex differences in dental health, it is necessary to remove the bias of age structure.

Caries

The proportion of caries in the deciduous and permanent teeth in the subadults from both samples was very low and similar between the samples (ranging from 0 to 2.2%) (Table 9.6). In the adult samples, the total proportion of carious lesions was also very similar between samples, with 4.5% for Ban Lum Khao and 4.8% for Noen U-Loke (Table 9.7 and Fig. 9.2). No significant differences were found.

Table 9.6. *The proportion of deciduous and permanent teeth with caries in Ban Lum Khao and Noen U-Loke subadults*

	Affected/total observed teeth (%)		Site difference p value[a]
	Ban Lum Khao	Noen U-Loke	
Deciduous teeth	4/182 (2.2)	1/89 (1.1)	>0.99
Permanent teeth	1/92 (1.1)	0/52 (0)	>0.99

[a]Fisher's exact test.

Table 9.7. *Summary of the proportion of teeth or tooth positions affected by the listed dental conditions in Ban Lum Khao and Noen U-Loke adults*

	Affected/total observed teeth (%)		Site difference p value[a]
	Ban Lum Khao	Noen U-Loke	
Advanced attrition	104/861 (12.1)	152/977 (15.6)	0.04
Caries	39/874 (4.5)	46/956 (4.8)	0.74
Periapical cavities	15/1138 (1.3)	34/571 (6.0)	<0.01
Antemortem tooth loss	59/1138 (5.2)	69/1334 (5.2)	>0.99

[a]Fisher's exact test.

Advanced attrition

The level of advanced attrition was significantly lower in the Ban Lum Khao sample, who had 12.1% of teeth affected compared with 15.6% in the Noen U-Loke sample (Table 9.7 and Fig. 9.2). This difference between the samples may have affected the observation of other dental conditions. As mentioned above, attrition can wear away the evidence of LEH; it could also eliminate some of the carious lesions. This effect would have been greater in the Noen U-Loke adults than in the Ban Lum Khao adults.

Tooth infection

The proportion of tooth positions with periapical cavities indicated significantly lower proportions of tooth infection in the Ban Lum Khao sample (1.3%) compared with the Noen U-Loke teeth (6.0%) (FET $p<0.01$) (Table 9.7 and Fig. 9.2). The Noen U-Loke teeth are well preserved but individuals frequently lacked alveolar bone. The inequality in bone preservation between the samples may have influenced this result by reducing the observable sites (that is, the denominator) for periapical

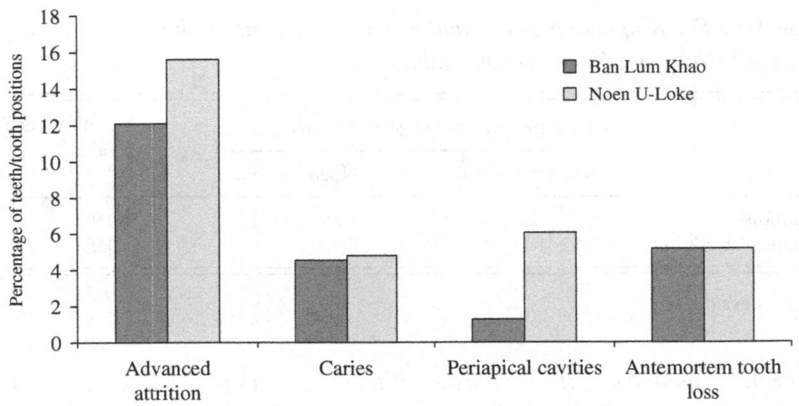

Figure 9.2. Percentage of teeth or tooth positions affected in the total adult sample for Ban Lum Khao and Noen U-Loke.

cavities in the Noen U-Loke sample and artificially inflating the prevalence in the small sample.

Antemortem tooth loss

The two samples have the same percentage of antemortem tooth loss (5.2%) (Table 9.7 and Fig. 9.2). However, the pathological cause of the loss may have been different. From these results, subject to the sample size caveat already referred to, it is possible that the higher level of advanced attrition in the Noen U-Loke sample exposed the pulp cavity and led to infection and eventual tooth loss. It is not as clear what caused the tooth loss in the Ban Lum Khao sample. It was not possible to compare the condition of the periodontal tissues, as the Ban Lum Khao alveolar bone tended to be obscured with the calcareous coating and the Noen U-Loke alveolar bone was poorly preserved.

Discussion

Infants and children are known to be particularly vulnerable to cultural and environmental stresses (Goodman and Armelagos 1989). The health status of infants and children can, therefore, be particularly illuminating in the assessment of the entire community from which they came. Evidence of infant and child mortality, growth disturbance and attained stature in adulthood can be integrated in an overall assessment of childhood health.

The bronze age at Ban Lum Khao, in comparison with the iron age at Noen U-Loke, was characterised by a lower infant mortality but a higher mortality of older children, particularly those aged 5–9 years (Table 9.1 and Fig. 9.1). In these subadults, there is evidence of higher levels of growth disruption (LEH) in comparison with Noen U-Loke subadults. This would imply that children dying prematurely in the Ban Lum Khao community died of chronic problems as there had been time for stress to affect the dentition. Alternatively, Ban Lum Khao children were weakened by chronic ill-health and were fatally affected by acute disease at a later age. Noen U-Loke subadults were more likely to die during the first year of life, with over 30% of the total sample dying as infants, in comparison with 19% in the Ban Lum Khao sample. It is possible that the infants in the Ban Lum Khao cemetery were under-represented (see comment below).

The mortality data illustrate the significance of the difference in the age structure of the two sites for the demographic status of the populations (Table 9.1). The JA and MCM statistics (Table 9.2), separately and when plotted together, consistently suggest a high fertility rate and rapid population growth at Ban Lum Khao compared with Noen U-Loke. The statistics from four other sites in northeast Thailand (Ban Chiang, Non Nok Tha, Non Pa Kluay, Ban Na Di) are most similar to those at Noen U-Loke (Pietrusewsky and Douglas, 2002b). Our data appear to suggest that Ban Lum Khao was anomalous for the region in having high fertility rates characteristic of a growing population, but this needs to be considered in the light of the dates of the other sites. At three of the sites (Ban Chiang, Non Nok Tha and Non Pa Kluay) the initial occupation predated the cemetery at Ban Lum Khao, but all included burials contemporary with Ban Lum Khao. The cemetery at Ban Chiang, in particular, was used over a long period of time but also included burials contemporary with Noen U-Loke. Over time, JA and MCM data from Ban Chiang (Pietrusewsky and Douglas 2002a) exhibit only minor changes compared with the differences seen at Ban Lum Khao. The cemetery at the fourth site (Ban Na Di) has been dated to the period of transition between Ban Lum Khao and Noen U-Loke (Higham 1996). A detailed comparison of evidence from different phases at these sites with the Ban Lum Khao and Noen U-Loke data would be ideal, but unfortunately the samples from the sites other than Ban Chiang are too small to be satisfactorily subdivided.

A decline in fertility, as experienced from the bronze age to the iron age at these two sites, is likely also to reduce the level of infant mortality (Jackes 1994). However, Noen U-Loke had a higher infant mortality rate than Ban Lum Khao. One implication of the high fertility rate and high growth rate suggested by the JA and MCM data is that the sample of infants in the

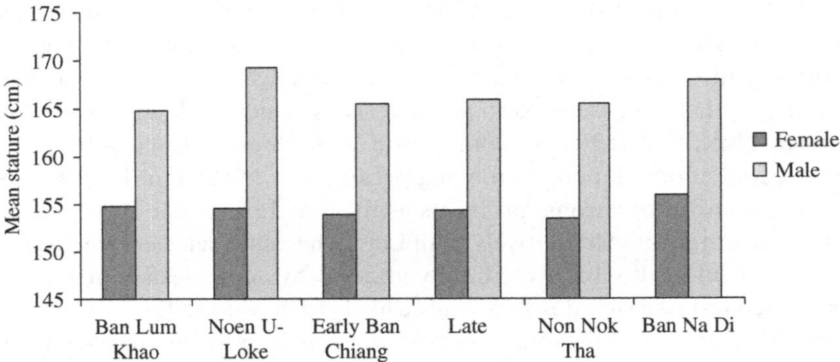

Figure 9.3. Mean stature for prehistoric northeastern Thai sites. Data for Ban Lum Khao and Ban Na Di from Domett (2001); For Ban Chiang from Pietrusewsky and Douglas (2002a); For Non Nok Tha from Douglas (1996); and for Noen U-Loke from this study.

skeletal sample from Ban Lum Khao should have been much larger. The under-representation could have been either because of poor survival of their skeletons or because of mortuary practices that resulted in them being interred elsewhere. Many infants in this cemetery were interred in burial jars ($n = 14$), ensuring their generally good preservation. The burial matrix was heavily disturbed by the activity of insects and small burrowing animals, so it would not be surprising if infant bones interred without the protection of burial jars did not survive. However, this does not appear to be the case, as a further 10 infants were not interred in burial jars and were successfully identified and excavated. Perhaps many infants were buried elsewhere.

Evidence of childhood growth and growth disturbance in the adult skeletons indicated that females between the sites were affected similarly, while males were not: higher LEH and shorter statures were identified in the Ban Lum Khao males. Given the potential, under appropriate conditions such as good nutrition, for a child to undergo a period of 'catch-up' growth after a period of stress, and assuming a degree of genetic homogeneity between the sites, this could indicate that the Ban Lum Khao males who survived to adulthood were more likely to have been affected severely by childhood illness and experienced growth disturbances. As a consequence, they did not attain their genetic potential for height. In contrast, the Noen U-Loke males, although showing signs of growth disturbance, had taller mean statures, suggesting that they were able to recover well from childhood stressors. The wide range of male statures at Ban Lum Khao suggests that there was considerable variation in health, implying inequality in access to resources in this subgroup. Comparison of the statures with those at the other northeast Thai sites (Fig. 9.3) shows that the Ban Lum Khao males were, on average, shorter than those at all other sites. The shortest individuals were shorter than at all other sites but the tallest were in the range of the other sites. In contrast the Noen U-Loke males were taller than at all other sites and all individuals were within the upper part of the ranges from those sites; however, the sample size was quite small.

Although the males at Noen U-Loke seem to have been more advantaged than those from Ban Lum Khao (and all other northeast Thai sites), the females have similar statures at all sites (Fig. 9.3). Stini (1985) suggested that males are more affected by stress than females and, therefore, sexual dimorphism tends to be lower in populations facing difficult conditions. The level of sexual dimorphism in the Ban Lum Khao sample was similar to the 6–7% at the other four sites (Pietrusewsky and Douglas 2002a) while it was higher at Noen U-Loke (8.7%). This is consistent with the suggested favourable conditions experienced by the iron age males.

Overall, the evidence of health during subadult life shows that, although there was high fertility and population growth during the bronze age, the population at Ban Lum Khao was experiencing stress that impacted on a number of the males in particular. The relatively wide range of statures suggests a hierarchical society with inequality, although there is no archaeological evidence of this (O'Reilly 2001). In comparison, by the iron age, the population had a lower growth rate demographically but appeared to be experiencing a better quality of life. Noen U-Loke was perhaps advantaged by more favourable technology and environmental conditions. It is possible that the people of Noen U-Loke, facing increasing population density, made a conscious decision to limit the number of children born to each woman. However, this is not a cultural feature of modern Thailand. The apparent contradiction in the relationship between fertility and health needs a more thorough and detailed examination on a larger sample.

The evidence of dental pathology indicates that the Ban Lum Khao and Noen U-Loke people had similar levels of caries and antemortem tooth loss but the Noen U-Loke sample had significantly higher proportions of periapical cavities and advanced attrition (Table 9.7). Bearing in mind the potential biases in the data, as outlined previously, what can these results tell us about each community's diet and dental health?

The higher prevalence of advanced attrition in the Noen U-Loke sample is most likely indicative of a more abrasive diet than at Ban Lum Khao, although the influence of genetic factors in tooth structure cannot be entirely eliminated. The low-to-moderate caries prevalence in both samples suggests their diet did not have a high proportion of cariogenic foods. The increase in the reliance on rice and less emphasis on hunter–gatherer foods would be expected to be one of the key differences between the bronze and iron ages in this region, although plants would provide the main carbohydrate in both these communities. However, despite archaeological evidence for increasing rice production, the caries rates between Ban Lum Khao and Noen U-Loke were very similar. They do not show a concomitant increase as is commonly reported for other carbohydrate staples (Turner 1979, Larsen 1997). As a previous study has indicated (Tayles *et al.* 2000), rice may not be a highly cariogenic carbohydrate compared with other carbohydrate staples such as maize. (Note: the Noen U-Loke and Ban Lum Khao caries data in Tayles *et al.* (2000) differ slightly from those presented here as a reassessment of the material has been carried out since that publication.)

Dental pathologies are generally highly interrelated and the construction of a dental pathology profile (Lukacs 1989) can provide a useful comparison among samples. Table 9.8 provides the dental pathology profiles from available prehistoric northeastern Thai skeletal samples. Good representation of teeth, compared with the poor representation of bones, has permitted the subdivision of some samples by time period. The terms low, moderate, and high are relative and appropriate only to the samples used here. The profiles do not show any marked patterns associated with cultural (temporal) changes. With few exceptions (Non Nok Tha Early Period and Ban Chiang Late Period), caries rates have remained stable in the northeast from the late neolithic through to the iron age. Antemortem tooth loss was also comparatively stable overall, with the exception of the high rate in the Late Period at Non Nok Tha. The most variation was seen in advanced attrition and tooth infection data, although there is no consistent pattern with time.

The dental pathology profile does not correlate with the evidence of better health at Noen U-Loke than at Ban Lum Khao. There appears to have been more tooth infection, associated with a more fibrous or gritty diet, although this may be a dental example of the 'osteological paradox', where better health in the iron age permitted these people to survive with their higher level of tooth infection than the inhabitants of Ban Lum Khao. However, the Noen U-Loke periapical bone was generally not well preserved and may be affecting these results.

Conclusions

Our results have indicated some key differences in the health status between bronze age Ban Lum Khao and iron age Noen U-Loke. There was evidence for significant population growth in the sample from Ban Lum Khao, while the Noen U-Loke population appears to have been experiencing a lower

Table 9.8. *Dental pathology profiles for prehistoric northeastern Thai sites*

Dental pathology	Site and cultural period, degree of pathology[a] (% teeth affected)						
	Non Nok Tha (Early): neolithic–bronze age[b]	Ban Chiang (Early): neolithic–bronze age[b]	Ban Lum Khao bronze age	Ban Chiang (Late): bronze–iron age[b]	Non Nok Tha (Late): early iron age[b]	Ban Na Di: early iron age[c]	Noen U-Loke: iron age
Advanced attrition	Low (5.5)	High (14.5)	Moderate (12.1)	High (17.0)	Low (7.6)	Moderate (12.0)	High (15.6)
Caries	Low (1.7)	Moderate–high (6.5)	Moderate (4.5)	High (7.7)	Moderate (4.1)	Moderate (4.7)	Moderate (4.8)
Periapical cavities	Low (1.4)	High (5.6)	Low (1.3)	High (6.6)	Moderate (2.7)	Low (2.1)	High (6.0)
Antemortem tooth loss	Moderate (5.0)	Moderate (6.9)	Moderate (5.2)	Moderate (6.9)	High (10.4)	Moderate (5.4)	Moderate (5.2)

[a]Terms low, moderate and high are relative and only appropriate to the samples used here.
[b]Data from Douglas (1996) and Pietrusewsky and Douglas (2002a).
[c]Data from Domett (2001).

growth rate. In the Ban Lum Khao sample, the high levels of child mortality were not matched by high levels of infant mortality, as would be expected in a rapidly growing population. This could possibly reflect under-representation of infants in the sample. Higher levels of mortality in young adult females at Ban Lum Khao is consistent with the suggestion of greater pressure on population growth, as it is possible that a proportion of these deaths could be attributed to problems at the time of birth (Domett 2001).

Higher levels of growth disruption in children at Ban Lum Khao suggest that they suffered multiple insults during their short lives compared with the few children who died at Noen U-Loke. In those who survived to adulthood, growth disruption was experienced to a similar level in females at both sites, whereas Ban Lum Khao males had significantly higher LEH levels than their Noen U-Loke counterparts. In addition, the lower mean stature in men at Ban Lum Khao indicates that poorer health was experienced by a subgroup from this population.

The greater range of stature in males at Ban Lum Khao is an enigma, as it suggests greater inequality during the bronze age. This is inconsistent with archaeological data, which suggests an increase in social complexity from the bronze to the iron age. The burial of infants may also reflect social inequality. It is possible that those infants buried in large jars were from families of higher status, while infants from families of lower status were interred directly in the ground. The suggestion of inequality is deserving of further investigation.

The evidence of dental pathology presents an equally difficult picture to interpret, partly again possibly because of doubts raised by preservation problems. Nevertheless, similar caries rates and antemortem tooth loss at both sites but an increase in attrition and dental infection over time is possibly reflective of a dietary change, with increasing fibrous, but not cariogenic, foods.

There is not a clear pattern of health change over time, but nor is there evidence of a general decline, as has been suggested for certain other parts of the world with intensification of agriculture. As Pietrusewsky and Douglas (2002b) found with their comparison of late and early inhabitants of Ban Chiang, evidence of health changes with the increasing reliance on agriculture in the northeast of Thailand does not follow that seen elsewhere. We suggest that the higher nutritional value of rice is a contributor, together with an undoubtedly complex set of as yet undefined social and environmental factors. In particular, the nature of any genetic relationships between these two populations is a factor that may significantly impact on their relative health status.

Until now, there has been no evidence for neolithic occupation in the Mun River valley, but recent excavations at the site of Ban Non Wat is providing evidence of habitation and a cemetery beginning in the second millennium BC and continuing into the early historic period (Higham, personal communication). With this new evidence of a continuous occupation of a single site, including earlier inhabitants, it is possible that this significant skeletal collection will enable more detailed analysis of the health status of the inhabitants of prehistoric northeast Thailand and of the Mun River valley in particular.

References

Boyd W. E. and McGrath R. J. 2001a. The geoarchaeology of the prehistoric ditched sites of the upper Mae Nam Mun valley, northeast Thailand, III: late Holocene vegetation history. *Palaeo* **171**: 307–328.

2001b. Iron age vegetation dynamics and human impacts on the vegetation of the upper Mun River floodplain, northeast Thailand. *New Zealand Geographer* **57**: 8–19.

Cohen M. N. and Armelagos G. J. (eds.). 1984. *Paleopathology at the Origins of Agriculture*. Orlando, FL: Academic Press.

Domett K. M. 2001. *British Archaeological Reports International Series*, No. 946. *Health in Late Prehistoric Thailand.* Oxford: Archaeopress.

Douglas M. T. 1996. Paleopathology in human skeletal remains from the pre-metal, bronze and iron ages, northeastern Thailand. Ph.D. Thesis, University of Hawaii.

Goodman A. H. and Armelagos G. J. 1989. Infant and child morbidity and mortality risks in archaeological populations. *World Archaeology* **21**: 225–243.

Goodman A. H., Armelagos G. J. and Rose J. C. 1980. Enamel hypoplasia as indicators of stress in three prehistoric populations from Illinois. *Human Biology* **52**: 515–528.

Higham C. F. W. 1996. *The Bronze Age of Southeast Asia.* Cambridge, UK: Cambridge University Press.

2002. *Early Cultures of Mainland Southeast Asia.* Bangkok: River Books.

Higham C. F. W. and Thosarat R. 1998. *Prehistoric Thailand. From Early Settlement to Sukhotai.* Bangkok: River Books.

Jackes M. 1992. Paleodemography: problems and techniques. In Katzenberg M. A. and Saunders S., eds., *Skeletal Biology of Past Peoples: Research Methods.* New York: Wiley–Liss, pp. 189–224.

1994. Birth rates and bones. In Chan L., ed., *Strength in Diversity: A Reader in Physical Anthropology.* Toronto: Canadian Scholars' Press, pp. 155–185.

Juliano B. O. 1993. *FAO Food and Nutrition Series*, No. 26: *Rice in Human Nutrition.* Geneva: UN Food and Agriculture Organization.

Larsen C. S. 1997. *Bioarchaeology: Interpreting behavior from the Human Skeleton.* Cambridge, UK: Cambridge University Press.

Lukacs J. R. 1989. Dental paleopathology: methods for reconstructing dietary patterns. In Iscan M. Y. and Kennedy K. A. R., eds., *Reconstruction of Life from the Skeleton.* New York: Liss, pp. 261–286.

McCaw M. 2000. The faunal remains: major variables. In Higham C. F. W., ed., *The Origins of the Civilisation of Angkor. The Final Report to the National Research Council of Thailand.* Bankok: National Research Council, pp. 188–199.

Molnar S. 1992. *Human Variation. Races, Types, and Ethnic Groups.* Englewood cliffs, NJ: Prentice Hall.

Nelsen K. M. 1999. The dental health of the people from Noen U-Loke. M.Sc. thesis, University of Otago, Dunedin.

O'Reilly D. J. W. 2001. From the bronze age to the iron age in Thailand: applying the heterarchical approach. *Asian Perspectives* **39**: 1–19.

Pechenkina E. A., Benfer R. A. and Zhijun W. 2002. Diet and health changes at the end of the Chinese Neolithic: the Yangshao/Longshan transition in Shaaxi Province. *American Journal of Physical Anthropology* **117**: 15–36.

Pietrusewsky M. and Douglas M. T. 2002a. Intensification of agriculture at Ban Chiang: is there evidence from the skeletons? *Asian Perspectives* **40**: 157–178.

2002b. *University Museum Monograph 111. Ban Chiang: A Prehistoric Village Site in Northeast Thailand. I: The Human Skeletal Remains.* Philadelphia, PA: Museum of Archaeology and Anthropology, University of Pennsylvania.

Steckel R. H. and Rose J. C. 2002. *The Backbone of History, Health and Nutrition in the Western Hemisphere.* Cambridge, UK: Cambridge University Press.

Stini W. A. 1985. Growth rates and sexual dimorphism in evolutionary perspective. In Gilbert R. I. and Mielke J. H., eds., *The Analysis of Prehistoric Diets* Orlando, FL: Academic Press, pp. 191–226.

Stodder A. L. W. and Douglas M. T. 2003. Childhood stress and determinants of adult sexual size dimorphism in late prehistoric skeletal assemblages from Guam, Mariana Islands. *American Journal of Physical Anthropology Supplement* **36**: 201.

Tayles N. G. 1999. *Report of the Research Committe* LXI. *The Excavation of Khok Phanom Di, a Prehistoric Site in Central Thailand,* Vol. V: *The People.* London: Society of Antiquaries.

Tayles N., Nelsen K., Buckley H. and Domett K. 1998. The Iron-age people of Noen U-Loke, Northeast Thailand: a preliminary review of the first two field seasons. *Bulletin of the Indo-Pacific Prehistory Association* **17**: 75.

Tayles N., Domett K. and Nelsen K. 2000. Agriculture and dental caries? The case of rice in prehistoric Southeast Asia. *World Archaeology* **32**: 68–83.

Thosarat, R. 2000. The fish remains. In Higham C. F. W., ed., *The Origins of the Civilisation of Angkor. The Final Report to the National Research Council of Thailand.* Bangkok: National Research Council, pp. 213–217.

Turner C. G. 1979. Dental anthropological indication of agriculture among the Jomon people of central Japan. X. Peopling of the Pacific. *American Journal of Physical Anthropology* **51**: 619–636.

Voelker J. 2002. From site specific to regional synthesis: using ceramics to bridge the gap. In *The 17th Indo-Pacific Prehistory Association Congress*, September, Taipei.

Welch D. J. 1985. Adaptation to environmental unpredictability: intensive agriculture and regional exchange at Late Prehistoric centres in the Phimai region in Thailand. Ph.D. thesis, University of Hawaii, Manoa.

10 *Palaeodietary change among pre-state metal age societies in northeast Thailand: a study using bone stable isotopes*

CHRISTOPHER A. KING
University of Hawaii at Manoa, Hawaii, USA

LYNETTE NORR
University of Florida, Gainesville, USA

Introduction

The use of stable isotopic analysis for reconstructing the diet of past populations has become common practice in the Americas, Europe, Africa, Japan and in the Pacific (e.g. Roksandic *et al.* 1988, Ambrose *et al.* 1997, Ambrose 2000, Sealy and Pfeiffer 2000, Larsen *et al.* 2001, White *et al.* 2001, Bocherens and Drucker 2003). Previous isotopic work to assess palaeodietary change among prehistoric human populations in Southeast Asia has been limited to Peninsular Malaysia and Sarawak (Krigbaum 2001). The project described here provides the first stable isotopic data from mainland Southeast Asia, offering a greater understanding of the interaction between environment and culture in changing food consumption patterns in prehistoric Southeast Asia.

King performed a study of human skeletal material ($n = 33$) from the archaeological site of Ban Chiang in northeast Thailand. The objectives of the study were to determine whether (a) carbon (from collagen and apatite) and nitrogen (from collagen) isotopic data could be collected from archaeological human remains from mainland Southeast Asia; (b) temporal changes in diet could be detected; and (c) sex differences in isotopic values could be detected. The Ban Chiang archeological site is an ideal place to start an in-depth case study examining synchronic and diachronic dietary patterns in an important and ecologically distinctive region of mainland Southeast Asia. The human skeletal remains from this important mortuary site were chosen for ease of accessibility and because the mortuary

Bioarchaeology of Southeast Asia. Marc Oxenham and Nancy Tayles.
Published by Cambridge University Press.

Table 10.1. *The Ban Chiang mortuary sequence*

Area period	Period (phase)	Resource base	Adults ($n = 102$)	Dates (calibrated years BP)
Pre-state metal age (bronze and iron)	Late (IX–X)	Wet-rice agriculture and iron continue	9	*c.* 2,300–1,800
Metal age (bronze/ early iron)	Middle (VI–VIII)	First appearance of iron (by MP VII) and water buffalo; forest recovery; possible commitment to wet rice cultivation	26	*c.* 2,900–2,300
Pre-metal and early bronze periods	Early (I–V)	Hunter–gatherer–cultivator economy; early bronze (EPIII) and domesticates including rice (from EPI)	67	*c.* 4,100–2,900

Modified from Pietrusewsky and Douglas (2002b, p. 158, Table 1).

sequence spans 2300 years (1,800 to 100 BP) that encompassed both the pre-metal and pre-state metal age periods (bronze and iron) of prehistoric Thailand (Table 10.1) (Pietrusewsky and Douglas 2002a). Moreover, the morphometric variability and palaeopathology of the collection have been comprehensively published (Pietrusewsky and Douglas 2002a).

Typically, the first line of evidence in palaeodietary reconstruction is the documentation of the presence or absence of particular plant and animal groups, referred to as the *menu* of prehistoric peoples (Bumsted 1984). By studying the frequency of occurrence of recovered plant and animal remains, archaeologists are able to provide a picture of what the overall subsistence base for the whole community was like over time. Questions about a community's resource consumption (diet), however, can be answered with a finer method of analysis through bone chemistry. Diet refers to the foods eaten by individuals in communities. Isotopic studies at an individual level provide a biocultural analysis of food selection, which can be coordinated with macroscopic studies of diet. Such studies allow a better understanding of the transition from horticulture to agriculture, inland versus coastal dietary differences, changes in subsistence patterns over time, within-group differences according to sex, status, age and weaning age, as indicators of population health and the relationship between diet and health (Norr 1995).

Environmental setting

The northern portion of northeast Thailand is subtropical monsoonal; as a result, there are two basic seasons: wet and dry. The area receives an average annual rainfall of 1400 mm, most of which falls between May and October (Penny 2001). Studies of pollen and phytoliths from Lake Kumphawapi in northeast Thailand (Fig. 1.1, p. 5), particularly from the sediment core 3KUM (Kealhofer 1996, Penny *et al.* 1996, Kealhofer and Penny 1998, Penny 1999, White *et al.* 2003), are of particular interest to this research. The sediment record from Lake Kumphawapi indicates the presence during the Holocene of predominantly deciduous tropical forests with some evergreen elements. This vegetation range can still be identified in the region today (White 1995). The pollen and phytolith records from the 3KUM Kumphawapi core suggest various vegetation changes during the early and middle Holocene including possible anthropogenic impacts. However, the archaeological record from the Kumphawapi catchment, including the Ban Chiang site, correlates with the top metre of the 3KUM core, in which there is pollen evidence but no phytolith record.

The 3KUM pollen evidence (Penny 1999) indicates that the region experienced significant forest disturbance beginning during the middle Holocene (from *c.* 6400 BP). With the beginning of the late Holocene (*c.* 4000 BP), about the time of the beginning of the Ban Chiang mortuary sequence, evidence for burning abruptly declines, but the most significant trend in forest recovery does not occur until *c.* 2900 BP. These changes associated with the late Holocene are interpreted as reflecting shifts in land use that might suggest changes in subsistence regimen. It is important to remember, however, that changes in burning patterns are not necessarily indicative of land clearance for agriculture *per se*. In Thailand, as in the rest of Southeast Asia, burning is ethnographically and ethnohistorically documented as a means to increase grazing fields, encourage the growth of particular species of wild plants and clear undergrowth, as well as for aesthetics (White 1995, Kealhofer and Penny 1998).

Subsistence change

Previous research on prehistoric subsistence in northeast Thailand has included studies of faunal remains (Higham 1975a, 1977, 1979, 1998, Higham and Kijngam 1979, Wilen 1987, Higham *et al.* 1998), human dental remains (Tayles *et al.* 2000, Domett 2001, Pietrusewsky and Douglas 2002a,b) and ethnographic analogy (White 1984, 1989, 1995).

The faunal spectrum is broad and exhibits a relatively low frequency of aggregation for both open and dense forest species (Wilen 1987, p. 76). Fauna typically recovered from archaeological sites in northeast Thailand has included mostly wild and domestic cattle, dog, water buffalo, pig, deer, civet, mongoose, rabbit, rat, mice, fish, shellfish, frog and lizard (Higham 1975b, 1979, 1998, Higham and Kijngam 1979, 1984, Wilen 1987). However, owing to issues of differential preservation and methods of archaeological excavation and analysis, the importance of various food categories cannot be determined from these studies (Dufour 1997).

Faunal evidence excavated from the site of Ban Chiang (Higham and Kijngam 1979) augments the pollen evidence for late Holocene environmental changes. Discriminating cultural from environmental factors in the range and changes of species over time at the site is not a simple task. Moreover, the faunal evidence needs reinterpretation in the light of the current Ban Chiang chronology and the pollen evidence. Nevertheless, the presence of grassland and open forest fauna and the absence of indigenous dense forest species (e.g. langurs and macaques) support the pollen evidence for open habitats in the region from the beginning of the mortuary sequence. One particularly interesting change in the Ban Chiang faunal spectrum over time, which may relate to the pollen evidence for forest regeneration beginning *c.* 2900 BP, is the probably concurrent decline of small open-habitat fauna (e.g. civets, hares, mongooses) and increased prevalence of open-forest herbivores (e.g. cervids, water buffalo).

Although food procurement practices do not necessarily remain the same over time, ethnographic data report dense clustering of wild resources in modern-day Ban Chiang (White 1984, 1989, 1995). White noted that wild rice was growing along the lake and stream edges and suggested that early residents of Ban Chiang may have ‘assisted’ the wild rice by expanding its habitat (White 1989). Additionally, wild yams, which overlap the life cycle of rice and grow in differing microhabitats, could have been harvested as another major source of starch (White 1989). All these resources can be exploited without significant clearing of forests and are not dependent on water control.

Diet and health at Ban Chiang

There has been substantial progress in understanding health and diet using human dental remains from Ban Chiang. The archaeological site of Ban Chiang is located in the northern part of northeast Thailand on a large

Table 10.2. *Dental pathological conditions in permanent teeth from Ban Chiang adults (over 18 years of age)*

Dental pathology	Affected/observed teeth (%)[a]		
	Male	Female	Total
Premortem loss	50/681 (7.3)	37/598 (6.2)	87/1279 (6.8)
Caries	45/527 (8.5)	29/489 (5.9)	74/1016 (7.3)
Abscessing	43/568 (7.6)	24/487 (4.9)	67/1055 (6.4)
Alveolar resorption[b]	79/451 (17.6)	36/418 (8.6)	115/869 (13.2)
Calculus[b]	142/458 (31.0)	111/435 (25.5)	253/893 (28.3)
Attrition[b]	118/531 (22.2)	51/484 (10.5)	169/1015 (16.7)

[a]Sex difference are statistically significant at $p \leqslant 0.10$.
[b]Advanced observations only.
From: Pietrusewsky and Douglas (2002a: Appendix B).

mound that is overlain by the modern-day village bearing the same name. The area is located near the junction of three small streams and is surrounded by rolling lowlands.

Two field seasons in 1974 and 1975 resulted in the recovery of 142 primary or disturbed inhumations with no secondary burials (Pietrusewsky and Douglas 2002a, p. 14). The skeletal collection has recently been completely reanalysed; the resulting monograph includes the most current assessments of biological and cultural change resulting from sedentism, agricultural dependency and rank and sex differences in health and disease (Pietrusewsky and Douglas 2002a,b).

As with all types of human skeletal analysis, the assessment of diet through macroscopic analyses has limitations based on environmental factors and sample representation. The study of dental pathological conditions has important implications for understanding health and disease, particularly in societies undergoing transitions in subsistence systems (Goodman *et al.* 1984, Lukacs 1989, Larsen and Ruff 1994, Larsen 1995). Overall, Ban Chiang dental health suggests that there was a mixed subsistence economy. The adult permanent teeth recovered have revealed low-to-moderate levels (<10%) of dental infection (premortem tooth loss, carious teeth and abscessing), suggesting that soft carbohydrate food as well as tough, fibrous food formed part of the subsistence economy (Table 10.2). An alternative explanation is the low cariogenicity of rice or even a certain amount of coarseness from unpolished rice consumption (Tayles *et al.* 2000). The latter is also supported by the level of advanced alveolar resorption (13.2%), suggesting a relatively coarse diet throughout the mortuary sequence. This would be in contradiction to the suggestion that rice, particularly wet-rice, agriculture increased the rate of carious lesions over time. The diet continued to be diverse throughout the mortuary sequence (Pietrusewsky and Douglas 2002b, p. 172).

In an examination of changes over time, no statistically significant change in caries frequency was found in the Ban Chiang dental series. However, statistically significant sex differences (males greater than females; $p \leqslant 0.10$) were found in the rates of carious lesions, abscessing, calculus and alveolar resorption (Pietrusewsky and Douglas 2002a, p. 61). These significant differences by sex are interpreted as possible differential access to food resources; different cultural activities, such as betel nut chewing; dental hygiene; or differential age structure (Pietrusewsky and Douglas 2002a, p. 79).

Curiously, with a seemingly mixed diet of soft and coarse foods, the relatively high frequency of calculus (28.3%) may actually reflect cultural activities such as betel nut chewing (Pietrusewsky and Douglas 2002a, p. 61). Pietrusewsky and Douglas (2002a, p. 351) noted dental staining in 102/1078 (9.5%) of all teeth at Ban Chiang; the typical thick, dark red-brown staining that is usually associated with betel nut chewing was not present. Whether betel nut chewing was a cultural activity at any point during Ban Chiang's mortuary sequence is uncertain, as no study as been performed to verify that any dark dental stains are actually from that source.

Data collected from a variety of sources from Ban Chiang indicate a landscape that was rich in resources for the entire 2300 years of the mortuary sequence. Staple foods such as rice (and most likely yams) and domestic animals were consumed in conjunction with gathered wild plants and diverse wild animals (e.g. fish, mammals and birds). Overall, the skeletal series from the Ban Chiang mortuary sequence have suggested good dental health representative of a widely diverse diet of soft and coarse foods even after apparent technological and subsistence changes in crop production. Dental evidence also suggests cultural rules for differential access to resources between males and females.

Stable isotopic analysis

The use of stable isotopic analysis of human bones and teeth is an established technique for further understanding palaeodietary changes in prehistoric populations (Ambrose and Katzenberg 2000, Katzenberg 2000,

Larsen 2002). Stable isotopic analysis as a research tool stems from the basic fact that the isotopic components of food eaten by an individual varies with the dietary food type and are incorporated into the skeletal system and can be used to interpret dietary patterns. Plant categories refer to the photosynthetic pathway used and are C_3 (usually temperate, tropical and boreal plants), C_4 (usually grassland plants) and CAM (crassulacean acid metabolism; usually desert but some tropical). Isotopic analysis can also differentiate proteins of terrestrial or aquatic origin. These dietary results may be used to investigate group differences over time or differences among groups in the same time period.

The three components of the skeleton used in palaeodietary reconstruction are the protein portion of bone (collagen), the non-protein portion of bone (hydroxyapatite), and dental enamel. These components offer the opportunity to develop a quantitative estimate of diet at the level of the individual, differences within or between subgroups of populations, and changes within a regional population over time. With samples from a large number of individuals from more than one site, community and regional characteristics can be compared (Schoeninger and Moore 1992, p. 252). Regional and methodological perspectives on isotopic analysis for palaeodietary reconstruction have demonstrated the usefulness of stable isotopes for interpreting human palaeodiet (reviewed by Keegan 1989, Katzenberg 1992, Ambrose 1993, Norr 1995, Pate 1997, Ambrose and Katzenberg 2000, Katzenberg and Saunders 2000). While describing the specific principles of carbon and nitrogen isotopic fractionation is beyond the scope of this chapter, some good reviews include those by DeNiro and Epstein (1978), O'Leary (1981, 1988, 1993) and for carbon and those by Farquhar *et al.* (1989a,b), Wada (1980), Schoeninger and DeNiro (1984), Dufour *et al.* (1999), Katzenberg and Weber (1999) and Ambrose (2000) for nitrogen.

Interpretive models

A review of palaeodietary reconstruction methods and interpretive models used here can be found in Norr (1995). In order to understand dietary changes that took place among prehistoric populations, it is essential to have first established the isotopic composition of the local or regional food web so that the human bone isotope results can be interpreted more accurately (Keegan and DeNiro 1988, Norr 1995). Consideration of the carbon-13 (δ^{13}C) and nitrogen-15 (δ^{15}N) composition of collagen yields information on the consumption of C_3, C_4 and CAM plant groups, and, in the absence of C_4 plants, marine versus terrestrial food groups (Schoeninger and DeNiro 1984, Chisholm 1989). The $\delta^{15}N$ values can distinguished between legume and non-leguminous plants, and aquatic versus terrestrial food groups when aquatic resources are consumed with C_4 plants (Minagawa and Wada 1984, Schoeninger and DeNiro 1984). Once isotopic differences in the local food resources have been noted, these results can be applied to interpretations of isotopic analyses of archaeological human bone in order to reconstruct the palaeodiet.

Although isotopic assessment of the local flora and fauna remains to be done, a general knowledge of isotopic signatures in the region allows provisional interpretation of the Ban Chiang results. In Southeast Asia, all edible plants are C_3 except for sugar cane, some types of millet, sorghum and Job's tears, which are C_4. Common edible plants that use the CAM photosynthetic pathways, including pineapple and vanilla, are New World domesticates and would not be a factor in prehistoric Southeast Asian cuisine. The discovery of rice remains in pottery from the earliest levels of Ban Chiang has led to the assumption that rice was a predominant plant in the diet of the early inhabitants. However, Mudar (1995) has provided some tantalising ideas for the possible use of C_4 crops at Ban Chiang. In her discussion of the alternative farming strategies in central Thailand, Mudar pointed out that millet has been part of the northern Chinese and island Southeast Asia agricultural repertoire since the eighth and fourth millennium BCE uncal., respectively (Mudar 1995, pp. 183–184). Thus, we can hypothesize that both rice (C_3) and millet (C_4) were used as food resources from the very beginning of the Ban Chiang mortuary sequence.

Stable isotopic values of carbon and nitrogen are used independently or in tandem to interpret dietary change, depending on the isotopic complexity of the environment and the food resources. By comparing the $\delta^{13}C$ values from inorganic (apatite carbonate) and organic (collagen) portions of human bone, patterns of dietary intake between populations and within populations over time can also be assessed. Results from laboratory experiments using rat bone have shown that bone collagen largely reflects dietary protein while bone apatite carbonate is more representative of an individual's whole diet (Ambrose and Norr 1993). The difference between the $\delta^{13}C$ values of carbonate and collagen in an individual, referred to as the apatite–collagen spacing ($\Delta^{13}C_{ca-co}$), will vary predictably with the isotopic composition of the dietary protein and energy sources. The resulting differences can then be used to gauge protein and carbohydrate intake (Lee-Thorp 1989, Ambrose and Norr 1993, Norr 1995).

Materials and methods

Using a sample of 33 prehistoric human remains from Ban Chiang, King conducted a study to determine if isotopic data could be collected successfully and, if so, whether changes in the diet could be detected in northeast Thailand. The results confirmed that the methodology is viable and suggested that dietary changes occurred over time (Tables 10.3–10.5).

Chemical extraction of bone collagen and apatite was conducted at the University of Florida, Gainesville, under the supervision of Norr. Mass spectrometry of the collagen and apatite was also performed at the University of Florida, Department of Geological Sciences. Procedures followed for cleaning the bone, chemical extraction and quality control for both collagen and apatite were as outlined in Ambrose (1993) and Lee-Thorp (1989). Bone collagen and apatite extraction protocols as well as standards for assessing sample quality for both bone collagen and apatite followed those outlined in Ambrose (1993). Non-diagnostic cortical bone fragments were preferentially selected, because the mineral and organic phases of the bone are the least susceptible to degradation and contamination (Lambert *et al.* 1982, 1985, 1991). In cases where no cortical bone fragment of a long limb bone was available, rib fragments were selected. Carbon and nitrogen isotopic composition varied only slightly when comparing sections of the same bone: same individual (mean difference $0.1 \pm 0.06‰$) and different skeletal elements from the same individual ($0.2 \pm 0.14‰$) (Chisholm 1989, p. 20).

Results

The analyses confirmed that human bone samples from Ban Chiang dating approximately 4,000 BP were adequately preserved (Table 10.3). Of the original 33 samples used, all individuals produced results for apatite data; three individuals (BC30, BC27, BC01) did not yield results for collagen, probably because of poor bone preservation.

In order to examine temporal changes, as well as differences between males and females, the skeletal series was divided by sex and by time periods (pre-metal was Ban Chiang Early Period (EP) I–II; bronze was Ban Chiang EPIII to Middle Period (MP) VI, and iron was Ban Chiang MPII to Late Period (LP) X) (Tables 10.4 and 10.5). Graphical and statistical analyses for each of the isotopic parameters ($\delta^{13}C_{co}$, $\delta^{15}N$, $\delta^{13}C_{ca}$, $\Delta^{13}C_{ca\text{–}co}$, where co indicates collagen, ca indicates carbonate) are presented in order to investigate trends in diet within and between

Table 10.3. *Individual stable carbon and nitrogen isotope results of collagen and apatite carbonate from Ban Chiang*

Burial[a]	Period	Age (years)	Sex	Collagen (wt. %)	Collagen composition					Apatite (wt. %)	Apatite carbonate composition		$\Delta^{13}C_{ca-co}$
					C (wt. %)	N (wt. %)	C:N ratio	^{15}N (‰)	^{13}C (‰)		C (wt. %)	^{13}C (‰)	
Pre-metal (n = 8)													
BC44	EPI	40–45	M	4.5	25.1	8.9	3.3	10.0	−18.8	54.4	1.35	−9.89	8.94
BC24	EPII	50+	F	6.2	34.6	12.3	3.3	9.1	−19.1	45.3	1.19	−13.44	5.63
BC33	EPII	45–50	F	3.9	29.6	10.1	3.4	9.5	−18.5	50.4	1.73	−12.04	6.50
BC34	EPII	40–45	F	3.7	15.0	5.4	3.3	10.0	−18.8	55.7	0.92	−12.65	6.15
BC45	EPII	18–22	F	3.8	32.7	11.1	3.5	9.8	−19.3	49.5	1.01	−13.37	5.93
BC43	EPII	35–40	M	3.3	27.8	9.4	3.5	10.1	−19.0	52.5	3.49	−11.64	7.40
BC47	EPII	25–30	M	2.2	20.9	6.9	3.5	10.1	−19.5	63.7	1.05	−12.47	7.07
BCES72	EPII–EPIII	35–40	M	7.2	31.9	11.4	3.3	9.9	−19.1	38.4	1.30	−12.96	6.10
Bronze age (n = 18)													
BCES49	EPIII	35–40	M	6.3	34.5	12.3	3.3	10.1	−18.5	36.5	1.50	−13.29	5.18
BCES76	EPIII	25–30	M	3.1	30.4	10.3	3.5	10.0	−18.9	42.2	1.22	−12.58	6.31
BCES69	EPIV	25–30	F	7.8	34.9	12.5	3.3	9.7	−19.2	35.8	1.06	−13.38	5.82
BC31	EPIV	50–60	F	6.3	29.3	10.4	3.3	8.5	−19.4	50.9	1.08	−12.50	6.96
BCES34	EPIV	25–30	F	4.9	34.8	12.4	3.3	9.2	−19.2	31.5	1.20	−12.97	6.23
BC26	EPIV	35–50	M	5.2	31.0	11.1	3.3	9.4	−19.3	44.5	0.91	−13.53	5.73
BCES55	EPIV	14–16	M	5.3	35.3	12.1	3.4	9.7	−18.9	42.0	0.96	−12.96	5.94
BCES31	EPIV	45–50	M	4.2	28.8	9.6	3.5	9.7	−18.4	43.5	1.15	−13.01	5.40
BC30	EPV	20+	F	1.1	3.2	0.7	5.2	—	—	49.5	1.37	−12.30	—
BCES32	EPV	35–50	M	8.3	36.5	12.4	3.4	9.5	−18.8	37.4	1.34	−12.56	6.28
BC12	EPV	35–50	M	3.0	32.7	11.2	3.4	11.0	−19.0	50.4	0.11	−12.10	6.95
BCES56	EPV	45–50	M	4.0	28.8	9.7	3.5	10.2	−18.7	39.8	1.08	−13.42	5.24
BCES53	EPV	20+	M	8.5	31.2	11.2	3.3	9.8	−18.9	45.9	1.22	−13.14	5.75

BC39	EPV	35–40	M	3.0	27.2	9.1	3.5	10.5	−18.6	60.7	1.02	−12.82	5.81
BCES36	EPV	45–50	M	3.1	30.4	10.3	3.4	10.2	−18.7	45.8	1.39	−12.54	6.17
BCES42	EPV	40–45	M	8.8	40.4	13.9	3.4	11.2	−19.2	33.8	3.22	−13.05	6.17
BC27	MPVI	25–30	F	1.1	2.1	0.4	6.1	—	—	45.4	1.57	−12.48	—
BCES25	MPVI	35–50	M	11.4	39.3	14.0	3.3	10.0	−18.6	30.1	2.28	−13.60	5.08
Iron age (n = 7)													
BCES75	MPVII	20–35	F	4.6	33.3	11.4	3.4	9.7	−19.4	42.5	1.41	−11.76	7.60
BCES20	MPVII	35–40	F	5.5	34.4	11.8	3.4	10.6	−18.9	51.1	2.55	−12.62	6.25
BCES24	MPVII	30–35	M	6.8	37.9	12.9	3.4	10.5	−18.7	41.1	1.41	−12.50	6.20
BCES03	LPX	35–50	F	2.7	28.2	9.5	3.5	10.0	−20.5	48.7	1.07	−12.31	8.20
BC01	LPX	20 +	F	1.6	1.8	0.4	5.9	—	—	51.9	2.63	−10.79	—
BCES07	LPX	45–55	M	4.5	33.7	11.5	3.4	10.4	−18.0	51.8	1.09	−12.60	5.44
BCES02	LPX	35–40	M	3.4	30.0	10.2	3.4	9.9	−18.6	47.2	1.58	−10.93	7.65

$\Delta^{13}C_{ca-co}$, difference between carbonate and collagen isotope values; EP, Early Period; MP, Middle Period; LP, Late Period; M, male; F, female; wt., weight.

[a]BC indicates the 1974 excavation and BCES the 1975 excavation.

Table 10.4. *Descriptive statistics and results from the general linear model for differences over time by sex for the isotopic composition of bone collagen and apatite carbonate from Ban Chiang*

Isotope analysis	Time period	No.	Mean (SD)	Minimum	Maximum
Collagen ^{13}C composition (‰)					
Female	Pre-metal	4	−18.93 (0.33)	−19.30	−18.54
	Bronze	3	−19.28 (0.14)	−19.44	−19.20
	Iron	3	−19.58 (0.84)	−20.51	−18.87
Male	Pre-metal*	4	−19.12 (0.30)	−19.54	−18.83
	Bronze*	13	−18.81 (0.26)	−19.26	−18.41
	Iron	3	−18.43 (0.37)	−18.70	−18.00
Collagen ^{15}N composition (‰)					
Female	Pre-metal	4	9.56 (0.39)	9.06	9.97
	Bronze	3	9.15 (0.61)	8.50	9.72
	Iron	3	10.10 (0.44)	9.71	10.58
Male	Pre-metal	4	10.02 (0.10)	9.87	10.10
	Bronze	13	10.08 (0.54)	9.38	11.19
	Iron	3	10.28 (0.30)	9.94	10.50
Carbonate ^{13}C composition (‰)					
Female	Pre-metal	4	−12.88 (0.66)	−13.44	−12.04
	Bronze	5	−12.72 (0.45)	−13.38	−12.29
	Iron	4	−11.87 (0.80)	−12.62	−10.79
Male	Pre-metal	4	−11.74 (1.35)	−12.96	−9.89
	Bronze	13	−12.96 (0.44)	−13.56	−12.06
	Iron	3	−12.00 (0.92)	−12.56	−10.93
$\Delta^{13}C_{ca-co}$					
Female	Pre-metal	4	6.05 (0.37)	5.63	6.50
	Bronze	3	6.34 (0.58)	5.82	6.96
	Iron	3	7.35 (1.00)	6.25	8.20
Male	Pre-metal**	4	7.38 (1.18)	6.10	8.94
	Bronze**	13	5.85 (0.54)	5.08	6.95
	Iron	3	6.43 (1.12)	5.44	7.65

$\Delta^{13}C_{ca-co}$, difference between carbonate and collagen isotope values; SD, standard deviation. Significant differences: $*p \leqslant 0.05$; $**p \leqslant 0.016$.

time periods. As a cautionary note, any significant relationships may either represent actual trends or may be the result of statistical noise owing to the small sample sizes (Fig. 10.1 and Tables 10.4 and 10.5).

The broader range of $\delta^{13}C$ values for carbonate (Fig. 10.1b) than for collagen (Fig. 10.1a) (graphed against same $\delta^{15}N$) indicates there is a broader range in carbon values for carbohydrates consumed than for proteins consumed. This would suggest that plant food consumption at

Table 10.5. *Descriptive statistics and results from the general linear model for differences between the sexes within each time period for the isotopic composition of bone collagen and apatite carbonate from Ban Chiang*

Isotopy analysis for time period	Sex	No.	Mean (SD)	Minimum	Maximum
Collagen ^{13}*C composition (‰)*					
Pre-metal	Female	4	−18.93 (0.33)	−19.30	−18.54
	Male	4	−19.12 (0.30)	−19.54	−18.83
Bronze	Female	3	−19.28 (0.14)	−19.44	−19.20
	Male	13	−18.81 (0.26)	−19.26	−18.41
Iron	Female*	3	−19.58 (0.84)	−20.51	−18.87
	Male*	3	−18.43 (0.37)	−18.70	−18.00
Collagen ^{15}*N composition (‰)*					
Pre-metal	Female	4	9.56 (0.39)	9.06	9.97
	Male	4	10.02 (0.10)	9.87	10.10
Bronze	Female**	3	9.15 (0.61)	8.50	9.72
	Male**	13	10.08 (0.54)	9.38	11.19
Iron	Female	3	10.10 (0.44)	9.71	10.58
	Male	3	10.28 (0.30)	9.94	10.50
Carbonate ^{13}*C composition (‰)*					
Pre-metal	Female	4	−12.88 (0.66)	−13.44	−12.04
	Male	4	−11.74 (1.35)	−12.96	−9.89
Bronze	Female	5	−12.72 (0.45)	−13.38	−12.29
	Male	13	−12.96 (0.44)	−13.56	−12.06
Iron	Female	4	−11.87 (0.80)	−12.62	−10.79
	Male	3	−12.00 (0.92)	−12.56	−10.93
$\Delta^{13}C_{ca\text{–}co}$					
Pre-metal	Female	4	6.05 (0.37)	5.63	6.50
	Male	4	7.38 (1.18)	6.10	8.94
Bronze	Female	3	6.34 (0.58)	5.82	6.96
	Male	13	5.85 (0.54)	5.08	6.95
Iron	Female	3	7.35 (1.00)	6.25	8.20
	Male	3	6.43 (1.12)	5.44	7.65

$\Delta^{13}C_{ca\text{–}co}$, difference between carbonate and collagen isotope values; SD, Standard deviation. Significant differences: *$p \leqslant 0.008$; **$p \leqslant 0.049$.

Ban Chiang was not dominated by one major crop during the site's mortuary sequence. The narrow variation in proteins consumed is ambiguous in that it can suggest that the animals (wild and domestic) being consumed had diets similar to the humans or that only selected species were being consumed.

In Fig. 10.1c, $\Delta^{13}C_{ca\text{–}co}$ values between 4 and 7 are produced from the consumption of protein and carbohydrates that are isotopically similar, and this indicated that C_3 plants were the dominant plant group consumed

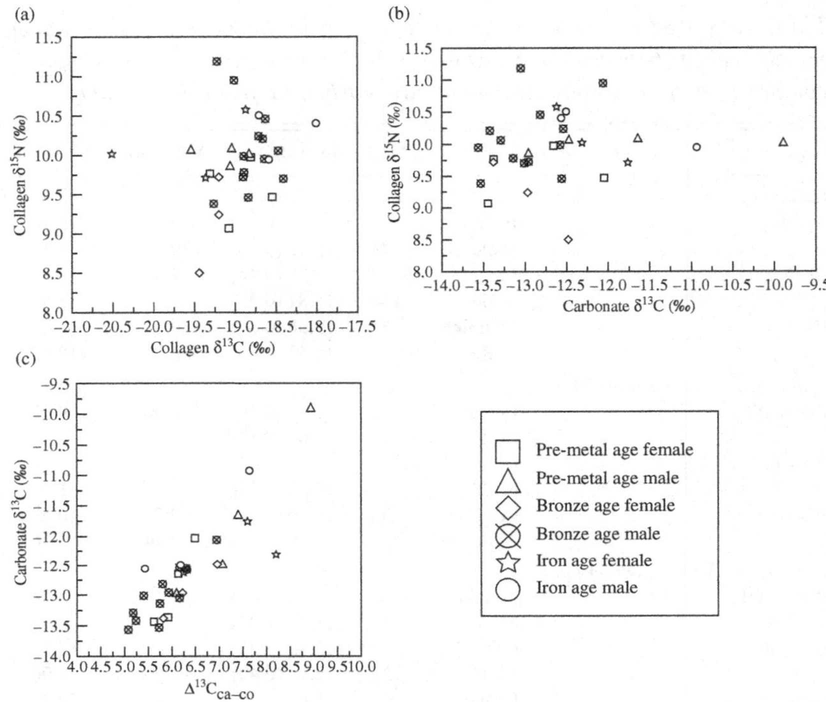

Figure 10.1. Comparison of stable isotopic values for $\delta^{13}C$ and $\delta^{15}N$ in bone collagen and carbonate. (a) Collagen $\delta^{15}N$ and $\delta^{13}C$ composition; (b) collagen $\delta^{15}N$ and carbonate $\delta^{13}C$; (c) carbonate $\delta^{13}C$ and the difference between collagen and carbonate ^{13}C ($\Delta^{13}C_{ca\text{–}co}$).

by both humans and animals. Values for $\Delta^{13}C_{ca\text{–}co}$ larger than 7 demonstrate carbohydrates that are more C_4-like, implying consumption of such foods as sugar cane, C_4-type millets, and/or Job's tears in their diets.

The general linear model (GLM) statistical procedure (Moore and McCabe 2003) was chosen to compare variances of the dependent variables – in this case time period and/or sex – each separately and then combined. If the overall F-test for the analysis of variance was statistically significant then *post hoc* comparison of means was made using the Tukey–Kramer procedure (Moore and McCabe 2003). For all tests, the level of significance was set at $p \leqslant 0.05$ for a significant difference over time and/or between the sexes.

The GLM procedure for all the samples combined (not controlling for time or sex) showed significant differences for $\delta^{13}C_{co}$ ($F_{2,24} = 5.962$; $p \leqslant 0.008$; $r^2 = 0.464$) and $\Delta^{13}C_{ca\text{-}co}$ ($F_{2,24} = 4.871$; $p \leqslant 0.017$; $r^2 = 452$), where the subscripts to F indicate the degrees of freedom. To explain these differences, the GLM procedure was applied controlling for time and sex. When controlling for sex (that is, sex combined) only $\delta^{13}C_{ca}$ varied significantly over the three periods ($F_{2,27} = 4.155$; $p \leqslant 0.027$). More specifically, the Tukey–Kramer procedure suggested that the only significant changes over time occurred during the transition from the bronze to iron ages ($p \leqslant 0.013$). As for differences between males and females, significant results were found for $\delta^{13}C_{co}$ ($F_{1,24} = 9.977$; $p \leqslant 0.004$) and $\delta^{15}N$ ($F_{1,24} = 7.080$; $p \leqslant 0.014$), suggesting differential access to food resources.

While these GLM results demonstrate significant differences over time and between sexes, only half of the variation in these variables can be explained when time and sex are analysed together. In other words, while these results may demonstrate statistically significant variability among the samples, variability within each subsample was high, which is most likely the result of small sample sizes; this reduces the overall explanatory power of the results.

A one-way analysis of variance with the Tukey–Kramer multiple comparison procedure (Moore and McCabe 2003) was used to examine further specific interactions between time periods and sex (Tables 10.4 and 10.5). The results indicated that temporal differences in food types were not the same among the three groups. More specifically, pre-metal males and bronze age males, bronze age females and bronze age males, and iron age females and iron age males were significantly different in at least one stable isotopic variable (Table 10.4). There were statistically significant differences ($\delta^{13}C_{ca}$, $p \leqslant 0.05$ and $\Delta^{13}C_{ca-co}$, $p \leqslant 0.016$) between pre-metal males and bronze age males. Table 10.5 indicates possible differential access to food resources between males and females within a time period. The $\delta^{15}N$ values in bronze age females were significantly lower ($p \leqslant 0.049$) than the values in bronze age males. The $\delta^{13}C_{co}$ values in iron age females were significantly different ($p \leqslant 0.008$) when compared with their iron age male counterparts. While these results may simply reflect the small sample sizes, they do suggest that dietary patterns among time periods and within single periods are detectable and warrant further study involving greater numbers of human samples, and most importantly, larger samples of different plant and animal species from this region.

Discussion

This study, using a sample of human remains from Ban Chiang, has demonstrated suitable preservation and varied isotopic results through time and between the sexes. The range of variation in carbon values in human bone collagen (−20.5‰ to −18.0‰) is expected from a region in which available food resources are almost exclusively based on C_3 plants. With the lack of isotopic data from known faunal samples, it is difficult to surmise the relative proportion of aquatic animals versus terrestrial animals (either wild or domestic).

The faunal evidence suggests an increase in domestic animal consumption in the later mortuary phases. This also corresponds to a decline in small browsing fauna associated with open habitats, such as civets and hares, and an increase in open-forest grazers, such as *Bos* spp. and cervids. A variety of cervids was recovered from Ban Chiang prehistoric remains, with some species indicating forest (e.g. *Cervus unicolor* and *Muntiacus muntjak*) and some open plain habitats (e.g. *Cervus porcinus* and *Cervus eldi*). These habitats potentially could provide isotopically very different food resources. All trees and shrubs are C_3 plants but grasses can be either C_3 or C_4. If some of the open-plains grazers were consuming C_4 grasses and then consumed directly by the Ban Chiang inhabitants, or indirectly through the consumption of a carnivore that ate the herbivores, this would explain some variation seen in the $\delta^{13}C_{co}$ and $\delta^{13}C_{ca}$ data (Fig. 10.1). The most probable explanation for the six individuals from two very different time periods with elevated $\Delta^{13}C_{ca\text{–}co}$ values could be their migration from another region. However, this result could also represent differential access to food resources, which may become apparent with increased sampling and inclusion of plant and animal isotopic data.

The isotopic values for both $\delta^{13}C_{co}$ and $\delta^{15}N$ showed a trend towards enrichment over time in the heavier isotopes (Fig. 10.1a). This shift may even suggest increased use of C_4 plants in the diet – perhaps millet. Significant changes over time ($\delta^{13}C_{ca}$ and $\Delta^{13}C_{ca\text{–}co}$) for the same sex (pre-metal males and bronze age males) or between sexes (iron age females and males ($\delta^{13}C_{co}$)) suggest the possibility of differential access to one or a combination of plant or animal foods according to sex, and/or temporal shifts in the types of food eaten (Tables 10.4 and 10.5). These results correlate with the osteological results shown in Table 10.2 from the dental palaeopathological analyses. However, the high variability among the samples, as demonstrated by the r^2 values from $\delta^{13}C_{co}$ (0.464) and $\Delta^{13}C_{ca\text{–}co}$ (0.452) reiterates the need to increase sample sizes to enhance the overall explanatory power of the results.

The variation shown in the nitrogen data could be tied to patterns in aquatic versus terrestrial animal consumption (Fig. 10.1a). Nitrogen values in fish will vary with differences in feeding habits between species, with origin of fish (e.g. river, lake, ocean), and even by age within a species. Fine screening for small faunal bones such as those of fish did not occur during the excavation of Ban Chiang, but evidence from the nearby site of Ban Na Di indicates a wide diversity of fish types and small game all through each of these mortuary sequences (Higham and Kijngam 1984).

The sex differences in δ^{15}N between pre-metal and bronze age ($p \leqslant 0.049$) males and females (Table 10.5) suggest the possibility of differential access to meat resources. Further, the wider range of δ^{15}N values in bronze age samples, coupled with the relatively narrow $\Delta^{13}C_{ca-co}$ values, suggest a wide range in the types of animals consumed, possibly as a result of landscape modifications from crop production. The higher values found in a couple of the male samples could be attributed to greater consumption of fish or carnivore meat by particularly skilled hunters or some other group of higher social status. Alternatively it may simply reflect the issues with sample size.

The δ^{15}N enrichment over time may also result from increased consumption of aquatic resources or carnivorous terrestrial animals, a shift toward wet-rice cultivation, or a combination of these. Wet rice can have higher δ^{15}N values than other plants because water nitrogen levels can be increased by bacterial activity or even by the use of manure as a fertiliser. An increased emphasis on wet-rice agriculture and domesticated animals would be in line with the current archaeological record, which suggests that these ancient inhabitants increased wet-rice agricultural production and domestic animal use during the iron age, as well as with the palaeoenvironmental record, which suggests agricultural intensification beginning in the early first millennium BCE.

Conclusions

This analysis achieved the original objectives of the study in that carbon (from collagen and apatite) and nitrogen were obtained, temporal changes in diet were detected, and sex differences in isotopic values were found to occur. While the study has demonstrated the viability of conducting stable isotopic research in Thailand, much more work needs to be done. Future work will expand this current dataset to include more individuals from Ban Chiang, as well as prehistoric human bone samples from other regions of northeast Thailand and the collection of modern plants and archaeological faunal remains. Future analyses will provide the first body of stable isotopic evidence from prehistoric human remains and dietary resources from mainland Southeast Asia. From these increased sample sizes, a clearer picture will emerge in understanding food production and subsistence patterns in northeast Thailand.

Integration of evidence from multiple prehistoric sites will stimulate communication amongst researchers and begin the process of developing a pan-regional synthesis of what has until now been largely single-site research projects. Data from Thailand will contribute to ongoing stable isotope research in Malaysia and other parts of island Southeast Asia.

Acknowledgements

C. A. King would like to thank several people for helping this study to occur. Dr Pietrusewsky granted permission to use the Ban Chiang human skeletal remains and gave his professional and moral support. Dr Andrew Taylor helped with statistical analysis and interpretation. We also extend our appreciation to Dr David Hodell and Dr Jason Curtis of the University of Florida Department of Geological Sciences for their expert advice and for the isotopic determinations on the archaeological samples. The authors also acknowledge suggestions from readers of earlier drafts of this manuscript, including Dr Michele Douglas, Dr Michael Pietrusewsky, Dr Joyce White, and two anonymous reviewers.

References

Ambrose S. H. 1993. Isotopic analysis of paleodiets: methodological and interpretive considerations. In Sandford M. K., ed., *Investigations of Ancient Human Tissue: Chemical Analyses in Anthropology*. Langhorn, PA: Gordon and Breach, pp. 59–130.

2000. Controlled diet and climate experiments on nitrogen isotope ratios of rats. In Ambrose S. H. and Katzenberg M. A., eds., *Advances in Archaeological and Museum Science 5: Biogeochemical Approaches to Paleodietary Analysis*. New York: Kluwer, pp. 243–259.

Ambrose S. H. and Katzenberg M. A. (eds.). 2000. *Advances in Archaeological and Museum Science 5: Biogeochemical approaches to paleodietary analysis*. New York: Kluwer.

Ambrose S. H. and Norr L. 1993. Experimental evidence for the relationship of the carbon isotope ratios of whole diet and dietary protein to those of bone collagen and carbonate. In Lambert J. B. and Grupe G., eds., *Prehistoric Human Bone: Archaeology at the Molecular Level*. Berlin: Springer-Verlag, pp. 1–38.

Ambrose S. H., Butler B. M., Hanson D. B., Hunter-Anderson R. L. and Krueger H. W. 1997. Stable isotopic analysis of human diet in the Marianas

Archipelago western Pacific. *American Journal of Physical Anthropology* **104**: 343–361.

Bocherens H. and Drucker D. 2003. Trophic level isotopic enrichment of carbon and nitrogen in bone collagen: case studies from recent and ancient terrestrial ecosystems. *International Journal of Osteoarchaeology* **13**: 46–53.

Bumsted M. P. 1984. Human variation: δ^{13}C in adult bone collagen and the relation to diet in an isochronous C_4 (maize) archaeological population. Ph.D. thesis, University of Massachusetts, Boston.

Chisholm B. S. 1989. Variation in diet reconstruction based on stable carbon isotopic evidence. In Price T. D., ed., *The Chemistry of Prehistoric Human Bone*. Cambridge, UK: Cambridge University Press, pp. 10–37.

DeNiro M. J. and Epstein S. 1978. Influence of diet on the distribution of carbon isotopes in animals. *Geochimica et Cosmochimica Acta* **42**: 495–506.

Domett K. M. 2001. *British Archaeological Reports International Series* No. 946: *Health in Late Prehistoric Thailand.* Oxford: Archaeopress.

Dufour D. L. 1997. Nutrition, activity, and health in children. *Annual Review of Anthropology* **26**: 541–565.

Dufour E., Bocherens H. and Mariotti A. 1999. Palaeodietary implications of isotopic variability in Eurasian lacustrine fish. *Journal of Archaeological Science* **26**: 617– 627.

Farquhar G. D., Ehleringer J. R. and Hubick K. T. 1989a. Carbon isotope discrimination and photosynthesis. *Annual Review of Plant Physiology and Plant Molecular Biology* **40**: 503–537.

Farquhar G. D., Hubick K. T., Condon A. G. and Richards R. A. 1989b. Carbon isotope fractionation and plant water-use efficiency. In Rundel P., Ehleringer J. and Nagy K., eds., *Ecological Studies 68: Stable Isotopes in Ecological Research.* New York: Springer-Verlag, pp. 21–40.

Goodman A. H., Lallo J., Armelagos G. J. and Rose J. C. 1984. Health changes at Dickson Mounds, Illinois (AD 950–1300). In Cohen M. N. and Armelagos G. J., eds., *Paleopathology at the Origins of Agriculture.* Orlando, FL: Academic Press, pp. 271–305.

Higham C. F. W. 1975a. Aspects of economy and ritual in prehistoric northeast Thailand. *Journal of Archaeological Science* **2**: 245–288.

1975b. *University of Otago Studies in Prehistoric Anthropology*, Vol. 7: *Non Nok Tha, The Faunal Remains.* Dunedin: University of Otago Press.

1977. Economic change in prehistoric Thailand. In Reed C. A., ed., *Origins of Agriculture.* The Hague: Mouton, pp. 385–412.

1979. The economic basis of prehistoric Thailand. *American Scientist* **67**: 670–679.

1998. The faunal remains from mortuary contexts. In Higham C. F. W. and Thosarat R., eds., *University of Otago Studies in Prehistoric Anthropology*, Vol. 18: *The Excavation of Nong Nor, a Prehistoric Site in Central Thailand.* Dunedin: University of Otago Press, pp. 315–319.

Higham C.F.W. and Kijngam A. 1979. Ban Chiang and northeast Thailand: the palaeoenvironment and economy. *Journal of Archaeological Science* **6**: 211–233.

1984. The biological remains from Ban Na Di. In Higham C. F. W. and Kijgnam A., eds., *British Archaeological Reports International Series* 231, Part ii: *Prehistoric Investigations in Northeast Thailand.* Oxford: Archaeopress, pp. 353–390.

Higham C. F. W., Fordyce R. E. and O'Reilly D. J. W. 1998. The faunal remains and worked bone. In Higham C. F. W. and Thosarat R., eds., *University of Otago Studies in Prehistoric Anthropology*, Vol. 18: *The Excavation of Nong Nor, a Prehistoric Site in Central Thailand.* Dunedin: University of Otago Press, pp. 119–127.

Katzenberg M. A. 1992. Advances in stable isotope analysis of prehistoric bones. In Saunders S. R. and Katzenberg M. A., eds., *The Skeletal Biology of Past Peoples: Research Methods*. New York: Wiley, pp. 105–120.

2000. Stable isotope analysis: a tool for studying past diet, demography, and life history. In Katzenberg M. A. and Saunders S. R., eds., *Biological Anthropology of the Human Skeleton.* New York: Wiley-Liss, pp. 305–328.

Katzenberg M. A. and Saunders S. R. 2000. *Biological Anthropology of the Human Skeleton*. New York: Wiley-Liss.

Katzenberg M. A. and Weber A. 1999. Stable isotope ecology and palaeodiet in the Lake Baikal region in Siberia. *Journal of Archaeological Science* **26**: 651–659.

Kealhofer L. 1996. The human environment during the terminal Pleistocene and Holocene in northeastern Thailand: preliminary phytolith evidence from Lake Kumphawapi. *Asian Perspectives* **35**: 229–254.

Kealhofer L. and Penny D. 1998. A combined pollen and phytolith record for fourteen thousand years of vegetation change in northeastern Thailand. *Review of Palaeobotany and Palynology* **103**: 83–93.

Keegan W. F. 1989. Stable isotope analysis of prehistoric diet. In Iscan M. Y. and Kennedy K. A., eds., *Reconstruction of Life from the Human Skeleton*. New York: Liss, pp. 223–236.

Keegan W. F. and DeNiro M. J. 1988. Stable carbon- and nitrogen-isotope ratios of bone collagen used to study coral-reef and terrestrial components of the prehistoric Bahamian diet. *American Antiquity* **53**: 320–336.

Krigbaum J. S. 2001. Human paleodiet in tropical Southeast Asia: isotopic evidence from Niah Cave and Gua Cha. Ph.D. thesis, New York University.

Lambert J. B., Vlasak S. M., Thometz A. C. and Buikstra J. E. 1982. A comparative study of the chemical analysis of ribs and femurs in woodland populations. *American Journal of Physical Anthropology* **59**: 289–294.

Lambert J. B., Simpson S.V., Szpunar C. B. and Buikstra J. E. 1985. Bone diagenesis and dietary analysis. *Journal of Human Evolution* **14**: 477–482.

Lambert J. B., Xue L. and Buikstra J. E. 1991. Inorganic analysis of excavated human bone after surface removal. *Journal of Archaeological Science* **18**: 363–383.

Larsen C. S. 1995. Biological changes in human populations with agriculture. *Annual Review of Anthropology* **24**: 185–213.

2002. Bioarchaeology: the lives and lifestyles of past people. *Journal of Archaeological Research* **10**: 119–166.

Larsen C. S. and Ruff C. B. 1994. The stresses of conquest in Spanish Florida: structural adaptation and change before and after contact. In Larsen C. S. and Milner G. R., eds., *In the Wake of Contact: Biological Responses to Conquest.* New York: Wiley-Liss, pp. 21–34.

Larsen C. S., Griffin M. C., Hutchinson D. L. *et al.* 2001. Frontiers of contact: bioarchaeology of Spanish Florida. *Journal of World Prehistory* **15**: 69–123.

Lee-Thorp J. A. 1989. Stable carbon isotopes in deep time: the diets of fossil fauna and hominids. Ph.D. thesis, University of Cape Town.

Lukacs J. R. 1989. Dental paleopathology: methods for reconstructing dietary patterns. In Iscan M. Y. and Kennedy K. A., eds., *Reconstruction of Life from the Human Skeleton.* New York: Liss, pp. 261–286.

Minagawa M. and Wada E. 1984. Stepwise enrichment of ^{15}N along food chains: further evidence and the relation between δ^{15}N and animal age. *Geochimica et Cosmochimica Acta* **48**: 1135–1140.

Moore D. S. and McCabe G. P. 2003. *Introduction to the Practice of Statistics*, 4th edn. New York: Freeman.

Mudar K. M. 1995. Evidence for prehistoric dryland farming in mainland Southeast Asia: results of regional survey in Lopburi Province, Thailand. *Asian Perspectives* **34**: 157–191.

Norr L. 1995. Interpreting dietary maize from bone stable isotopes in the American tropics: the state of the art. In Stahl P. W., ed., *Archaeology in the Lowland American Tropics.* Cambridge, UK: Cambridge University Press, pp. 198–223.

O'Leary M. H. 1981. Carbon isotope fractionation in plants. *Phytochemistry* **20**: 553–567.

1988. Carbon isotopes in photosynthesis. *Bioscience* **38**: 328–336.

1993. Biochemical basis of carbon isotope fractionation. In Ehleringer J. R., Hall A. E. and Farquhar G. D., eds., *Stable Isotopes and Plant Carbon–Water Relations.* San Diego, C.A: Academic Press, pp. 19–28.

Pate F. D. 1997. Bone chemistry and paleodiet: reconstructing prehistoric subsistence–settlement systems in Australia. *Journal of Anthropological Archaeology* **16**: 103–120.

Penny D. 1999. Palaeoenvironmental analysis of the Sakon Nakhon Basin, Northeast Thailand: palynological perspectives on climate change and human occupation. *Bulletin of the Indo-Pacific Prehistory Association* **18**: 139–149.

2001. A 40,000 year palynological record from north-east Thailand; implications for biogeography and palaeo-environmental reconstruction. *Palaeogeography, Palaeoclimatology and Palaeoecology* **171**: 97–128.

Penny D., Grindrod J. and Bishop P. 1996. Holocene palaeoenvironmental reconstruction based on microfossil analysis of a lake sediment core, Nong Han Kumphawapi, Udon Thani, northeast Thailand. *Asian Perspectives* **35**: 209–228.

Pietrusewsky M. and Douglas M. T. 2002a. *University Museum Monograph 111. Ban Chiang, A Prehistoric Village Site in Northeast Thailand I: The Human Skeletal Remains.* Philadelphia: Museum of Archaeology and Anthropology, University of Pennsylvania.

2002b. Intensification of agriculture at Ban Chiang: is there evidence from the skeletons? *Asian Perspectives* **40**: 157–178.

Roksandic Z., Minagawa M. and Akazawa T. 1988. Comparative analysis of dietary habits between Jomon and Ainu hunter–gatherers from stable carbon isotopes of human bone. *Journal of the Anthropological Society of Nippon* **96**: 391–404.

Schoeninger M. J. and DeNiro M. J. 1984. Nitrogen and carbon isotopic composition of bone collagen from marine and terrestrial animals. *Geochimica et Cosmochimica Acta* **48**: 625–639.

Schoeninger M. J. and Moore K. M. 1992. Bone stable isotope studies in archaeology. *Journal of World Prehistory* **6**: 247–296.

Sealy J. and Pfeiffer S. 2000. Diet, body size, and landscape use among Holocene people in the Southern Cape, South Africa. *Current Anthropology* **41**: 642–655.

Tayles N., Domett K. and Nelsen K. 2000. Agriculture and dental caries? The case of rice in prehistoric Southeast Asia. *World Archaeology* **32**: 68–83.

Wada E. 1980. Nitrogen isotope fractionation and its significance in biogeochemical processes occurring in marine environments. In Goldberg E. D., Horibe Y. and Saruhaski K., eds., *Isotope Marine Chemistry*. Tokyo: Uchida Rokakuho, pp. 375–398.

White C. D., Pohl M. E. D., Schwarcz H. P. and Longstaffe F. J. 2001. Isotopic evidence for Maya patterns of deer and dog use at Preclassic Colha. *Journal of Archaeological Science* **28**: 89–107.

White J. C. 1984. Ethnoecology at Ban Chiang and the emergence of plant domestication in Southeast Asia. In Bayard D. T., ed., *University of Otago Studies in Prehistoric Anthropology*, Vol. 16. *Southeast Asia Archaeology at the XV Pacific Science Congress: The Origins of Agriculture, Metallurgy, and the State in Mainland Southeast Asia.* Dunedin: University of Otago Press, pp. 26–35.

1989. Ethnoecological observations on wild and cultivated rice and yams in northeastern Thailand. In Harris D. R. and Hillman G. C., eds., *One World Archaeology 13. Foraging and Farming: The Evolution of Plant Exploitation.* London: Unwin Hyman, pp. 152–158.

1995. Modeling the development of early rice agriculture: ethnoecological perspectives from northeast Thailand. *Asian Perspectives* **34**: 37–68.

White J. C., Penny D., Kealhofer L. and Maloney B. 2003. Vegetation changes from the late Pleistocene through the Holocene from three areas of archaeological significance in Thailand. *Quaternary International* **113**: 111–132.

Wilen R. N. 1987. The context of prehistoric food production in the Khorat Plateau piedmont, Northeast Thailand. Ph.D. thesis, University of Hawaii.

11 *The oral health consequences of the adoption and intensification of agriculture in Southeast Asia*

MARC OXENHAM
Australian National University, Canberra, Australia

NGUYEN LAN CUONG AND NGUYEN KIM THUY
Institute of Archaeology, Hanoi, Vietnam

Introduction

In 1979, Christy Turner put the Asia Pacific region on the global bioarchaeological radar with his highly cited paper characterising subsistence economies using mean caries frequencies. Collating dental caries data from some 64 samples, Turner (1979, p. 622) found a mean caries frequency for hunter–gatherers of 1.3%, for mixed economies of 4.8% and for agriculturalists of 10.4%. Rose *et al.* (1984) also posited a quantitative benchmark where a mean carious lesion rate per individual exceeding 2 indicated agricultural dependency. In fact, the contributions in the widely cited book by Cohen and Armelagos (1984) helped to formulate a consensus view that the frequency of carious lesions will increase with the move from foraging to agriculturally oriented subsistence bases, at least in the maize-based economies on the North American continent. Other contributions to that volume detailed the same correlation between increasing frequencies of lesions and intensification of millet-based agriculture in North Africa (Martin *et al.* 1984) and of barley- and wheat-based agriculture in the Levant (Smith *et al.* 1984). Lukacs (1992) has also documented the same trend in South Asia with the intensification of wheat and barley agriculture. With respect to Europe, Molnar and Molnar (1985) argued not so much an increase in carious lesions per se but rather for a high prevalence of root lesions with the increased use of unspecified domesticated grains. It is also worth pointing out that other types of grain have been associated with this apparent global trend. For instance, Rose *et al.* (1991) noted a major increase in carious lesions in communities living in the lower Mississippi valley (southeastern USA) that were increasingly

Bioarchaeology of Southeast Asia. Marc Oxenham and Nancy Tayles.
Published by Cambridge University Press.

targeting starchy seeds such as maygrass (*Phalaris caroliniana*), knotweed (*Polygonum erectum*) and goosefoot (*Chenopodium californicum*) well before the intensified use of maize at the end of the first millennium CE. While a globally fairly robust and general pattern characterised by declining oral health in the face of transitions to grain-based agricultural subsistence is fairly clear (Larsen 1995), there may be an important regional exception.

An analysis of an ostensibly pre-agricultural (*c.* 5,000 years BP) and an arguably rice-based agricultural sample (*c.* 2,500 to 1,700 years BP) in northern Vietnam (Oxenham *et al.* 2002a) found very low frequencies of carious lesions for both time periods. There are a number of potential explanations for this finding, including the low cariogenicity of rice-based diets (Sreebny 1983); the elevated component of marine foods in the diet, and associated reduced cariogenicity (Kelley *et al.* 1991, Larsen *et al.* 1991); the evidence for the use of betel nut (*Areca catechu*), which is also arguably cariostatic (although see Oxenham *et al.* 2002b); and finally the possibility that the grain growers were not the chief consumers of their produce but rather the Han (Chinese) who had administrative control of the rice-growing areas. Domett (2001) has recently examined the oral health, among other variables sensitive to health, of four skeletal samples from Thai sites, Khok Phanom Di, Nong Nor, Ban Na Di and Ban Lum Kao, which date to between 4,000 and 2,500 years BP. Domett (2001, p. 134, italics in original) concluded that 'in direct contrast to many other world trends the intensification of agriculture in prehistoric Thailand appears to have coincided with a *decrease* in the prevalence of caries, and a general improvement in dental health'. Pietrusewsky and Douglas (2002a,b) have also argued that there is no evidence for a decline in oral health, specifically with respect to the frequency of carious lesions, with the intensification in agriculture at Ban Chiang, Thailand, over time (the Ban Chiang skeletal material dates to between 4,100 and 1,800 years BP). Finally, Tayles *et al.* (2000) examined the frequency of carious lesions over time using three dental assemblages in Thailand (Khok Phanom Di, Ban Lum Khao and Noen U-Loke) spanning the period from *c.* 4,000 to *c.* 1,700 years BP. They have argued that there is clearly no evidence for an increase in carious lesion frequency over time.

The aim of this chapter is to review the evidence for oral disease in the majority of Holocene Southeast Asian assemblages investigated to date and to determine if there is a demonstrable secular change in oral health. To this end, the authors have reanalysed published and unpublished data from the seven largest skeletal assemblages in Thailand and three samples representing the majority of skeletal material in Vietnam. Given the number of dental collections examined here, the authors have focused on three pathological conditions that are particularly sensitive to oral health and diet in general, in addition to other oral behaviours: carious lesions, alveolar defects of pulpal origin (often incorrectly termed abscesses) and antemortem tooth loss (AMTL). In the process of addressing the main aim of this study, several related questions were examined. What does the oral pathology profile for Southeast Asia as a whole look like? Are there discernable Southeast Asia-wide (regional) oral health trends? If such trends occur, do they correlate with specific archaeologically described behaviours (e.g. subsistence)? Are these oral health profiles consistent with the archaeological evidence for lifestyles, including social structure, in the region?

Materials and methods

This is the first time the skeletal and dental materials from all of the main cemetery sites in mainland Southeast Asia have been collectively examined in terms of oral health. With the exception of human material recovered from Ban Lum Kao, Ban Na Di, Nong Nor and Noen U-Loke, the data in this study have been derived from the published literature or our own research. Additional unpublished raw data pertaining to the aforementioned assemblages were graciously supplied by Kathryn Domett and Nancy Tayles. A total of 10 assemblages from the region are reviewed. The sample is increased to 12 with the division of the temporally extensive samples from Non Nok Tha and Ban Chiang into early and late samples. This study includes some 548 adult individuals (the adult cut-off point differs from study to study but is always at a minimum 15 years or above) with 9033 permanent teeth and 12 469 tooth positions (alveoli), spanning approximately 3,500 to 4,000 years and extending from the mid Holocene through to the adoption of iron technology. The three dental conditions selected to develop a comparative oral pathology profile have been collected and reported on in a consistent manner: carious lesions, alveolar defects of pulpal origin and AMTL. While there are clearly correlations between the frequency, manifestation and distribution of all oral pathologies, these three conditions are commonly examined in the literature (Lukacs 1989) and are collectively representative of most of the more common pathologies seen in the oral cavity.

The overall frequency of each condition has been reported by tooth count in addition to the frequencies for the anterior (incisors and canines) and posterior (premolars and molars) dentitions separately. Further, for carious lesions, the proportion of lesions affecting the root (including the cementoenamel junction (CEJ)) was also reviewed, given the purported relationship between these lesions and diet (Molnar and Molnar 1985), tooth wear and/or age at death (Hillson 2001). None of the studies in this review included attrition-based pulp chamber exposure in the carious lesion count. A number of so-called caries-correction functions are available to account for carious teeth lost antemortem (e.g. Lukacs 1995, Erdal and Duyar 1999) or to attempt to make bioarchaeological indices of oral health more comparable with modern studies (Kelley *et al.* 1991, Saunders *et al.* 1997). However as Hillson (2001, p. 257) noted, these corrections 'make assumptions which are difficult to test'. The frequency of AMTL, upon which corrections are based, has an extremely complex aetiological history that will be very different from assemblage to assemblage and includes such factors as the degree of periodontal disease, carious lesions, tooth wear, extra oral activities (using the teeth as tools), tooth ablation, trauma and so forth. A caries-correction function has not been used on any of the data presented in this study.

The term abscess has been avoided here, accepting the argument of Dias and Tayles (1997) that the term abscess should be restricted to alveolar defects that present fistulae or drainage sinuses (active infection). However, the alternative term used by Dias and Tayles (1997), granuloma, was not adopted as this terminology, by definition, assumes that the lesion is benign, although this does not exclude the potential for these lesions to develop subsequently into sites of infection. Given that both lesion types are ultimately pulpal in origin, and the practical difficulty of differentiating between lesions that were the result of chronic cysts and those that resulted from infectious processes, the authors have used the inclusive and assumption-free term alveolar defect of pulpal origin. Finally, given that tooth wear plays an important aetiological role in the development and patterning of numerous oral pathologies, it would have been useful to compare a measure of this variable among the samples. However, marked differences in how this variable was identified and reported has meant that meaningful inter-assemblage comparisons of this variable could not be performed.

Statistical tests of inter- and intrasample significance were carried out for each oral pathology. While there is a great deal of consistency in data-collection protocols and reporting methods across the studies reviewed here (see references in Table 11.1), it was not possible to test for interobserver error. Intrasample differences by sex and tooth position were examined using standard chi-square tests of homogeneity. Because these tests cannot be used in multiple intersample comparisons, because of the

Table 11.1. *Dental assemblage sample sizes for adults (15 years or more)*

Assemblage	Date (years BP)	Physical location	Chief subsistence[a]	Teeth No.				Alveoli				Total indiv-iduals[b]	Source
				Total	Male	Female	$\overline{X}$	Total	Male	Female	$\overline{X}$		
Vietnam Da But	6,000–5,500	Estuary/ river	Fa/t	944	445	333	13.3	1430	691	540	20.1	71	This study
Khok Phanom Di	4,000–3,500	Estuary/ coast	Fa/t (A?)	1282	625	657	19.1	2047	944	1103	30.6	67	Tayles (1999)
Early Non Nok Tha	4,800–3,400	Inland/ river	M	666	284	382	17.5	900	391	509	23.7	38	Douglas (1996)
Early Ban Chiang	4,100–2,900	Inland/ river	M	534	310	198	19.8	667	390	252	24.7	27	Douglas (1996)
Late Non Nok Tha	3,400–2,200	Inland/ river	M	539	292	244	13.1	766	449	317	18.7	41	Douglas (1996)
Nong Nor	3,100–2,700	Coast/ river	A (Fa)	1044	527	456	15.4	1402	693	630	20.6	68	Tayles *et al.* (1998), Domett, personal communication
Ban Lum Khao	3,000–2,500	Inland/ river	M	874	397	477	20.3	1138	531	607	26.5	43	Domett 2001, Domett, personal communication
Late Ban Chiang	2,900–1,800	Inland/ river	M	560	244	316	15.1	698	323	375	18.9	37	Douglas (1996)
Ban Na Di	2,600–2,400	Inland/ river	A (F)	516	147	368	15.2	707	241	466	20.8	34	Domett 2001, Domett, personal communication

Table 11.1. (*cont.*)

Assemblage	Date (years BP)	Physical location	Chief subsistence[a]	Teeth No.				Alveoli				Total indiv-iduals[b]	Source
				Total	Male	Female	$\overline{X}$	Total	Male	Female	$\overline{X}$		
Vietnam Ma, Ca rivers	2,500–1,700	Inland/ river	M	828	435	294	18.4	938	545	302	20.8	45	This study
Noen U-Loke	2,300–1,700	Inland/ river	A/H	956	422	382	17.7	1250	533	519	23.1	54	Domett, personal communication
Vietnam Red River	2,200–1,700	Estuary/ river	M	290	51	168	12.6	526	122	288	22.9	23	This study
Total				9033	4179	4275	16.5	12 469	5853	5908	22.8	548	

F, foraging; M, foraging and agriculture mix; A, agriculture; H, cattle herding; a, aquatic; t, terrestrial.
$\overline{X}$, mean number per individual.
[a]Subsistence in parentheses indicates secondary importance.
[b]refers to individuals with assessable oral structures.

increasing likelihood of incorrect significant results, the Kruskall–Wallis one way analysis of variance (ANOVA) was used. Significant ANOVA values were examined using Dunn's multiple comparison test (Zarr 1974).

The assemblages

With the exceptions of the Vietnamese Da But samples, the Thai, Ban Lum Khao samples, and possibly the Thai Nong Nor samples, some level of social differentiation or hierarchy has been suggested for the communities represented by these human remains. Whether the material culture associated with burials at these sites represents social hierarchy or is better seen in terms of heterarchical models of social organisation is unclear at present (see discussion in White (1995)). Table 11.1 summarises the assemblages with respect to date, physical location (see also Fig. 1.1, p. 5), subsistence orientation and sample sizes.

Beginning with the temporally earliest assemblage, the Vietnamese Da But period series is represented by the dental remains of 71 adults from Con Co Ngua, which is close to the Ma river in Thanh Hoa province and approximately 30 km from the current coast. The inhabitants of the site are best described as sedentary foragers, with a range of vertebrates and shellfish being excavated from the site: buffalo, pig, turtle, fish, oyster, mussels and snails. Artefacts include numerous grinding stones, ground axes and mat-impressed pottery fragments (Oxenham *et al.* 2002a). A recent limited analysis of the dentin chemistry of a small sample of teeth indicates a marine-dependent diet (Y. Yasutomo *et al.*, unpublished data). The date of this post-Hoabinhian assemblage has been estimated by way of material culture associations to approximately 5,500 to 6,000 years BP, a date consistent with the nitrogen decay rate in the dentin samples (Y. Yasutomo *et al.*, unpublished data).

Khok Phanom Di, adjacent to the coast in the Gulf of Thailand for much of its occupation, is the earliest site to show evidence of some form of rice cultivation in prehistoric Thailand. The site was occupied between 4,000 and 3,500 years BP and of the remains of 154 individuals recovered, 67 adults with preserved dental remains are available for study. A great deal is known of the diet of these individuals because of the extraordinarily good preservation and meticulous recovery of biological remains at the site. Direct evidence for the consumption of fish and cultivated rice comes from preserved human stomach contents and coprolites (Tayles 1999). There is also evidence that the intensity of rice production was such that surpluses were being stored (Tayles 1999). The enormous quantity and variety of marine remains also attests to the reliance on such a diet. Tayles (1999) has also alluded to the probable availability of foods such as banana, yam and taro to the inhabitants of Khok Phanom Di. This site is pre-metal but shows evidence for a sophisticated ceramics industry and trading.

Non Nok Tha, south and west of Ban Chiang in northern Thailand, is a site with a contested chronology. The entire site spans general periods A to D as defined by Bayard (1984), although intact skeletal material is confined to periods A to C. We have used Douglas' (1996) division of the skeletal assemblage into an 'early group' (initial occupation of the site Early Period (EP) 1 to Middle Period (MP) 2 (EP1–MP2) with 38 adult dentitions) and a 'late group' ((MP3–MP8), with 41 adult dentitions). Douglas (1996) noted that it would have been natural to divide the assemblage at between EP3/MP1 and MP2, where a clear cultural change occurs, including the appearance of bronze. However, the resultant sample size would have been prohibitively small. Moreover, a further cultural change occurs with the advent of MP3, with the cessation of burial mounding and the appearance of bronze ornaments, in distinction to more functional or utilitarian bronze artefacts (Douglas 1996, p. 56). The dating controversy surrounding this site is too complex to deal with here. We have taken a conservative position and placed the temporal extent of the site with skeletal remains between 4,800 and 2,200 years BP, with the split between early and late group Non Nok Tha occurring at approximately 3,400 years BP.

Ban Chiang, located just north of Ban Na Di, is perhaps one of the best-known archaeological sites in Thailand. Dated to between 4,100 and 1,800 years BP, it extended from the pre-metal period through to the adoption of iron. The remains of 142 individuals, including 64 adults with assessable dental material, have been recovered. The sample has been divided into early (burial phases EPI to EPIV, including 27 adult dentitions) and late (EPV to Late Period (LP) X, including 37 adult dentitions) subsamples following Douglas (1996). The division between late and early groups occurs at approximately 2,900 years BP (M. T. Douglas, personal communication). Remains from a range of domesticated animals have been excavated from the site, including dogs, chickens, cattle and pigs. While, in general, a mixed agriculture and foraging economy is indicated, over time the site saw increasing amounts of forest clearance. This and the domestication of the buffalo indicate 'the beginning of bunded [embanked], wet-rice agriculture' (Douglas 1996, p. 65).

The following characterisations of Nong Nor, Ban Lum Kao, Ban Na Di and Noen U-Loke are based on recent work by Domett (2001) and Higham (1996). Nong Nor is located only 14 km from Khok Phanom Di and during its occupation was situated by a mangrove-fringed coast in the Gulf of Siam (Higham and Thosarat 1998, p. 523). Skeletal material dating to between 3,100 and 2,700 years BP, representing 155 individuals, was recovered from the site (Tayles *et al.* 1998), of which 68 adults had assessable dental remains. Little detailed information exists for the cemetery phase of the site, although a marine orientation and the cultivation of rice seems certain (Higham 1996). In fact, the area would have been ideal for seasonally flooded rice cultivation. Faunal evidence also indicated the presence of chickens, suids, cattle and domesticated dogs (Higham 1998).

Ban Lum Khao is located in the southern Khorat plateau in the Mun River valley, and the skeletal assemblage has been dated to between 2,500 and 3,000 years BP (Domett 2001). Of 110 sets of human remains recovered, 43 adults with assessable dental remains have been included in this study. The faunal remains recovered from this site are similar to those from Ban Na Di, although there is a suggestion that Ban Lum Khao was in an area being newly exploited by humans. Domett (2001) argued that rice was a component of the diet of those represented by the skeletal remains. This was a community that had a sophisticated pottery industry, knowledge of bronze technology and was involved in some level of trading (O'Reilly 2001).

The Ban Na Di assemblage was originally studied by Wiriyaromp (1984) and subsequently reanalysed by Domett (1999, 2001). The 78 individuals recovered have been dated to between 2,600 and 2,400 years BP and 34 of these are adults with assessable dental remains. Environmentally, the site is on the Khorat plateau, close to Ban Chiang, and is located by several small streams. Numerous domestic and wild animals were recovered from the site in addition to many species of fish from diverse habitats. Cultivated rice remains and the site's ideal rice-growing location have suggested to researchers that 'rice agriculture would have been a predominant subsistence activity' (Domett 2001, p. 23). Technologically, Ban Na Di has shown evidence of pottery, bronze, bone, glass and shell working and an extensive trading network.

Noen U-Loke is situated in the Mun River valley west of Ban Lum Khao. Skeletal remains of 120 individuals were recovered from the site, with 54 adults having preserved dental remains. The assemblage is unique in being the first major skeletal sample from the iron period in Thailand, which saw large increases in population density and village sites. While details on the subsistence base of the site are not available, it would appear that rice was intensively cultivated, given the practice of filling in graves with this grain (Higham and Thosarat 1998) and palynological evidence for extensive forest clearing (Boyd and McGrath 2001). In addition to evidence for fish exploitation, cultivated rice and remnants of other plants, there are the butchered remains of water buffalo, cattle and suids. An exceptionally rich material culture has been identified at the site, including bronze, iron, glass, ceramic and agate artefacts.

Finally the Vietnamese metal period, the most developed phase being well known as the Dong Son period, is represented by the dental remains of 68 adult individuals recovered from 11 separate sites in the Red, Ma and Ca river areas. Because of the small individual sample sizes, the 11 assemblages have been aggregated into two samples for the purposes of analysis. One sample named Vietnam Red River includes 23 individuals recovered from sites in the Red River delta and has been dated to between 2,200 and 1,700 years BP. The other sample comprised 45 individuals principally from the Ma river and has been dated to 2,500 and 1,700 years BP. The Ma river sites are in the same geographic area as the mid Holocene Vietnam Da But assemblage. Higham (1989, p. 30) has described the Vietnamese Dong Son as one of several Southeast Asian examples demonstrating social differentiation and a move 'from village autonomy towards centralised chiefdoms'. Apart from the diverse range and technological sophistication of material culture objects, evidence has been found for marked craft specialisation, a complex ritual life, the development of an aristocratic and centralised elite, maritime trade and sophisticated military skills and equipment (Oxenham *et al.* 2002a).

Results

Adult age-at-death distributions

Because of the age-accumulative nature of dental disease, it is necessary to review and compare the adult age distributions of each assemblage. As can be seen from Table 11.2, there is some variation in the age profiles of the samples. Simply looking at the composition of younger (first two age groups) to older (latter two groups), the Vietnam Da But, early and late Non Nok Tha and early Ban Chiang samples have the greatest number of older individuals. Nong Nor, Ban Na Di and the late Ban Chiang sample also have relatively more aged individuals, while Khok Phanom Di, Ban Lum Khao, Vietnam Red River, Vietnam Ma and Ca river and Noen U-Loke are fairly evenly divided between older and younger cases.

Oral health profiles for Southeast Asia

Carious lesions

Table 11.3 summarises the oral pathology profile for each assemblage. There is no clear secular trend in caries frequency, with one of the lowest

Table 11.2. *Dental sample adult (15 years or over) age-at-death and sex distributions*

Assemblage[a]	Age category (years)				Male:female ratio	Percentage with details available for	
	15–19	20–29	30–39	40+		Age	Sex
Vietnam Da But	11.7	18.2	28.6	41.6	1.4:1	89.4	85.0
Khok Phanom Di	11.8	33.8	36.8	17.6	1:1.1	100.0	100.0
Early Non Nok Tha	5.1	17.9	25.6	51.3	1:1.3	72.2	98.1
Early Ban Chiang	6.9	20.7	20.7	51.7	1.3:1	76.3	94.7
Late Non Nok Tha	2.4	14.3	35.7	47.6	1:1.1	60.0	97.1
Nong Nor	2.4	30.6	50.6	16.5	1.1:1	69.7	76.2
Ban Lum Khao	8.5	42.4	28.8	20.3	1:1.2	100.0	98.3
Late Ban Chiang	15.2	18.2	33.3	33.3	1:1.0	51.6	96.9
Ban Na Di	0.0	36.1	36.1	27.8	1.7:1	72.0	70.0
Vietnam Ma, Ca rivers	11.9	38.1	23.8	26.2	1.1:1	93.3	84.4
Noen U-Loke	7.4	44.4	25.9	22.2	1.2:1	80.6	71.6
Vietnam Red River	13.6	36.4	27.3	22.7	1:2.2	95.7	72.7

[a]See Table 11.1 for data sources.

frequencies occurring in the earliest sample (Vietnam Da But, 1.5%) and the highest frequency occurring in the next earliest sample (Khok Phanom Di, 10.9%). There is significant (Kruskall–Wallis one-way ANOVA, $H = 143$, $p = 0.000$) intersample heterogeneity for the distribution of carious lesions (Table 11.4). Khok Phanom Di had significantly more lesions (Dunn's *post hoc* multiple comparison test, $p < 0.05$) than any other assemblage, with the exception of late Ban Chiang. Additionally, early and late Ban Chiang, Nong Nor, Ban Lum Khao and Noen U-Loke samples showed significantly more carious lesions than Vietnam Da But. The frequency of carious lesions was high at Khok Phanom Di despite the relative youth of the sample. The older age of samples from early and late Ban Chiang and Nong Nor also needs to be considered when interpreting their relatively elevated caries rates, given the age-cumulative nature of this disease. However, Vietnam Da But and early Non Nok Tha had some of the lowest rates of caries and the greatest percentage of older (>30 years) individuals, suggesting that longevity may not be one of the most important factors in the aetiology of caries in Southeast Asia. There is also no obvious correlation between the frequency of carious lesions and putative subsistence orientation. However, for the two samples divided into early and late subsamples,

Table 11.3. *Sample oral pathology profiles*[a]

Assemblage[a]	Chief subsistence[b]	CAR	ADP	AMTL	CAR		ADP		AMTL		CAR				ADP		AMTL	
					M[c]	F[c]	M	F	M	F	Other	R/CEJ	Ant.	Post.	Ant.	Post.	Ant.	Post.
Vietnam Da But	Fa/t	1.5	1.5	4.8	1.6	2.1	10.0	9.0	4.8	6.5	64.3	35.7	0.0	1.9*	0.4	2.2*	7.5*	3.4
Khok Phanom Di	Fa/t (A?)	10.9	6.0	17.2	6.9	14.6*	4.7	7.1*	11.7	21.9*	92.8	7.2	5.9	12.7*	5.2	6.4	6.2	10.6*
Early Non Nok Tha	M	1.7	1.2	5.0	2.1	1.3	2.0	0.6	7.4*	3.1	100	0.0	0.5	2.1	0.0	1.9*	4.0	5.6
Early Ban Chiang	M	6.2	4.6	6.6	7.7	4.5	4.4	5.6	6.7	7.1	97.0	3.0	1.7	8.7*	1.6	6.4*	4.5	7.9
Late Non Nok Tha	M	4.1	2.1	10.4	2.7	5.7	2.4	1.6	11.1	9.5	100	0.0	1.1	5.6*	1.0	2.7	8.2	11.8
Nong Nor	A (Fa)	6.5	0.8	4.2	4.7	9.4*	1.2	0.5	2.6	6.3*	98.5	1.5	3.5	8.0*	0.0	1.2*	2.0	5.3*
Ban Lum Khao	M	4.5	1.3	5.3	1.8	6.7*	1.7	1.0	5.3	5.1	94.9	5.1	0.7	6.5*	1.4	1.3	4.0	6.1
Late Ban Chiang	M	7.7	5.2	6.9	8.6	7.0	8.0*	2.7	8.4	5.6	86.0	14.0	5.7	8.7	4.8	5.3	2.0	9.6*
Ban Na Di	A (F)	4.7	2.1	5.4	8.2*	3.3	5.4*	0.4	2.9	6.7*	95.8	4.2	0.0	6.9*	3.1	1.6	2.3	7.2*
Vietnam Ma, Ca rivers	M	2.8	3.1	2.9	1.6	4.8*	2.6	4.0	2.0	3.0	69.6	30.4	0.4	3.2*	3.0	3.2	0.9	4.0*
Noen U-Loke	A/H	4.8	2.7	5.5	4.5	5.8	3.0	2.9	3.0	8.9*	100	0.0	1.6	6.3*	2.3	2.9	2.6	7.0*
Vietnam Red River	M	1.4	3.2	3.6	0.0	1.8	0.8	3.1	2.5	5.6	100	0.0	0.0	1.4	4.2	2.7	6.3*	2.1

CAR, carious lesions; ADP, alveolar defects of pulpal origin; AMTL, antemortem tooth loss; R/CEJ, root/cementoenamel junction; Ant., anterior teeth; Post., posterior teeth.
[a]See Table 11.1 for data sources.
[b]See Table 11.1 for definitions.
[c]Male, M; Female, F.
*Intrasample difference significant at $p < 0.05$ by chi-square test.

Table 11.4. *Results of Kruskall–Wallis one-way analysis of variance for caries, alveolar defects and antemortem*[a] *tooth loss*[b]

	KPD	NNTE	BCE	NNTL	NN	BLK	BCL	BND	VMC	NUL	VR
VDB	1,2,3	*	1,2	3	1	1	1,2	*	*	1	*
KPD		1,2,3	1	1,2	1,2,3	1,2,3	*	1,2	1,2,3	1,2,3	1,3
NNTE			*	3	*	*	2	*	*	*	*
BCE				*	*	*	*	*	*	*	*
NNTL					*	*	*	*	3	*	*
NN						*	2	*	*	*	*
BLK							*	*	*	*	*
BCL								*	*	*	*
BND									*	*	*
VMC										*	*
NUL											*
VR											

VDB, Vietnam Da But; KPD, Khok Phanom Di; NNTE, early Non Nok Tha; BCE, early Ban Chiang; NNTL, late Non Nok Tha; NN, Nong Nor; BLK, Ban Lum Khao; BCL, late Ban Chiang; BND, Ban Na Di; VMC, Vietnam Ma, Ca rivers; NUL, Noen U-Loke; VR, Vietnam Red River.

[a]Khok Phanom Di antemortem tooth loss was calculated minus contribution from ablated teeth.

[b]Kruskall–Wallis result: caries, $H = 143.0$, $p = 0.000$; alveolar defects of pulpal origin, $H = 133.3$, $p = 0.000$; antemortem tooth loss, $H = 85.2$, $p = 0.000$. *Post hoc* multiple comparisons using Dunn's test ($p < 0.05$) showed a significant difference in caries (1), in alveolar defects of pulpal origin (2), in antemortem tooth loss (3).

*Indicates significant difference.

there are observable changes over time. The rate of lesions increased at Non Nok Tha from 1.7% to 4.1%, albeit not a significant difference. For Ban Chiang, there was also a non-significant increase in the rate of carious lesions from 6.2% to 7.7%.

Females displayed a higher rate of carious lesions in 8/12 samples and the difference was significant in the samples from Khok Phanom Di, Nong Nor, Ban Lum Khao and Vietnam Ma and Ca rivers (Table 11.3). Again, there was no clear association between the level of female lesions and time, subsistence orientation and location.

In examining the distribution of carious lesions by tooth position (Table 11.3), the posterior teeth were more affected in every instance, although in three samples this difference was not significantly different. The rate of anterior tooth lesions was relatively elevated for Khok Phanom Di (5.9%), Nong Nor (3.5%) and late Ban Chiang (5.7%). An intersample comparison indicated heterogeneity in the distribution of anterior ($H = 56.8, p = 0.000$) and posterior lesions ($H = 99.3, p = 0.000$). Khok Phanom Di had a significantly higher anterior rate (Dunn's test, $p < 0.05$) than Vietnam Da But, early and late Non Nok Tha, Ban Lum Khao, Ban Na Di, Vietnam Ma Ca and rivers and Noen U-Loke. Further, Nong Nor and late Ban Chiang had significantly higher rates of anterior lesions than Vietnam Da But. The frequency of anterior lesions also increased from early (1.7%) to late Ban Chiang (5.7%) and early (0.5%) to late Non Nok Tha (1.1%), albeit not significantly. The situation is a little different for the distribution of posterior caries. Khok Phanom Di had a significantly higher rate (Dunn's test, $p < 0.05$) than Vietnam Da But, early and late Non Nok Tha, Ban Lum Khao, Noen U-Loke and Vietnam Red River. All samples, with the exception of early and late Non Nok Tha and Vietnam Red River, had significantly more posterior caries than Vietnam Da But.

The proportion of carious lesions affecting the root and/or CEJ was relatively low for most samples. Only in the Vietnam Ma and Ca Rivers and Vietnam Da But samples did it account for a considerable proportion of all lesion types. There was also a statistically non-significant ($\chi^2 = 2.7$, $p > 0.5 < 0.2$) increase in the proportion of root and/or CEJ lesions from early Ban Chiang (3.0%) to late Ban Chiang (14.0%).

Alveolar defects of pulpal origin

As was the case for carious lesions, there was no clear trend, secular or otherwise, in the distribution of alveolar defects of pulpal origin (Table 11.3). Khok Phanom Di had the highest rate of alveolar defects (6.0%) and Nong Nor the lowest (0.8%). There was significant intersample heterogeneity ($H = 133.3$, $p = 0.000$) in the frequency of alveolar defects (Table 11.4). Khok Phanom Di had significantly more alveolar defects (Dunn's test, $p < 0.05$) than all samples with the exception of early and late Ban Chiang and Vietnam Red River. Early and late Ban Chiang samples had significantly more defects than Vietnam Da But, while late Ban Chiang also had significantly more defects than early Non Nok Tha and Nong Nor. For the two cases where intrasample comparisons could be made, there were slight, statistically non-significant, changes over time, with alveolar defects increasing from 4.6% to 5.2% at Ban Chiang and from 1.2% to 2.1% at Non Nok Tha. As was the situation with carious lesions, there was no consistent association between the age profile of the sample and the frequency of alveolar defects.

Males displayed a greater rate of alveolar defects than females in 8/12 samples and this was a statistically significant difference at late Ban Chiang and Ban Na Di (Table 11.3). While the sex distribution was even for late Ban Chiang, it was, as noted above, highly skewed toward males at Ban Na Di. Khok Phanom Di was the only sample with significantly more female than male alveolar defects. Moreover, females were slightly better represented at Khok Phanom Di. The distribution of defects by sex does not appear to be associated with assemblage age, location or sample subsistence orientation.

The distribution of alveolar defects of pulpal origin (Table 11.3) was very different from that seen for carious lesions. In only four instances was the frequency of posterior defects significantly greater than anterior defects, occurring at Khok Phanom Di, early Non Nok Tha, early Ban Chiang and Nong Nor. An intersample examination of the frequency of posterior defects indicated a degree of heterogeneity ($H = 89.0$, $p = 0.000$). Khok Phanom Di had significantly more posterior defects (Dunn's test, $p < 0.05$) than all samples with the exception of early and late Ban Chiang and Vietnam Red River. Early and late Ban Chiang samples also had significantly more posterior defects than Vietnam Da But. With respect to anterior defects, there was also a degree of intersample heterogeneity ($H = 67.9$, $p = 0.000$). Khok Phanom Di had significantly more anterior defects (Dunn's test, $p < 0.05$) than early and late Non Nok Tha, Nong Nor, Ban Lum Khao and Vietnam Da But. Late Ban Chiang and Vietnam Red River samples had significantly more defects than Vietnam Da But. In terms of the temporal distribution of anterior alveolar defects, samples with elevated rates tended to occupy the later period while those with relatively low frequencies of defects were found in the samples from the earlier period of the region.

Antemortem tooth loss

There was no consistent trend in the distribution of AMTL between assemblages (Table 11.3). Khok Phanom Di had the highest rate (17.2%) and Vietnam Ma and Ca rivers had the lowest (2.9%). The high rate of AMTL at Khok Phanom Di was confounded by extreme levels of deliberate anterior tooth ablation and taking this into account reduced the frequency of AMTL to 8.9%. In order to make inter- and intrasample comparisons more meaningful, only AMTL attributable to pathological loss at Khok Phanom Di, and not deliberate ablation, was considered for the analyses below. The distribution of AMTL was distributed heterogeneously ($H = 84.9, p = 0.000$) among the assemblages (Table 11.4). Late Non Nok Tha and Khok Phanom Di had a significantly higher rate of AMTL (Dunn's test, $p < 0.05$) than early Non Nok Tha, Vietnam Da But and Vietnam Ma and Ca rivers. Additionally, the rate was significantly higher at Khok Phanom Di than Nong Nor, Ban Lum Khao, Noen U-Loke and Vietnam Red River. For the two samples where intrasample comparisons were possible, Ban Chiang showed a slight, non-significant increase from 6.6% to 6.9% while Non Nok Tha had a greater, but still non-significant, increase in AMTL of 5.0% to 10.4%. There was no consistent relationship between the rate of AMTL and sample age profile.

Females displayed a higher rate of AMTL in 8/12 samples and the difference was significant for Khok Phanom Di, Nong Nor, Ban Na Di and Noen U-Loke (Table 11.3). With the exception of Khok Phanom Di, the samples with significantly more female AMTL all had a good representation of males, particularly Ban Na Di, suggesting that these differences are not simply statistical artefacts. In the one instance where males displayed significantly more AMTL than females, early Non Nok Tha, there was a large sex imbalance in favour of females.

Looking at the distribution of AMTL by tooth position (Table 11.3), it can be seen that the posterior teeth were affected to a greater degree than the anterior teeth in 10/12 samples and significantly so in six instances. Only Vietnam Da But and Vietnam Red River displayed significantly more anterior tooth loss. Both anterior ($H = 50.8, p = 0.000$) and posterior ($H = 82.9, p = 0.000$) AMTL was distributed heterogeneously among the assemblages (Table 11.4). Vietnam Da But had a significantly greater rate of AMTL (Dunn's test, $p < 0.05$) than Nong Nor, late Ban Chiang, Ban Na Di, Vietnam Ma and Ca rivers and Noen U-Loke. Khok Phanom Di had a significantly greater frequency of anterior AMTL than Vietnam Ma and Ca rivers. In a reversal of the trend seen for alveolar defects of pulpal origin, the temporally later assemblages tended for the most part to have relatively lower rates of anterior AMTL, while the earlier samples tended to have relatively higher rates.

In terms of posterior AMTL, Khok Phanom Di had a significantly higher frequency than the three Vietnamese samples, early Non Nok Tha, Nong Nor and Ban Lum Khao. Further, early and late Ban Chiang, late Non Nok Tha and Noen U-Loke samples displayed a significantly greater rate of posterior AMTL than Vietnam Da But. Late Non Nok Tha also displayed a significantly higher rate than Vietnam Red River. There is no clear trend in the distribution of posterior AMTL by time, subsistence orientation or site location.

Intra-sample comparisons

As noted above, because of the large temporal extent of the Ban Chiang and Non Nok Tha samples, they were divided (after Douglas 1996) into early and late subsamples. It was found that the frequency of carious lesions, alveolar defects of pulpal origin and AMTL increased, albeit not in a significant manner based on the Kruskall–Wallis one-way ANOVA results, within each sample from early to late periods. The changes at Ban Chiang did not increase in a significant manner over time, while the increases in carious lesions ($\chi^2 = 6.2$, $p < 0.03$) and AMTL ($\chi^2 = 15.2$, $p < 0.001$) at Non Nok Tha were statistically significant.

Discussion

The chief impetus for this review arose from the need to examine critically the oral health of samples from Southeast Asia spanning the transition to and intensification of agriculture. A number of the most prominent and active researchers into Southeast Asian palaeohealth have argued for improvements occurring in one or more variables determining oral health over time in the region, either locally (Pietrusewsky and Douglas 2002a,b) or regionally (Tayles *et al.* 2000; Domett 2001). These findings are not consistent with what seems to be a generalised decline in oral health globally (see contributions in Cohen and Armelagos (1984) and Larsen (1997)). The following discussion will demonstrate that oral health was remarkably homogeneous over time in the region and will provide a number of explanations why the dramatic oral health costs associated with the adoption and/or intensification of grain agriculture in other regions of the world did not occur in Southeast Asia.

Patterns in Southeast Asian oral health

Carious lesions

Overall, the frequency of carious lesions in these Southeast Asian samples was very low, with a range between 1.4% and 10.9% and a mean of 5.2%, which reduces to 4.2% if Khok Phanom Di is treated as an outlier. Despite the significant intersample heterogeneity, there was no clear pattern in the distribution of what are for the most part low frequencies of carious lesions. However, where intrasample comparisons were possible, an increase in the rate of carious lesions could be detected at both Ban Chiang and Non Nok Tha, in particular, over time, which could reflect intensification in the agricultural component of their otherwise mixed subsistence economies.

There is some support for the observation that oral health was poorer in females than in males, at least in terms of carious lesions, in Southeast Asia. However, there is no apparent consistent association between the relatively higher rate of female carious lesions and time, age composition or sex composition of the samples. Poorer female oral health, as measured by the frequency of carious lesions, is not an unexpected finding and numerous explanations have been offered in the literature, including earlier female tooth eruption and hence increased exposure to cariogenic processes, pregnancy, differential male and female diets and/or frequency of eating and so forth (see Larsen *et al.* 1991). It is worth noting that greater rates of female carious lesions do not necessarily reflect differential social status within the communities these samples represent, as poorer female oral health has been observed in modern egalitarian societies (Walker and Hewlett 1990).

The finding that the posterior dentition is preferentially disposed to carious lesions in every sample is not unexpected. Hillson (2001) has summarised the differential susceptibility of teeth by position to carious lesions. With an increasing frequency of carious lesions, tooth positions become progressively affected in the following order: occlusal sites in first molars, occlusal sites in second molars, approximal sites of molar crowns, premolar occlusal sites, approximal upper incisors, other premolar sites and finally the canines (Hillson 2001, pp. 251–252). In samples with quite low rates of carious lesions, as is the case for these Southeast Asian dental assemblages, it would be expected that incisor and canine lesions would be rare. Further, as expected, relatively higher rates of anterior tooth lesions are only seen in samples with overall higher rates of carious lesions.

A contributing factor in the elevated rate of anterior lesions may be particular dietary regimens. For instance, while carious lesions are relatively rare in apes (Kilgore 1989, Lovell 1990), Lovell (1991) found higher rates of anterior, as compared with posterior, lesions in orang-utans. While free-ranging orang-utans have a diverse diet, they prefer fruits when available (Rodman 1988) and are among those primates that use their anterior teeth extensively in food preparation and processing (Ungar 1994). The anterior dentition is also used extensively by humans in the consumption of food stuffs such as taro, yams and bananas. All of these foods were postulated as having been present at Khok Phanom Di (Tayles 1999) and may have contributed to the overall high rate of lesions in this sample and specifically have been responsible for elevated anterior tooth decay.

Root lesions were examined because of the purported association between an increase in such lesions with an increase in the grain component of the diet (Molnar and Molnar 1985, Moore 1993). There is no pattern of association between more agriculturally oriented samples and the proportion of root and/or CEJ lesions in this study. Notwithstanding this, a lack of support for the hypothesis that agricultural intensification may lead to an increase in root and/or CEJ lesions is informative in itself. It may well be that agricultural intensification, as it is generally understood on the global stage, was not occurring in bronze age Southeast Asia, or that rice is less cariogenic than other cereals (a point examined in more detail later in the chapter). The relatively high proportion of root and/or CEJ lesions seen in the late Ban Chiang, Vietnamese Da But and Vietnamese metal phase Ma and Ca River samples would appear to be unusual compared with the other Southeast Asian assemblages. It has been argued that samples with older age profiles will show higher rates of root lesions as a result of the effects of continuous tooth eruption and exposure of the root to acidogenic bacteria (Hillson 2001), in addition to periodontal disease in general. Both late Ban Chiang and Vietnam Da But assemblages had relatively high proportions of older individuals, although so did a number of other samples that did not exhibit high proportions of root and/or CEJ lesions.

Alveolar defects of pulpal origin

As is the case for carious lesions, the frequency of alveolar defects of pulpal origin was low and ranged between 0.8% and 6.0%, with a mean of 2.9%. Again, although there was significant intersample heterogeneity, there was no clear pattern in the distribution of these defects. In a reversal of the trend seen for carious lesions, males had higher rates of alveolar defects than females in a majority of samples, and in only four instances was the frequency of posterior defects significantly greater than anterior defects. Both Khok Phanom Di and late Ban Chiang displayed elevated rates of anterior caries and anterior alveolar defects, suggesting an association between these two conditions. Another observation that needs to be explained is the low frequency of anterior defects in the temporally earlier samples, with the exception of Khok Phanom Di, and despite the overlap in sample date ranges.

The aetiology of alveolar defects is complex and includes the effects of carious lesions, tooth trauma, tooth wear, use of the dentition as tools and periodontal disease. As mentioned above, different protocols used to measure and record tooth wear and periodontal disease has meant that these variables could not be assessed and controlled for in this study. However, one potential contributing factor to elevated levels of anterior alveolar defects at Khok Phanom Di, late Ban Chiang and the metal period Vietnamese sites is the evidence for the use of the dentition as tools at each of these sites (Tayles 1999, Oxenham *et al.* 2002a, Pietrusewsky and Douglas 2002a). It may well be that a major behavioural shift occurred in mid to late Holocene Southeast Asia, with Khok Phanom Di being the earliest instance, involving a move to or intensification of the use of the dentition as tools, although the specific tasks involved remain unclear. Further, it appears that it may have been males, for the most part, that were more intensively involved in this apparent behavioural shift.

Behavioural considerations aside, it is difficult to assess the health implications of alveolar defects of pulpal origin. Dias and Tayles (1997) argued that only abscesses are harmful to human health and defined such abscesses as alveolar defects with fistulae or drainage sinuses. They suggested that most other defects are relatively benign and should be classified by size as either periapical granuloma (2–3 mm in diameter) or the apical periodontal cyst (>3 mm diameter), which follows. They go on to argue that the presence of these lesions indicates the host's ability to contain these centres of infection over the long term. However, it is not clear that all lesions lacking fistulae are simply benign and, with the exception of Tayles (1999), the data-recording protocols of researchers in Southeast Asia have not included criteria to assist in differentiating between clearly benign and clearly infectious alveolar lesions. Given the extreme difficulty in differentiating between cavities that are the result of chronic cysts (benign) and those that were caused by infectious processes (inimical to health), the health implications for these samples are unclear.

Antemortem tooth loss

The frequency of AMTL ranged between 2.9% and 10.4%, with a mean of 5.9%, making it the most common oral pathology. There was significant intersample heterogeneity but no clear pattern in the distribution of tooth loss. Not unlike the trend seen for carious lesions, a majority of female samples displayed higher rates of tooth loss. This is not surprising given that the frequency of AMTL will be partially dependent, to an unknown degree in each sample, on the frequency of carious lesions. However, the frequency of AMTL is also dependent on a host of other factors, including the aetiological repertoire of alveolar defects of pulpal origin. While AMTL was more common in the posterior dentition in 10/12 samples, it was significantly more so in only six of these. Interestingly, the Vietnamese Da But and Red River samples had significantly more anterior tooth loss than posterior. Both samples displayed evidence of the use of the dentition as tools (Oxenham *et al.* 2002a), although the case is much stronger for the Red River sample. It has been suggested that anterior tooth ablation was possibly practiced at Viet Nam Da But (Oxenham *et al.* 2002a) and this is certainly consistent with the elevated anterior AMTL rate. While Khok Phanom Di was characterised by significantly more posterior tooth loss, anterior tooth loss combined with ablated teeth increased to 28.3% (Tayles 1997). While anterior tooth ablation has been clearly demonstrated for Khok Phanom Di (Tayles 1997), it is unclear what the chief contributing factors were to the very high rate of anterior AMTL at late Non Nok Tha. Given the low overall rate of caries and alveolar defects at Non Nok Tha, in addition to the lack of evidence for using the teeth as tools, deliberate tooth ablation may be indicated.

In summary, the oral health profile for Southeast Asia is more complex than that described in the global literature, which demonstrates generally clear declines in oral health with the adoption or intensification of agriculture. While there are certainly important differences in the frequency and distribution of the oral pathologies examined here, the overall pattern is of general homogeneity over time. There is certainly no observable trend that could be interpreted as an improvement or decline in oral health from the mid Holocene through to early metal period in Southeast Asia.

The question, then, is why is the observed pattern of oral health seen in Southeast Asia so inconsistent with the often dramatic declines seen in other regions of the globe? Were the effects on human health of the adoption or intensification of agriculture in Southeast Asia substantively different? Is rice, the principle crop associated with agriculture in this region, less detrimental in terms of oral health than cereals such as wheat, millet or maize, which have all been associated with declining oral health in other regions of the globe?

The consequences of agriculture in Southeast Asia

There are two separate issues that need to be dealt with here. The first is the cariogenicity of rice itself and the significance of its role in oral disease. The second issue relates to the particular nature of agricultural subsistence economies, and social structure, in Southeast Asia. In other words, patterns in oral health may be reflecting a unique subsistence–social system in this region of the world. In terms of the relationship between the consumption of rice and oral health, at least one study has indicated that rice consumption, as opposed to wheat consumption, does not appear to be correlated with caries prevalence *per se* (Sreebny 1983), although high levels of rice consumption may be correlated with relatively higher caries prevalence. Krasse (1985) has also pointed out that the caries-inducing potential of rice is very low. Tayles *et al.* (2000) have recently reviewed a number of studies that examined sub-adult oral health in modern Asian countries with rice-based diets. In all cases, except those samples drawn from more recent urban settings, oral health was extremely good. Tayles *et al.* (2000) also argued that until recent times rice would have been consumed unpolished, thus reducing its cariogenic potential further through stimulating saliva flow and abrading the teeth and contributing to the loss of favoured lesion-formation sites. In the opinion of Tayles *et al.* (2000, p. 79), 'there is no obvious reason for the development and intensification of rice agriculture to have increased caries rates in the populations affected'.

Southeast Asia in the late pre-metal to metal period was not the only region and time to show a less clear-cut change in oral health with the adoption of grain foods. For example, while Lubell *et al.* (1994, p. 213) demonstrated differences in the rates of oral pathology between the mesolithic and neolithic (*c.* 7,000 years BP) periods in Portugal, they conclude that 'dental pathology is not universally a good marker of the shift to agriculture ... it is evident that European domesticates were either not as cariogenic as maize proved to be in North America or that such cariogenic sources were less completely integrated into the diet'.

Regarding the second issue, it is proposed here that the generally accepted model for declining oral health with the adoption and/or intensification of agriculture may not be appropriately tested using the currently available samples from late pre-metal and early metal period Southeast Asia. It is becoming clear that the early metal period in Southeast Asia cannot be modelled upon the hierarchical pattern favoured for complex societies in other regions of the globe. White (1995) has presented powerful arguments for viewing early metal period Southeast Asia as structurally heterarchical for the most part. Among a range of characteristics of such a structure immediately relevant to oral health, the focus of (agricultural) production being based on individual households and a general lack of centralisation of production and distribution may be significant (White 1995, p. 104). O'Connor (1995, p. 972) has characterised such pre-modern agriculturalists as 'flood-managing garden farmers'. The term refers to both garden-based horticulture and rice growing. Such a system is contrasted to the intensive wet-rice system that came to dominate the region beginning some thousand years or more ago.

It is not until the introduction of iron in the region, *c.* 2,500 years BP, that a hierarchical structure developed and became entrenched. For instance, O'Reilly (2001) has suggested that an increase in warfare is indicated by the presence of weapons in cemeteries. During this period, substantial villages developed with apparent 'moats' surrounding them. Regardless of the purpose of such earthworks, they are seen as representing large public works and the concomitant social organisation necessary for such efforts. Site density and site size also increased significantly, with some settlements attaining up to 2000 inhabitants, compared with an estimated maximum of 500 in the bronze period. O'Reilly (2001, p. 7) argued that the manner, number and nature of grave goods in the iron period indicate the development of social ranking and attendant centralisation of social structures. Trade also increased and may have been under some form of centralised control. Importantly, there is also evidence for an increase in agricultural intensification and agricultural surpluses, based on an increase in agricultural grave goods and the volume of rice found in iron period sites (O'Reilly 2001). It would, therefore, appear that the major change in Southeast Asia, in terms of human health in general and oral health specifically, should be sought at the transition from the bronze to the iron period.

Many of these arguments are perhaps not relevant in the case of northern Vietnam during the same period, and alternative answers are required. It has been suggested (Oxenham 2000), based on archaeological and historical data, that initial rice agricultural intensification in northern Vietnam was associated with Han (Chinese) moves to develop the region into a rice bowl. If such was the case, much of the agricultural output of the Vietnamese would not have been consumed locally, thus contributing to the low observed rate of carious lesions. In fact, the lowest frequency of carious lesions in Southeast Asia to date is seen in the Vietnam Red River sample. This sample, and not the Vietnam Ma and Ca rivers sample, occupies the epicentre of Han control at this time. There is also archaeological evidence and some hints from stable isotope studies (Y. Yasutomo *et al.*, unpublished data) that the Vietnamese diet in the bronze period was quite varied and not reliant on rice.

Conclusions

The chief aim of this chapter was to reevaluate recent claims for an improvement or at least the maintenance of the status quo in terms of oral health in Southeast Asia with the adoption of rice agriculture. The oral health profiles of every major dental assemblage in the region were examined and a complex albeit essentially homogeneous, for the most part, pattern across all oral health markers was demonstrated. This pattern is not what would be expected based on the often dramatic deteriorations in oral health seen with the adoption and/or intensification of agriculture in other regions of the globe. The principal reason for this apparent anomaly is argued to be related to both the intrinsic low cariogenicity of rice itself and the observation that the marked social and subsistence changes generally associated with declines in health do not occur in Southeast Asia until the adoption of iron, a period relatively poorly sampled bioarchaeologically in this region to date. Clearly, oral health profiles are sensitive to diet and lifestyles in general. The oral health of the inhabitants of Southeast Asia, as sampled in this study, is consistent with the archaeological model for socio-political heterarchy in the bronze period. Further excavation and assessment of iron period human remains will enable the hypothesis for demonstrably clearer signs of worsening health in general and oral health specifically to be tested.

Acknowledgements

We would like to thank Professor Ha Van Tan, Director of the Institute of Archaeology, Hanoi, Vietnam for permission to study curated human skeletal and dental material. We would also like to thank three anonymous reviewers for useful comments and corrections.

References

Bayard D. T. 1984. A tentative regional phase chronology for Northeast Thailand. In *Otago University Studies in Prehistoric Anthropology*, Vol. 16: *Southeast Asian Archaeology at the XV Pacific Science Congress*. Dunedin: University of Otago Press, pp. 87–128.

Boyd W. E. and McGrath R. 2001. Iron Age vegetation dynamics and human impacts on the vegetation of Upper Mun River floodplain, NE Thailand. *New Zealand Geographer* **57**: 21–32.

Cohen M. N. and Armelagos G. J. (eds.). 1984. *Paleopathology at the Origins of Agriculture*. Orlando, FL: Academic Press.

Dias G. and Tayles N. 1997. 'Abscess cavity': a misnomer. *International Journal of Osteoarchaeology* **7**: 548–554.

Domett K. M. 1999. Health in late prehistoric Thailand. Ph.D. thesis, University of Otago, Dunedin.

2001. *British Archaeological Reports International Series* No. 946: *Health in Late Prehistoric Thailand.* Oxford: Archaeopress.

Douglas M. T. 1996. Paleopathology in human skeletal remains from the pre-metal, bronze and iron ages, Northeastern Thailand. Ph.D. thesis, University of Hawaii.

Erdal Y. S. and Duyar I. 1999. A new correction procedure for calibrating dental caries frequency. *American Journal of Physical Anthropology* **108**: 237–240.

Higham C. F. W. 1989. *The Archaeology of Mainland Southeast Asia.* Cambridge, UK: Cambridge University Press.

1996. *The Bronze Age of Southeast Asia.* Cambridge, UK: Cambridge University Press.

1998. The faunal remains from mortuary contexts. In Higham C. F. W. and Thosarat R., eds., *University of Otago Studies in Prehistoric Anthropology*, Vol. 18: *The Excavation of Nong Nor, a Prehistoric Site in Central Thailand.* Dunedin: University of Otago Press, pp. 315–320.

Higham C. F. W. and Thosarat R. 1998. *Prehistoric Thailand from Early Settlement to Sukhothai.* Bangkok: River Books.

Hillson S. 2001. Recording dental caries in archaeological human remains. *International Journal of Osteoarchaeology* **11**: 249–289.

Kelley M. A., Levesque D. R. and Weidl E. 1991. Contrasting patterns of dental disease in five early northern Chilean groups. In Kelley M. A. and Larsen C. S., eds., *Advances in Dental Anthropology.* New York: Wiley-Liss, pp. 203–213.

Kilgore L. 1989. Dental pathologies in ten free-ranging chimpanzees from Gombe National Park, Tanzania. *American Journal of Physical Anthropology* **80**: 219–227.

Krasse B. 1985. The cariogenic potential of foods: a critical review of current methods. *International Dental Journal* **35**: 36–42.

Larsen C. S. 1995. Biological changes in human populations with agriculture. *Annual Review of Anthropology* **24**: 185–213.

1997. *Bioarchaeology: Interpreting Behavior from the Human Skeleton.* Cambridge, UK: Cambridge University Press.

Larsen C. S., Shavit R. and Griffin M. C. 1991. Dental caries evidence for dietary change: an archaeological context. In Kelley M. A. and Larsen C. S., eds., *Advances in Dental Anthropology.* New York: Wiley-Liss, pp. 179–202.

Lovell N. C. 1990. Skeletal and dental pathology of free-ranging mountain gorillas. *American Journal of Physical Anthropology* **81**: 399–412.

1991. *Patterns of Injury and Illness in the Great Apes: A Skeletal Analysis.* Washington, DC: Smithsonian Institution Press.

Lubell D., Jackes M., Schwarcz H., Knyf M. and Meiklejohn C. 1994. The Mesolithic–Neolithic transition in Portugal: isotopic and dental evidence of diet. *Journal of Archaeological Science* **21**: 201–216.

Lukacs J. R. 1989. Dental paleopathology: methods for reconstructing dietary patterns. In Iscan M. Y. and Kennedy K. A. R., eds., *Reconstruction of Life from the Skeleton*. New York: Wiley-Liss, pp. 261–286.

1992. Dental paleopathology and agricultural intensification in South Asia: new evidence from Bronze Age Harappa. *American Journal of Physical Anthropology* **87**: 133–150.

1995. The 'caries correction factor': a new method of calibrating dental caries rates to compensate for ante-mortem loss of teeth. *International Journal of Osteoarchaeology* **5**: 151–156.

Martin D. L., Armelagos G. J., Goodman A. H. and van Gerven D. P. 1984. The effects of socioeconomic change in prehistoric Africa: Sudanese Nubia as a case study. In Cohen M. N. and Armelagos G. J., eds., *Paleopathology at the Origins of Agriculture*. Orlando, FL: Academic Press, pp. 193–214.

Molnar S. and Molnar I. 1985. Observations of dental diseases among prehistoric populations of Hungary. *American Journal of Physical Anthropology* **67**: 51–63.

Moore W. F. 1993. Dental caries in Britain from Roman times to the nineteenth century. In Geissler C. G. and Oddy D. F., eds., *Food, Diet and Economic Change Past and Present*. Leicester, UK: Leicester University Press, pp. 50–61.

Oxenham M. F. 2000. Health and behaviour during the mid-Holocene and metal period of Northern Viet Nam. Ph.D. thesis, Northern Territory University, Darwin, Australia.

Oxenham M. F., Nguyen L. C. and Nguyen K. T. 2002a. Oral health in Northern Viet Nam: neolithic through metal periods. *Indo-Pacific Prehistory Association Bulletin* **22**: 121–134.

Oxenham M. F., Locher C., Nguyen L. C. and Nguyen K. T. 2002b. Identification of *Areca catechu* (betel nut) residues on the dentitions of Bronze Age inhabitants of Nui Nap, Northern Vietnam. *Journal of Archaeological Science* **29**: 909–915.

O'Connor R. A. 1995. Agricultural change and ethnic succession in Southeast Asian studies: a case for regional anthropology. *Journal of Asian Studies* **54**: 968–996.

O'Reilly D. J. W. 2001. From the Bronze Age to the Iron Age in Thailand: applying the heterarchical approach. *Asian Perspectives* **39**: 1–19.

Pietrusewsky M. and Douglas M. T. 2002a. *University Museum Monograph 111: Ban Chiang, a Prehistoric Site in Northeast Thailand. I: the Human Skeletal Remains*. Philadelphia, PA: University of Pennsylvania Press.

2002b. Intensification of agriculture at Ban Chiang: is there evidence from the skeletons? *Asian Perspectives* **40**: 157–178.

Rodman P. S. 1988. Diversity and consistency in ecology and behavior. In Schwartz J. H., ed., *Orang-utan Biology*. New York: Oxford University Press, pp. 31–51.

Rose J. C., Burnett B. A., Nassaney M. and Blaeuer M. W. 1984. Paleopathology and the origins of maize agriculture in the lower Mississippi Valley and Caddoan culture areas. In Cohen M. N. and Armelagos G. J., eds.,

Paleopathology at the Origins of Agriculture. Orlando, FL: Academic Press, pp. 393–424.

Rose J. C., Marks M. K. and Tieszen L. L. 1991. Bioarchaeology and subsistence in the central and lower portions of the Mississippi Valley. In Powell M. L., Bridges P. S. and Mires A. M. W., eds., *What Mean These Bones: Studies in Southeastern Bioarchaeology*. Tuscaloosa, AL: University of Alabama Press, pp. 7–21.

Saunders S., DeVito C. and Katzenberg A. 1997. Dental caries in nineteenth century Upper Canada. *American Journal of Physical Anthropology* **104**: 71–88.

Smith P., Bar-Yosef O. and Sillen A. 1984. Archaeological and skeletal evidence for dietary change during the late Pleistocene/early Holocene in the Levant. In Cohen M. N. and Armelagos G. J., eds., *Paleopathology at the Origins of Agriculture*. Orlando, FL: Academic Press, pp. 101–136.

Sreebny L. M. 1983. Cereal availability and dental caries. *Community Dentistry and Oral Epidemiology* **11**: 148–155.

Tayles N. 1997. Tooth ablation in prehistoric Southeast Asia. *International Journal of Osteoarchaeology* **6**: 333–345.

1999. *Report of the Research Committee* LXI. *The Excavation of Khok Phanom Di, a Prehistoric Site in Central Thailand*, Vol. V: *The People*. London: Society of Antiquaries.

Tayles N., Domett K. and Hunt V. 1998. The people of Nong Nor. In Higham C. and Thosarat R., eds., *University of Otago Studies in Prehistoric Anthropology*, No. 18: *The Excavation of Nong Nor, a Prehistoric Site in Central Thailand.* Dunedin: University of Otago Press, pp. 321–368.

Tayles N. Domett K. and Nelsen K. 2000. Agriculture and dental caries? the case of rice in prehistoric Southeast Asia. *World Archaeology* **32**: 68–83.

Turner C. G. II. 1979. Dental anthropological implications of agriculture among the Jomon people of central Japan. *American Journal of Physical Anthropology* **51**: 619–636.

Ungar P. S. 1994. Incisor microwear of Sumatran anthropoid primates. *American Journal of Physical Anthropology* **94**: 339–363.

Walker P. L. and Hewlett B. S. 1990. Dental health, diet and social status among central African foragers and farmers. *American Anthropologist* **92**: 383–398.

Wiriyaromp W. 1984. The human skeletal remains from Ban Na Di: patterns of birth, health and death in prehistoric North East Thailand. M.A. thesis, University of Otago, Dunedin.

White J. C. 1995. Incorporating heterarchy into theory on socio-political development: the case from Southeast Asia. In Ehrenreich R. M., Crumley C. L. and Levy J. E., eds., *Archaeological Papers of the American Anthropological Association*, No. 6 *Heterarchy and the Analysis of Complex Societies*. Arlington, VA: American Anthropological Association, pp. 101–123.

Zarr Z. H. 1974. *Biostatistical Analysis*. Englewood Cliffs, NJ: Prentice Hall.

12 *Cranial lesions on the late Pleistocene Indonesian* Homo erectus *Ngandong 7*

ETTY INDRIATI
Gadjah Mada University, Yogyakarta, Indonesia

Introduction

More than five decades have passed since Weidenreich's brief description of the cranial lesions on the Indonesian *Homo erectus* Solo skull VI, now known as Ngandong 7 (Ng7), in his monograph *The Morphology of Solo Man* (1951). The Ng7 skull does not have a mandible or associated post-cranial bones. Re-examination of the calvarial lesions on Ng7 invites further explanation, as Weidenreich (1951, p. 238) only briefly described the lesion on the left parietal bone without further analysis or a differential diagnosis.

> The lesion consists of very irregular tuberosities and, in two places, of deep erosions, none of which perforate the bone. The entire lesion area is circumvallated by a smooth, rounded bone wall. All these details suggests that the lesion was originally an inflamed wound with considerable consumption of bone substance; the smooth surrounding wall indicates that healing had begun at the time of death.

The Ng7 skull was one of 12 Solo skulls discovered between September 1931 and November 1933, at Ngandong, a small area along the left bank of the Solo river, about 6 miles north of Ngawi, East Java, Indonesia. Weidenreich (1951) reported that the hominid remains from Ngandong were irregularly distributed throughout the whole site, rather than being from a particular area or a special layer of excavation. The Ngandong fauna was discovered in a terrace about 20 m above the Solo river. They consisted of 13 vertebrate genera including the hominid under discussion here. Artefacts commonly found at Ngandong included perfectly round unpolished stone balls made of andesite, varying from 10 to 22 cm in diameter or even larger. This type of stone ball is also found in lower and middle Pleistocene remains at Sangiran (Jacob 1973, 1975). Ng7 was listed as excavation number 9975 and was discovered on 13 June 1931. The skulls were first described by Dubois (1937). The function of the stone balls remains unclear but they may have been used as tools, for hunting or even for playing a game.

Bioarchaeology of Southeast Asia. Marc Oxenham and Nancy Tayles.
Published by Cambridge University Press.

The Ngandong remains were initially found to be about 100,000 years old, which would date them to the late Pleistocene era (von Koenigswald 1951). However, more recent investigations using electron spin resonance and mass spectrometric U-series of fossilised bovid teeth collected from the hominid-bearing levels at Ngandong have resulted in significantly younger dates, in the order of 53,000–27,000 years BP (Swisher *et al.* 1996). According to Jacob (in Oakley *et al.* 1975), Ng7 was a young female and the frontal, parietals, occipitals, temporals and sphenoids were preserved. The skull has been reconstructed and is one of the most complete skulls in the Ngandong series (Indriati 2003a). On the frontal surface, the left supraorbital aspect is broken; and on the sphenoid, the right pterygoid plate is missing. The external table shows erosion and cracks. Previous studies considered Ngandong hominids as Neandertal and/or Neandertal variants (Oppenoorth 1937, Brace 1967, Brose and Wolpoff 1971, Mann and Trinkaus 1974) but more recently, anthropologists have predominantly clustered Ngandong hominids under *H. erectus* (e.g. Santa Luca 1980, Rightmire 1993, Antón 1999). Compared with Indonesian hominids from other sites such as Sangiran and Modjokerto, the Ngandong hominids were the most morphologically advanced (Antón 1999).

Methods and results

The cranium was scanned with a three-dimensional UMAX scanner in order to map the size and location of the lesions. The scanned picture allowed mapping of the outline of the lesions and each lesion was assigned a number for the purposes of discussion (Fig. 12.1). Observation of the cranium of Ng7 revealed the multifocal nature of the lesions, which are located on the parietal and occipital bones. The lesions on the parietals were designated numbers 1–4 (Figs. 12.1 and 12.2; Table 12.1), and a lesion on the occipital was designated 5 (Fig. 12.3 and Table 12.1) (Indriati 2002). Right lateral and frontal views of the whole skull (Fig. 12.4) assist in showing the locations and depths of the lesions on the vault.

Lesion 1 (Figs. 12.1 and 12.2) is a large discrete circular area with an anteroposterior diameter of 55.98 mm and a mediolateral diameter of 51.25 mm, thus affecting almost one-third of the left parietal bone surface. The lesion is at the sagittal suture, with a perpendicular distance of the lesion to the coronal suture of 31.26 mm, to the lambdoid suture of 12.35 mm and to the temporal line of 15.23 mm. This lesion consists of abnormal bone loss mixed with abnormal bone formation, with clear marginal definition. The combination of abnormal bone loss and bone

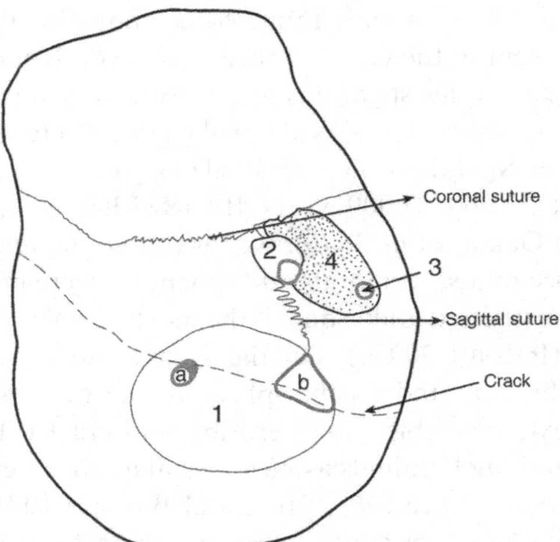

Figure 12.1. Schematic diagram of lesions on the cranium of Ngandong 7.

formation has created undulations on the base of the circular lesion. These irregular, 'wrinkled' undulations might have been produced by woven bone indicative of reaction to infection. Postmortem damage impeded observation of the lesion character. The thinning of the outer table created a slight depression, but there is no endocranial indentation. The bone loss affected the external table and the diploë but did not penetrate the internal table. However, the computed tomographic (CT) scan shows the external table, diploë and internal table were all affected, as evidenced by the homogeneity of the mineralisation, with the tables and diploë appearing as one layer. The external table, diploë and internal table appear very dense overall. The reactive bone formation is smooth and undulating. The border of the circular lesion is smooth, striated and well defined. Within the circle, there are two depressions: one ovoid (a) and one triangular (b) (Fig. 12.1). The ovoid depression is located on the anterolateral part of the circle, and the triangular depression is located in the medial wall. The ovoid depression is 8.48 mm and 10.70 mm in mediolateral and antero-posterior diameters, respectively. The depth of the ovoid depression is 1.89 mm. The sagittal side of the triangular depression is 11.53 mm long, the anterior side is 8.95 mm long, and the posterior side is 9.51 mm long. The depression is 2.11 mm in depth.

Lesion 2 (Figs. 12.1 and 12.2) is an ovoid lesion, smaller than lesion 1. It is located about 4.10 mm posterior to bregma and straddles the sagittal suture, thus affecting the suture interdigitation posterior to bregma. The

(a)

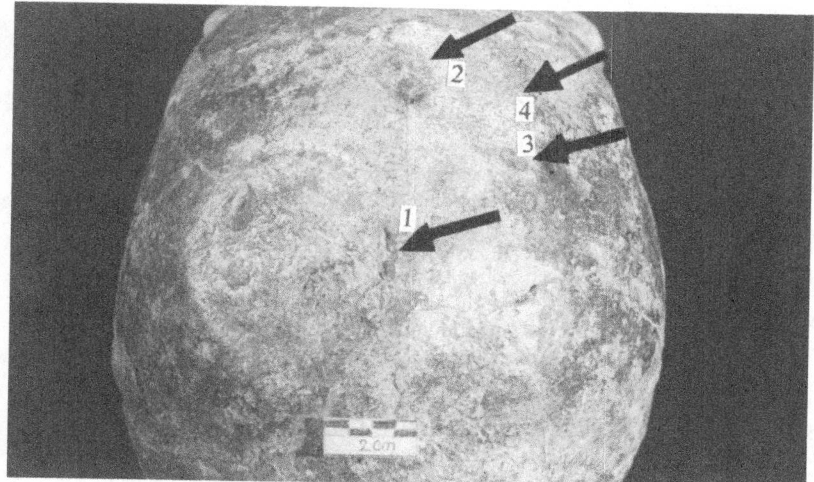

(b)

Figure 12.2. Parietal lesions on the cranium of Ngandong 7. (a) Superior view of lesions 1–4. (b) The same view in close up; note the circularity of lesion 1 (arrow).

anteroposterior length of this lesion is 19.31 mm and the mediolateral width is 16.41 mm. The greatest depth is 2.29 mm, and the base resembles a cyst. An elevated undulation running anterolaterally has divided the depression into two areas. The border of lesion 2 is clearly defined, and

Table 12.1. *Lesion character on Indonesian* Homo erectus *Ngandong 7 calvarium*

Lesion	Size (mm)	Location	Character
1	51 × 56	Left parietal	Margin is well defined. Base is undulating, non-perforating, and circular. Wall is smooth. Lesion shows mixed abnormal bone formation and loss; woven bone reaction is irregular (appears 'wrinkled'). External table, diploë and internal table are affected (CT shows homogenous mineralisation). No endocranial indentation. No raised tissue surrounding lesion
2	16 × 19	Posterior to bregma	Margin is well defined and elliptic. Base is rough, without perforation, similar in appearance to multiple cysts. External table, diploë and internal table are affected (CT shows homogenous mineralisation), thinning of external table, slight depression of external table. No endocranial indentation, no raised structure surrounding lesion
3	7 × 9	Right parietal	Margin is diffuse. Base shows erosion mixed with hypervascularity, only surface of external table affected, no perforation
4		Right parietal	Margin is diffuse. Base shows no perforation; erosion mixed with hypervascularity. Surface of external table only affected. Lesion is in the form of a continuous depression connecting lesions 2 and 3
5	9.43 × 7.69	Left occipital	Non-pathological lesion. Margin is round and well defined. Base is perforated. Wall is smooth. No evidence of reactive bone

CT, computed tomography.

the wall is smooth. A CT scan of lesion 2 indicated that the internal aspect of the sagittal and coronal sutures were not completely fused (Balzeau *et al.* 2003), concordant with the external aspect. The width of the coronal suture is approximately 2 mm. Zimmerman (1999) reported that 2 mm is the normal width of this suture, with traumatic separation when a fracture (injury) extends into the suture (diastasis of suture) only indicated when the sutural width is more than 3 mm. Figure 12.4 shows the slight depression of the external tables, with no raised margins, of lesions 1 and 2.

Lesion 3 (Figs. 12.1 and 12.2) was probably an extension of lesion 2. It is located posterolaterally to lesion 2 and approximately 23.46 mm perpendicular to the sagittal suture. Lesion 3 is smaller than either lesion 1 or 2. Its anteroposterior diameter reaches 7.11 mm, and its mediolateral diameter reaches 8.67 mm, with a depth of 0.63 mm. The border of lesion 3 is not clearly defined, but it is smooth and ovoid shaped anterolaterally. Lesion 3

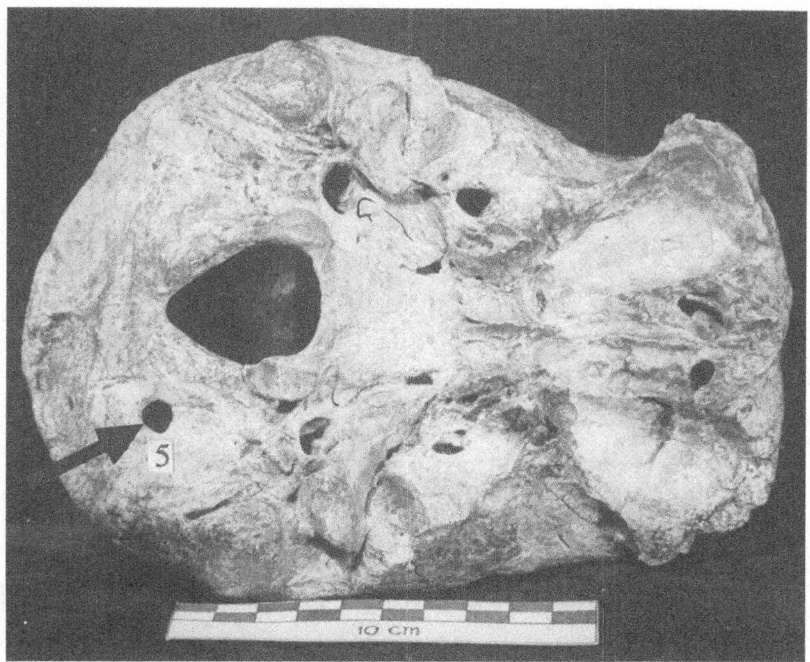

Figure 12.3. Inferior view of the cranium of Ngandong 7, showing lesion 5 (indicated by the arrow) on the left occipital bone; this is probably an artificial lesion.

can be characterised as a slight depression with porosity and abnormal bone loss. Although there is some postmortem damage, lesion 3 looks as if it was in the early stages of formation. The inferred continuation of lesion 2 into lesion 3 has been based on palpation: as the lesion was palpated by running bare fingers from lesion 2 to lesion 3 (posterolaterally), a continuous depression connecting both lesions could be identified.

On Fig. 12.1, the shaded area depicts the slight depression area that connects lesion 2 to lesion 3, and so it was labelled as lesion 4 (see also Fig. 12.2b). Lesion 4 is basically an area of porosity with slight depression. The porosity suggests hypervascularity in response to inflammation, which is blurred by taphonomic erosion. The indistinct boundary of lesion 4 is ovoid shaped posterolaterally.

Another lesion (5) is located on the occipital bone and is very likely to be artificial (non-pathological) (Fig. 12.3). The lesion is 9.43 mm (anteroposterior) and 7.69 mm (mediolateral) in diameter. The location of this lesion is 16.55 mm perpendicular to the foramen magnum, 28.08 mm from

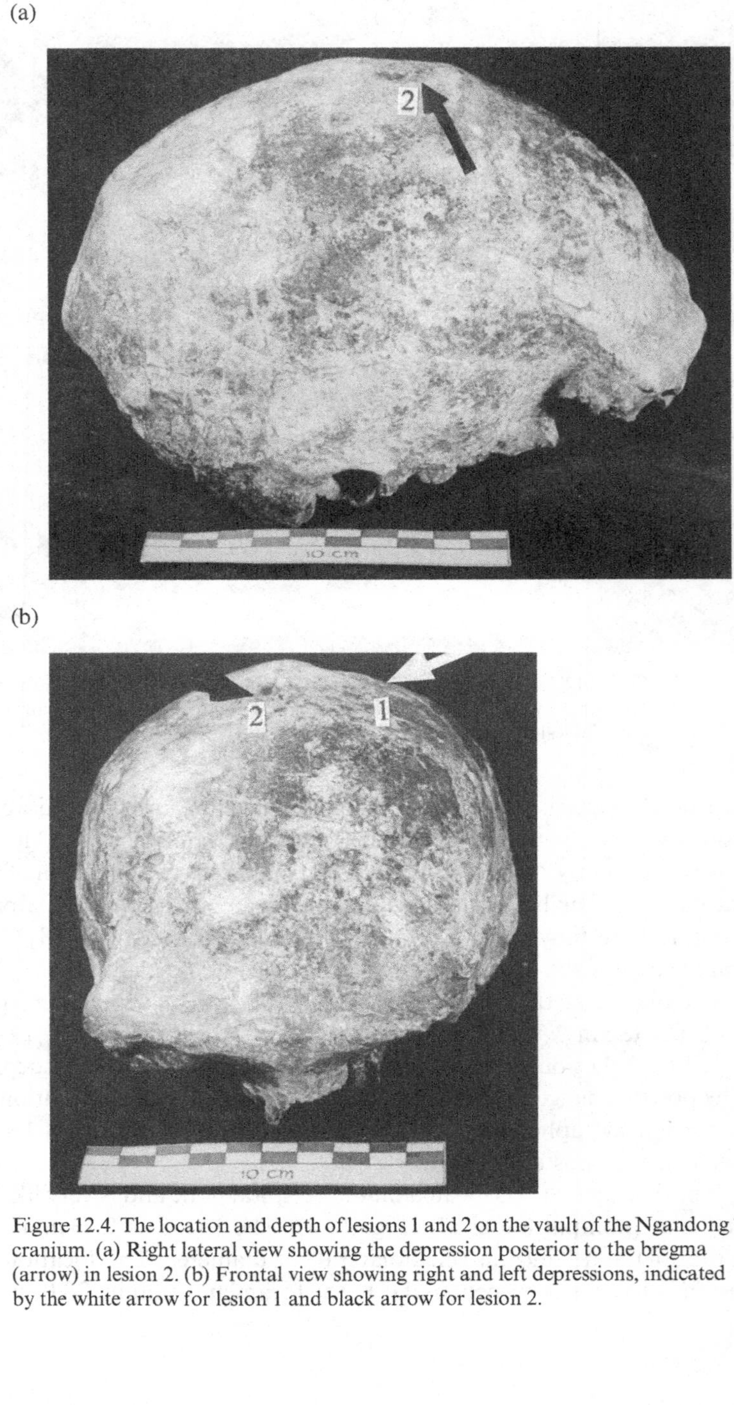

(a)

(b)

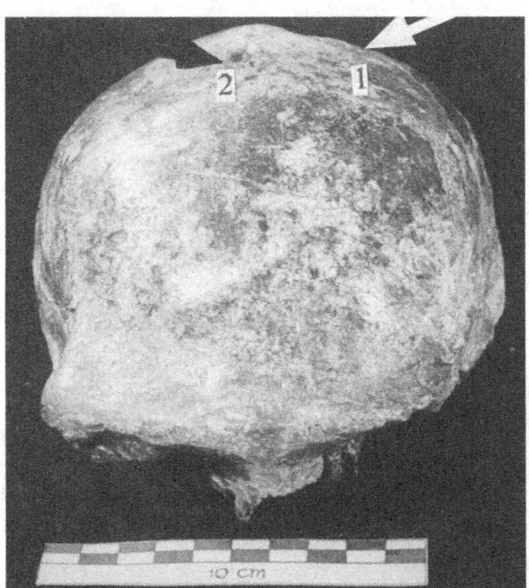

Figure 12.4. The location and depth of lesions 1 and 2 on the vault of the Ngandong 7 cranium. (a) Right lateral view showing the depression posterior to the bregma (arrow) in lesion 2. (b) Frontal view showing right and left depressions, indicated by the white arrow for lesion 1 and black arrow for lesion 2.

the mastoid process and 47.33 mm from the external occipital protuberance. The hole has penetrated the bone, with a smooth wall on the lateral and anterior area. This hole does not meet the pathological criteria of holes in the calvarium (Kaufman *et al.* 1997), and there was no evidence of bone reaction.

Discussion

Lesion interpretation

To arrive at a conclusion about the cause of cranial lesions, we must first ask what types of disease or trauma manifest in skulls. We must then ask how cranial bones react to disease and trauma. This section discusses these two questions in order to interpret the cranial lesions evident on Ng7.

First, it must be noted that the bones of the calvarium have different reactions to cranial trauma and disease. For instance, calvaria react differently to neoplastic and infectious diseases. Neoplasms often cause general disruptions, such as irregular bone growth or large perforation on skulls, whereas infections often manifest themselves in a more specific manner, such as the appearance of stellate scars in venereal syphilis, and sequestra and involucra as a result of osteomyelitis (Steinbock 1976). Trauma on skulls often results in perforation or fracture, whereas metabolic disturbances could cause hyperdensity and thickening of skull vaults.

Because the subject of study here consists merely of a cranium with no mandible or postcranial bones, we will narrow the focus to diseases and trauma that affect skulls, and the reaction of the skull structure to these diseases and trauma. While recognising that the pattern of lesions in the whole skeleton is an important contributor to a differential diagnosis, in the absence of the rest of the skeleton this discussion will of necessity proceed based on cranial evidence. The discussion will focus on the cranial reaction to systemic conditions, neoplasms, trauma, infection and post-mortem alteration (taphonomy). Taphonomic effects will be discussed because of the fossilised state of the subject studied.

Cranial bone reaction to systemic conditions

The types of disease manifested in calvaria are generally classified into six categories (Olmsted 1981): tumours, infections, vascular diseases, metabolic (endocrinological) disorders, congenital disorders and idiopathic diseases.

Based on radiological examinations, Olmsted (1981) reported that the most common systemic diseases affecting the skull include fibrous dysplasia, hyperparathyroidism, Paget's disease, histiocytosis, sickle cell anaemia and thalassaemia. Clinical studies generally categorise cranial bone reaction to disease into pagetoid (56%), sclerotic (23%) and cystic (21%), based on radiological observation (Fries 1957). More recent clinical studies on cranial diseases using magnetic resonance imaging (MRI) and CT have described tumours, trauma, and infections affecting skulls (Lee *et al.* 1999). Olmsted (1981) use Fries (1957) categories to describe cranial bone reaction to diseases. The following radiographic evidence of calvarial bone reaction to systemic disease is summarised from Olmsted (1981).

- fibrous dysplasia is characterised by very dense sclerosis of the skull base
- hyperparathyroidism is shown by a classic 'salt and pepper' pattern and by the absence of a distinct outer table
- Paget's disease leads to sclerotic patches, which, when coalesced, yield a cotton-wool appearance, accompanied by thickening of the skull vault and basilar invagination
- sickle cell anaemia and thalassaemia commonly lead to diploic widening.

In contrast, fibrocystic diseases affect the skull by producing round or oval cystic spaces without sclerotic boundaries in the diploë. Fibrocystic diseases include eosinophilic granuloma, Hand–Schüller–Christian disease and Letterer–Siwe disease. As none of these reactions are present in the Ng7 calvarium, we must look elsewhere for a diagnosis.

Cranial bone reactions to neoplastic processes

Common neoplastic processes leading to skull lesions include congenital epidermal inclusion cysts, osteoma, meningioma, multiple myeloma and metastasised tumours (Ortner and Putschar 1981, Kaufman *et al.* 1997). Table 12.2 lists the condition, type and nature of lesions, and the usual lesion location of tumour manifestations on skulls. Neoplasms have been recorded in some ancient people and earlier hominid fossils (e.g. DuBoulay 1965, Strouhal 1976, Deeley 1983, Tkocz and Bierring 1984, Grupe 1988, Gladykowska-Rzeczycka 1991, Ortner *et al.* 1991, Roberts and Manchester 1997). None of the skeletal descriptions of neoplasms in these ancient remains resembles the lesion on Ng7.

Cranial CT scanning of the Ng7 calvarium showed that all three cranial layers (external table, diploë, and internal table) were affected at the sites of both lesion 1 and lesion 2. The three cranial layers had become

Table 12.2. *Tumour manifestations on skulls*[a]

Condition	Type	Nature of lesion	Location
1. Congenital epidermal inclusion cyst	Cyst	Round lytic, single, perforate	Skull vault
2. Osteoma	Blastic	Dense lamellar bone; single/ multiple	Frontal, parietal, ear canal
3. Meningioma	Arises from mesothelial lining cells of dura mater	Radiating spicules; hyperostosis; massive, reactive bone, but no internal table spiculae; as in osteosarcoma multiple lysis, perforated	Skull base
4. Osteosarcoma	Lytic to massive sclerotic	Cortical destruction; hypervascularity; dense radiating layers of reactive bone	Skull vault: rare
5. Chordoma	Lytic	Large lytic defect; no mineralised matrix	Skull base
6. Metastatic tumour	Blastic/lytic; breast origin is mostly lytic; prostate origin is mostly blastic; lung origin is mixed blastic and lytic, a fine spongy lytic process, multifocal	Strong osteoblastic response, new bone replacing old trabeculae; diffuse density on fast growth, or lytic on slow growth	Spine 80%; skull, pelvis 25%
7. Myeloma	Lytic, from plasma cell of the bone marrow	Multiple lytic, perforations with diameter 1–2 cm	Skull vault

[a]Sex and age predilection: 1, none; 2, none; 3, none; 4, male:female ratio 2:1; affects young adult/adolescents; 5, males >40 years and a male:female ratio 2:1 in young adult/adolescents; 6, none; 7, males and females equally; affects >40 years.

homogenously mineralised. As with the analysis above, none of the bone reactions similar to the neoplastic processes described in Table 12.2 was present in the skull of Ng7, meaning that neoplasm must also be excluded as an explanation for the cranial lesions.

Cranial bone reactions to traumatic injury

Lesions on crania from traumatic injury can either be perforated or non-perforated. When perforated, the trauma is usually the result of contact with a sharp object. When non-perforated, a depressed fracture is the most common result. The Ng7 calvarium has a slight depression without any endocranial indentation, but this might not have been caused primarily by trauma as commonly defined in palaeopathology. Although the categorisation of trauma in the literature varies among authors, we have used Steinbock's (1976) scheme. He divided trauma in palaeopathology into five categories, of which three are relevant to the skull: fractures, crushing injuries and bone wounds caused by sharp instruments (including arrow and spear wounds, scalping, trephination, sincipital T scarring and amputation). The cranial vault lesion on Ng7 does not exhibit evidence fitting any of these specific categories of trauma. However, accidental scalp trauma with consequent fatal infection was reported by Ortner and Putschar (1981) and exhibited undulating lesions indicative of bony response to granulation tissue and hypervascularisation that resembled those of Ng7. It is compelling to suggest that Ng7 might have suffered partial scalp trauma.

Cracks also occurred on the cranial vault of Ng7, running diagonally from the left to the right parietal, from anterolateral to posterolateral, across lesion 1 (Fig. 12.1, crack line). The frontal view of the CT scan shows the crack affected the external table, the diploë and the internal table. The occipital bone also has a crack apparent on the sagittal CT scan, superior to the external occipital protuberance. The jagged and torn edges and the lack of bone reaction in these cracks suggest that they occurred postmortem, most likely from soil compression postfossilisation. A less likely explanation is that they were caused through accidents in the transport of the skull across continents before it was returned to its homeland in 1976 (Indriati 2003) from other depositories in Europe and the USA. The disadvantage of a fossil being transported is apparent in the Sambungmacan 3 skull, which had more cracks after travelling from New York to Yogyakarta (Indriati 2003), perhaps caused by the airline engine's vibration, despite the fact that it was very well packed. The jagged fracture on Ng7 suggests that it occurred in dry bone rather than in green bone. Galloway *et al.* (1999) have reported that a clean cut is indicative of fresh bone fracture in forensic cases. Soil compression appears to have caused the crack on the large lesion 1, since the external table was thinned at this point and thus would have been more vulnerable to cracking. The lack of outbending of the external table on the periphery of the lesion suggests that the crack was not caused by a forceful impact to the area.

To further examine whether Ng7 experienced trauma, Walker's approach to cranial injuries was adopted. Walker (1989) used four criteria to identify traumatic injuries: (a) an absence of reactive bone indicative of infectious aetiology (also indicates fatal trauma); (b) a tendency for the injuries to be single, with a lack of lesions elsewhere in the skeleton suggestive of systemic infection; (c) a well-delineated circular or ellipsoidal shape of many of the lesions; and (d) the retention in some cases of fracture lines at the periphery of the depressed area. The Ng7 cranium shows one indicator of trauma matching Walker's criteria, that of the circular and ellipsoidal shape of lesions 1 and 2. The reactive bones indicative of infectious aetiology are present in the form of an irregularly undulating base. Lesions are multiple, and there are no fracture lines at the margins of the slightly depressed areas. The absence of fracture lines radiating out from the rounded indentation differentiate the lesions from the evidence of cranial injuries in the Zhoukoudian fossils described by Weidenreich (1939, 1943). In addition, Ng7 does not suffer the depressed fracture that commonly occurs in cranial injuries as described by Walker (1989) and Webb (1995). Only the circular and ellipsoidal shape of the calvarial lesions on Ng7 matches the criteria of cranial injury. In this case, the presence of reactive bone indicative of infectious aetiology led the author to examine the possibilities of infection as a more appropriate diagnosis of the lesions on the cranial vault.

Cranial bone reactions to infection

Diseases affecting the skeleton demonstrate a predilection toward manifestations in certain locations: some affect only the skull, some the spinal column, some the extremities and some multiple regions. Nevertheless, conclusions are complicated by the fact that a single disease can produce different effects on different bones in an individual. For instance, bone changes resulting from *Echinococcus* infection are different in the long bones than in the flat bones (Ortner and Putschar 1981).

Some diseases also predilect different geographical regions (latitude, climate), and according to nutritional intake, occupation, population density, sociocultural norms, sex and age. Table 12.3, summarised from Ortner and Putschar (1981), lists the nature of lesions on skulls in numerous infectious diseases affecting bones. I would warn against oversimplification and conclusions must be drawn cautiously as different diseases can lead to similar lesions on bones. However, the table is intended to narrow down the possibilities while attempting to identify lesions of infectious origin on the Ng7 skull.

Among the 11 categories of infectious diseases that can affect the skeleton (Ortner and Putschar 1981), the most apparent systemic

Table 12.3. *Lesions characteristic of infection affecting the skull*

Infection	Nature of lesion on skull
Bacterial infections	
Osteomyelitis	Rare on skull; perforation, sequestra, involucra, diffuse
Periostitis	Intact diploë; periosteal woven bone formation, diffuse
Glanders	Similar to tertiary syphilis and leprosy
Tuberculosis	Rare on skull except in young children; absence or limited bony reaction, perforation, irregular margin, active resorption, defect on inner table is larger than outer table
Leprosy	Rare in vault, often in nasal bones, hands and feet; bone resorption, irregular margin
Treponematosis	Multiple cavitations, sclerosis, reactive bone, diffuse, caries sicca, gumma, sequestra (venereal syphilis and yaws)
Actinomycosis	Mastoid process, petrous part of temporal bone; diffuse cortical perforation
Fungal infections	
Cryptococcosis	Well-circumscribed lytic lesion on cranial bones
Coccidioidomycosis	Lytic lesions with central cavitation and osteoporosis
Sporotrichosis	Periosteal lesion: focal or extensive periostitis
Viral infections	
Rubella	Skull may show fontanel enlargement and poor mineralisation
Parasitic infections	
Echinococcosis	In flat bones, cystic and the outer contour is often expanded by slow destruction of the old cortex and formation of new reticulated cortex
Sarcoidosis	In adolescents and young adults; nasal and maxillary bones; purely lytic, round 1–10 mm; minimal or no periosteal bone formation

From Ortner and Putschar (1981).

infections manifested in the skull are syphilis and osteomyelitis, with tuberculosis also showing a relatively high frequency (Table 12.3). The typical lesions resulting from syphilis, osteomyelitis and tuberculosis – such as stellate scars, sequestra and involucra, and perforation – are not apparent in the calvarial lesions of Ng7. Infections of the skull often have focal infections on teeth; for instance, cocci bacteria in caries leading to systemic infection. This is suggested in early clinical studies (Blum 1924, Wilensky 1932).

Another type of infection, not listed in Table 12.3, is that of the scalp, which can be manifested in the skull mainly in the form of reaction to granulation tissue. Ortner and Putschar (1981) included scalp infections as secondary to trauma, as does Steinbock (1976). Compiled from Reese (1940), Hamperl and Laughlin (1959), Hamperl (1967), Morse (1969),

Steinbock (1976) and Ortner and Putschar (1981), the sequence of skull bone changes in scalping includes: (a) deprivation of periosteal blood supply to the external table; (b) necrosis of the external table; (c) granulation and new bone formation occurring to start healing of the deeper layer; (d) granulation and separation of the new bone, shedding the necrotic part of the external table; (e) external table losing its regular thickness, becoming thin, but appearing dense because of new bone formation, while a slight depression of spongy bone occurs; (f) internal table often not perforated and no indentation of it is apparent; (g) bone surface shows irregular erosion/granulation/osteitis/necrosis on the affected area, often elliptic, oblong or round, where the margin can be sharply circumscribed or diffuse. The size of calvarial bone scarring on scalping measures about 7 cm × 10 cm according to Hamperl and Laughlin (1959), and 8 cm × 10 cm according to Morse (1969), indicating large areas are affected. Scalping can be accidental (e.g. the case of a factory worker who had her hair trapped in an industrial machine and survived for 8 months (Ortner and Putschar 1981)) or intentional (Hamperl 1967).

Lesion 1 on the Ng7 cranial vault might have been from a scalp infection following an open wound in the affected area. This is suggested by the nature of the lesion (Table 12.1). The 5.1 cm × 5.6 cm round lesion matches the skull bone changes in scalping, especially the irregularly wrinkled and undulated base, which might have been caused by formation of granulation tissue and separation of the new (woven) bone, shedding a necrotic part from the external table. Lesion 2, however, resembles cyst formation. Cyst formation is a common effect of infection by coccal bacteria such as *Echinococcus* spp. or by fungi (Ortner and Putschar 1981). The most likely explanation for these lesions on Ng7 is, therefore, traumatic injury causing an open wound in the scalp and subsequent infection by bacteria, fungi or parasites. The cause of the open wound trauma, however, remains unclear; it might have been caused by falling rock or any of the other many mishaps that can lead to scalp infection.

The andesite stone balls found at Ngandong might have been a possible cause of the injury, although lesion 1 on Ng7 skull also resembles an almost healed trephination. While tools to perform trephination, such as iron knives, were not available during the late Pleistocene, trephination can be performed using stone tools such as flaked stones, especially flint or obsidian (Lisowski 1967). Scraping the osseous tissue away using a stone flake will result in widely bevelled edges and powdery bones (Lisowski 1967). Lesion 1 does not have bevelled edges and diffuse indentations such as those described by Lisowski (1967, his Fig. 2), but the possibility of healed trephination in lesion 1 should not be dismissed as a differential diagnosis.

Another possible case of infectious disease in *H. erectus* in the late Pleistocene has been reported from Gongwangling, China, although the nature of lesion is completely dissimilar. The Chinese lesion is on the supraorbital region (Caspari 1997) and deformed the supraorbital area. Unfortunately, the pathology of *H. erectus* has not yet been widely studied, despite advances in research on their morphology and taxonomy (Rightmire 1993, 1998, Antón and Indriati 2002, Antón 2003, Baba *et al.* 2003).

Taphonomic effects on cranial bones

After death, biological remains undergo various changes, beginning with the decomposition of the soft tissue, possible transportation by scavengers, soil erosion impact on the bone surfaces, roots of plants damaging the bone surfaces and destruction by water seepage (Behrensmeyer 1975, Shipman 1981, Indriati 2002, 2003b). The fact that the Ng7 cranium is fossilised (the organic material in the bones is impregnated with minerals from sediment) makes the interpretation of possible taphonomic effects on the lesions more difficult. Lists of typical postmortem alterations in fossils include breakage, plastic deformation (causing asymmetry), erosion and distortion (Shipman 1981). The absence of distortion, plastic deformation and breakage suggests that the lesions of Ng7 are not the result of postmortem processes. The multiple, localised lesions on Ng7 were, therefore, not likely to have taphonomic causes. However, surface erosion of the external table does occur on the Ng7 vault, suggesting that to some extent the appearance of the lesions are the result of taphonomic erosion. Most Ngandong skulls were collected after they had weathered out of a river bank (Hooijer 1951). The Ng7 skull might have been sorted, exposed, transported and redeposited, as is suggested by the absence of postcranial remains. Fluvial transport is suggested by Behrensmeyer (1975), inferring that the river must have been responsible for the sorting of skeletal elements, as the sediment deposited with the fauna contained sands and not very-fine-grained particles. The taphonomic effect of fluvial transport in hominid fossils, in the form of erosion on the bone surface, is evident on the Ng7 skull. The CT scan suggested that the crack and erosion occurred after fossilisation (Balzeau *et al.* 2003).

Conclusions

It is suggested that the Indonesian *H. erectus* Ng7 might have suffered an infection of the calvarium following open wounds of the scalp. This is based on the undulating appearance of the bone in the base of the large and circular lesion, the lack of alterations to the margin of the lesions and the cystic appearance of the smaller lesion. Taphonomic processes appear to have also eroded the bone surface. The morphology of *H. erectus* has been widely studied but little attention has been given to its pathology; consequently, the research described here will contribute to the palaeopathology of this species.

Acknowledgements

I would like to thank Nancy Tayles and Marc Oxenham for their invitation to write this chapter and the anonymous reviewers for their helpful comments.

References

Antón S. C. 1999. Cranial growth in *Homo erectus*: how credible are Ngandong juveniles? *American Journal of Physical Anthropology* **108**: 223–236.

2003. Natural history of *Homo erectus*. *Yearbook of Physical Anthropology* **46**: 126–170.

Antón S. C. and Indriati E. 2002. Earliest Pleistocene *Homo* in Asia: comparison of Dmanisi and Sangiran. *American Journal of Physical Anthropology Supplement* **34**: 38.

Baba H., Aziz F., Kaifu Y. *et al.* 2003. *Homo erectus* calvarium from the Pleistocene of Java. *Science* **299**: 1384–1388.

Balzeau A., Indriati E., Grimaud-Hervé D. and Jacob T. 2003. Computer tomography scanning of *Homo erectus* crania Ngandong 7 from Java: internal structure and postmortem-history. [In Indonesian] *B. I. Ked (Journal of Medical Science)* **35**: 135–143.

Behrensmeyer A. K. 1975. The taphonomy and paleoecology of Plio-Pleistocene vertebrate assemblages east of Lake Rudolf, Kenya. *Bulletin of the Museum of Comparative Zoology* **146**: 473–578.

Blum T. 1924. Osteomyelitis of the mandible and maxilla. *Journal of the American Dental Association* **2**: 802–805.

Brace C. L. 1967. *The stages of human evolution*. Englewood Cliffs, NJ: Prentice Hall.

Brose D. S. and Wolpoff M. H. 1971. Early upper Paleolithic man and late middle Paleolithic tools. *American Anthropologist* **713**: 1156–1194.

Caspari R. 1997. Brief communication: evidence of pathology on the frontal bone from Gongwangling. *American Journal of Physical Anthropology* **102**: 565–568.

Deeley T. J. 1983. A brief history of cancer. *Clinical Radiography* **34**: 597–608.

DuBoulay G. H. 1965. *Principles of X-ray Diagnosis of the Skull*. New York: Butterworth.

Dubois E. 1937. On the fossil human skulls recently discovered in Java and *Pithecanthropus erectus*. *Man* **37**: 1–7.

Fries J. W. 1957. The roentgen features of fibrous dysplasia of the skull and facial bones. *American Journal of Roentgenology* **77**: 71–77.

Galloway A., Symes S. A., Haglund W. D. and France D. L. 1999. The role of forensic anthropology in trauma analysis. In Galloway A., ed., *Broken Bones: Anthropological Analysis of Blunt Force Trauma*. Springfield, IL: Charles C. Thomas, pp. 5–31.

Gladykowska-Rzeczycka J. 1991. Tumors in antiquity in East and Middle Europe. In Ortner D. J. and Aufderheide A. C., eds., *Human Paleopathology: Current Synthesis and Future Options*. Washington, DC: Smithsonian Institution Press, pp. 257–260.

Grupe G. 1988. Metastasizing carcinoma in a medieval skeleton: differential diagnosis and etiology. *American Journal of Physical Anthropology* **75**: 369–37.

Hamperl H. 1967. The osteological consequences of scalping. In Brothwell D. R. and Sandison A. T., eds., *Diseases in Antiquity*. Springfield, IL: Charles C. Thomas, pp. 630–634.

Hamperl H. and Laughlin W. S. 1959. Osteological consequences of scalping. *Human Biology* **31**: 80–89.

Hooijer D. A. 1951. The geological age of *Pithecanthropus, Meganthropus*, and *Gigantopithecus*. *American Journal of Physical Anthropology* **9**: 265–281.

Indriati E. 2002. Cranial lesion on Indonesian *Homo erectus* Ngandong 7. In *The Inter-Congress of the International Union of the Anthropological and Ethnological Sciences*, September, Tokyo.

2003. Indonesian hominid fossil discovery of 1889–2003: catalogue and problems. In Akiyami, S., Miyawaki, R., Kubodora, T. and Higuchi, M., eds., *Proceedings of the Fifth and Sixth Symposia on Collection Building and Natural History Studies in Asia and the Pacific Rim. National Science Museum Monographs* **24**, 163–177.

2003b. Mati: Tinjauan Klinis dan Antropologi Forensik. [In Indonesian] [Death: clinical and forensic anthropological perspectives.] B. I. Ked. (*Journal of Medical Science*) **35**, 1–7.

Jacob T. 1973. Paleoanthropological discoveries in Indonesia with special reference to the finds of the last two decades. *Journal of Human Evolution* **2**: 473–485.

1975. Morphology and paleoecology of early man in Java. In Tuttle R. H., ed., *Paleoanthropology, Morphology and Paleoecology*. The Hague: Mouton, pp. 311–325.

Kaufman M. H., Whitaker D. and McTavish J. 1997. Differential diagnosis of holes in the calvarium: application of modern clinical data to palaeopathology. *Journal of Archaeological Science* **24**: 193–218.

Lee S. H., Rao K. C. V. G. and Zimmerman R. A. (eds.). 1999. *Cranial MRI and CT*, 4th edn. New York: McGraw-Hill.

Lisowski F. P. 1967. Prehistoric and early historic trepanation. In Brothwell D. R. and Sandison A. T., eds., *Disease in Antiquity*. Springfield, IL: Charles C. Thomas, pp. 651–672.

Mann A. and Trinkaus E. 1974. Neandertal and Neandertal-like fossils from the Upper Pleistocene. *Yearbook of Physical Anthropology* **17**: 169–193.

Morse D. 1969. *Illinois State Museum Reports Investigations*, No. 15: *Ancient Disease in the Midwest*. Springfield, IL: Illinois state Museum.

Oakley K. P., Campbell B. G. T. and Molleson T. I. 1975. *Catalogue of Fossil Hominids*. Part III: *America, Asia, Australia*. London: British Museum.

Olmsted W. M. 1981. Some skeletogenic lesions with common calvarial manifestations. *Radiologic Clinics of North America* **19**: 703–713.

Oppenoorth W. F. F. 1937. The place of *Homo soloensis* among fossil men. In MacCurdy G. G., ed., *Early Man*. Philadelphia, PA: Lippincott, pp. 349–360.

Ortner D. J. and Putschar W. G. J. 1981. *Smithsonian Contributions to Anthropology*, No. 28: *Identification of Pathological Conditions in Human Skeletal Remains*. Washington, DC: Smithsonian Institution Press.

Ortner D. J. Manchester K. and Lee F. 1991. Metastatic carcinoma in a leper skeleton from a medieval cemetery in Chichester, England. *American Journal of Physical Anthropology*. **80**: 369–376.

Reese H. H. 1940. The history of scalping and its clinical aspects. *Yearbook of Neurology, Psychiatry, and Endocrinology*, pp. 3–19.

Rightmire G. P. 1993. *The Evolution of* Homo erectus, *Comparative Anatomical studies of an Extinct Human Species*. New York: Cambridge University Press.

1998. Evidence from facial morphology of similarities of Asian and African representatives of *Homo erectus*. *American Journal of Physical Anthropology*. **106**: 61–85.

Roberts C. and Manchester K. 1997. *The Archaeology of Disease*, 2nd edn. Ithaca, NY: Cornell University Press.

Santa Luca A. P. 1980. The Ngandong fossil hominids: a comparative study of a far eastern *Homo erectus* group. *Yale University Publications in Anthropology*, No. 78. New Haven, CT: Yale University Press.

Shipman P. 1981. *The History of a Fossil*. Cambridge, UK: Cambridge University Press.

Steinbock R. T. 1976. *Paleopathological Diagnosis and Interpretation*. Springfield, IL: Charles C Thomas.

Strouhal E. 1976. Tumors in the remains of ancient Egyptians. *American Journal of Physical Anthropology* **45**: 613–620.

Swisher C. C., Rink W. J., Anton S. C. *et al.* 1996. Latest *Homo erectus* of Java: potential contemporaneity with *Homo sapiens* in Southeast Asia. *Science* **274**: 1870–1874.

Tkocz I. and Bierring F. 1984. A medieval case of metastasizing carcinoma with multiple osteosclerotic bone lesions. *American Journal of Physical Anthropology* **65**: 373–380.

von Koenigswald G. H. R. 1951. Introduction. In Weidenreich F., ed., *Anthropological Papers of the American Museum of Natural History*, No. 43, part 3: *Morphology of Solo Man*. Washington, DC: Smithsonian Institution Press, pp. 211–221.

Walker P. L. 1989. Cranial injuries as evidence of violence in prehistoric southern California. *American Journal of Physical Anthropology* **80**: 313–323.

Webb S. 1995. *Paleopathology of Aboriginal Australians*. Cambridge, UK: Cambridge University Press.

Weidenreich F. 1939. Six lectures of *Sinanthropus pekinensis* and related problems. *Bulletin of the Geological Society of China* **19**: 1–110.

1943. The skull of *Sinanthropus pekinensis*: a comparative study of a primitive hominid skull. *Paleontologica Sinica New Series D* **10** (Whole series **27**): 1–184.

1951. *Anthropological Papers of the American Museum of Natural History*, No. 43, Part 3 *Morphology of Solo Man*. Washington, DC: Smithsonian Institution Press, pp. 205–290.

Wilensky A. O. 1932. Osteomyelitis of the jaws. *Archives of Surgery* **25**: 187–237.

Zimmerman R. A. 1999. Craniocerebral trauma. In Lee S. H., Rao K. C. V. G. and Zimmerman R. A., eds., *Cranial MRI and CT*, 4th edn. New York: McGraw-Hill, pp. 413–452.

13 'The predators within': investigating the relationship between malaria and health in the prehistoric Pacific Islands

HALLIE R. BUCKLEY
University of Otago, Dunedin, New Zealand

Introduction

The success of human settlement in tropical environments was undoubtedly influenced by the presence of infectious disease. Of the tropical diseases, malaria has caused more human suffering and death than any other infectious disease (Bloland *et al.* 1998). Considering the continued onslaught of malaria, despite intensive medical treatment and research, the cost to prehistoric populations must have been profound. The Pacific Islands provide an invaluable opportunity for examining the potential effects of malaria on prehistoric populations by the variable endemicity among western island groups and its complete absence east of Vanuatu.

Human settlement of the western Pacific Islands has been very long. Yet, despite the development of agriculture in the Holocene, populations in Near Oceania remained small and technological advancement was limited. (Near and Remote Oceania define geographical regions in the southwest Pacific Islands (Green, 1997) that have distinct ecological and geological influences on prehistoric human settlement (Fig. 13.1).) This is a markedly different pattern from the demographic shift seen in other regions of the world where agriculture was developed. Based on epidemiological, entomological and archaeological evidence, Groube (1993, p. 166) suggested that economic development and population increase in Near Oceania was retarded compared with Asian contemporaries because of '... the role in Sahuland prehistory of the "predators within", the parasites of infectious diseases, particularly the presence in Melanesia, and probably northern Australia as well, of that most ancient of human-specific diseases, malaria.'

Bioarchaeology of Southeast Asia. Marc Oxenham and Nancy Tayles.
Published by Cambridge University Press.

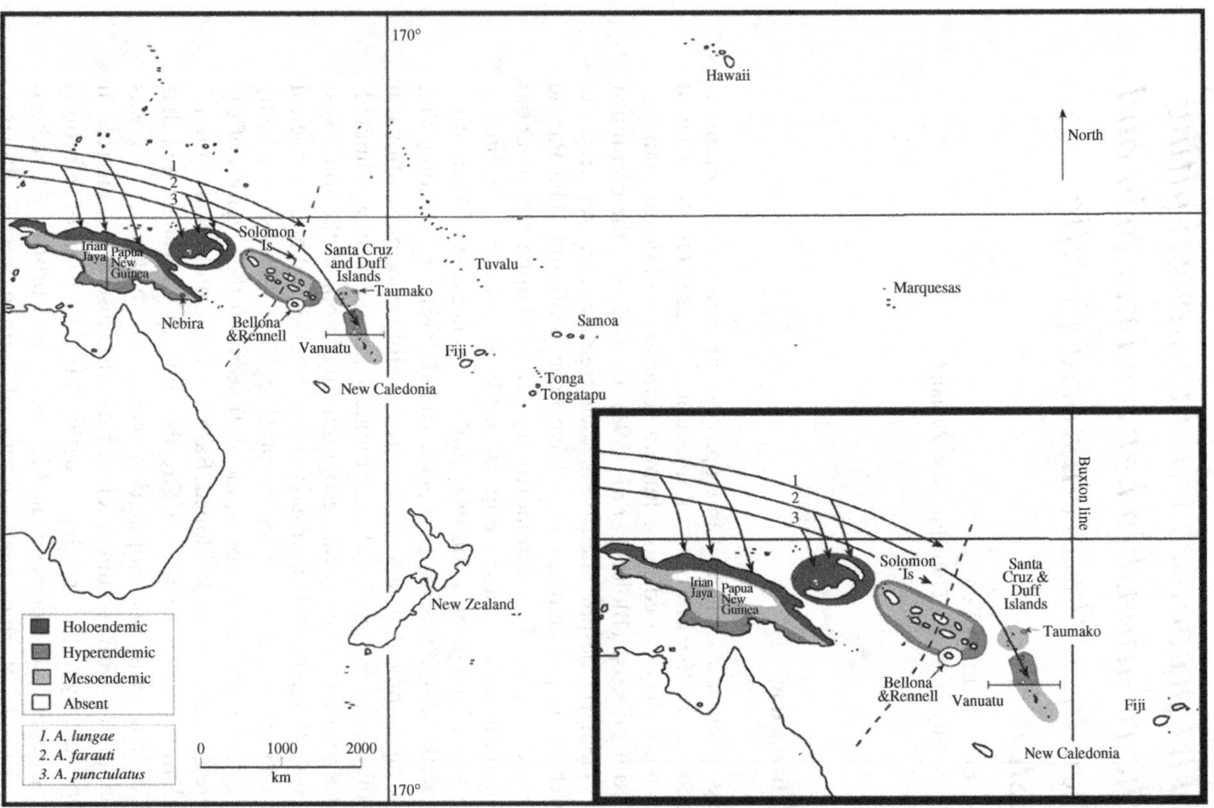

Figure 13.1. The southwest Pacific, showing the location of the skeletal samples used and illustrating the probable relationship between malaria vector distribution and variable endemicity. Remote Oceania is from the Santa Cruz and Duff Islands to the east, including New Caledonia. Near Oceania is from the Solomon Islands westwards.

By analysing settlement patterns throughout Oceania, Kirch (2000, p. 308) expanded Groube's hypothesis and argued that 'the demographic history of Oceania has been intimately linked with the history of disease, especially malaria'. He suggested that the absence of malaria and relatively low incidence of other infectious diseases in Polynesia allowed rapid population growth and social development, which led to the intensely hierarchical chiefdoms encountered by early European explorers.

As Groube and Kirch argued, the complete absence of malaria in Polynesia may have significantly influenced the demographic history of this region of the world. However, an investigation of the impact of malaria using the most direct avenue of evidence may lend clues to why people in this region died and, more importantly, how they lived. This may be achieved by palaeopathological studies of dental and skeletal remains of the prehistoric people themselves.

This chapter considers the epidemiological history of the Pacific Islands through the examination of palaeopathological parameters of child health in archaeological skeletal samples. It may be possible to test the validity of the hypothesis put forward by Groube (1993) and Kirch (2000) by assessing the skeletal and dental evidence of health and disease of populations living in malarial and non-malarial zones of the Pacific Islands. A review of literature concerning the ecology of malaria in the Pacific Islands is presented to provide a background on the possible effects of malaria on health in prehistory. Recent palaeopathological research using skeletal samples from malarial and non-malarial sites in the Pacific Islands is also discussed in the context of malaria ecology and distribution.

Malaria epidemiology

All four human malaria species are present in the Pacific Islands. *Plasmodium vivax*, *P. ovale*, *P. malariae* and *P. falciparum*. The life cycle of malaria parasites is essentially the same across all human-specific species. It comprises two phases of multiplication: one in the mosquito, acting as a vector, and one in the human host (Gilles and Warrell 1993).

Infection is transmitted when a female *Anopheles* mosquito inoculates the human host with plasmodial sporozoites during feeding. The clinical disease associated with malaria is characterised by fever, which is an inflammatory response to destruction of parasitised red blood cells (Wyler 1982). Other aspects of clinical disease are considered below in the context of maternal and infant health.

The particular pathogenesis of each malaria species influences the course of clinical disease. Infection with *P. falciparum* is continuous and more severe than the other species because the parasite multiplies in and destroys blood cells of all ages. By comparison, *P. vivax* and *P. ovale* are restricted to the invasion of young blood cells, while *P. malariae* attacks older cells (Wyler 1982). Both *P. ovale* and *P. malariae* infections are characterised by low parasitaemia and mild clinical disease (Taylor and Strickland 2000). The clinical literature has concentrated on *P. falciparum* and *P. vivax*, which have higher parasitaemia and more severe clinical disease. The latter can remain dormant in the liver cells for months or even years, accounting for the intermittent fevers that characterise its clinical expression (White 1996).

The ability of *P. vivax* to remain dormant supports the hypothesis put forward by Groube (1993), that it was carried into the Pacific Islands with the first human settlers. For *P. falciparum* to be transported into the western Pacific region, human settlement would have to be rapid. If not, the carrier may have died before the parasite could be transmitted to the vector. The maritime technology necessary for rapid settlement of these islands was not seen until the appearance of the Lapita culture in the western Pacific Islands approximately 3,500–3,400 years BP (Kirch 2000). It has been proposed that the more lethal *P. falciparum* was introduced by the Lapita people (Clark and Kelly 1993, Kirch 2000).

The severity of illness and threat of death from malaria varies with its endemicity. Traditionally, the degree of endemicity of malaria has been defined in terms of parasite rates in the spleens of children between two and nine years of age: *hypoendemic*, parasite rate 0–10%, considered unstable and contributing to epidemic malaria; *mesoendemic*, parasite rate 10–50%, unstable; *hyperendemic*, parasite rate 50–75%, stable; *holoendemic*, parasite rate over 75%, stable (White 1996). The degree of endemicity depends on whether infection rates are stable or unstable. In areas of stable (holoendemic and hyperendemic) endemicity, infection occurs repeatedly from the first year of life and this constant exposure aids in achieving a state of premunition or partial immunity in survivors (Hendrickse 1987). If partial immunity develops, clinical symptoms are minimal or completely absent (Taylor and Strickland 2000). Malaria can also be expressed as an epidemic disease. Epidemics are caused when new susceptible hosts migrate into the area and disrupt the fine balance achieved between the parasite and the host. Mortality during epidemics is generally high compared with that in holoendemic and hyperendemic areas (White 1996). If *P. falciparum* was a later arrival in the western Pacific Islands, epidemics of malaria amongst existing populations would have been a probable consequence of the arrival of Lapita peoples in Near Oceania.

Certain individuals are at risk of greater morbidity than others, such as pregnant women, the fetus carried by an infected woman, and young children between the ages of one and five years. A greater susceptibility to malaria is especially found in women who are pregnant with their first child, as acquired immunity diminishes during pregnancy (Taylor and Strickland 2000). Severe anaemia and depression of red blood cell production increase the risk of death in pregnant women and the fetus. Maternal illness may significantly affect fetal growth and may induce spontaneous abortion, still birth or prematurity (Lee 1988). Another clinical aspect of malaria and pregnancy is the parasitisation of the placenta, which compromises fetal health (Lee 1988). The ability of the parasite to transfer across the placenta has been reported in all *Plasmodium* spp. (Lee 1988).

The epidemiology of infant and childhood malaria is complex. In infants less than three months of age, parasitism does occur but the clinical symptoms of malaria are rare (White 1996). This is probably a result of maternal immunity and also the presence of fetal haemoglobin, which restricts the multiplication of parasites (Pasvol *et al.* 1976). After early infancy, parasite loads, with associated clinical symptoms, increase steadily until five years of age, when the incidence and severity of clinical malaria usually wanes as a result of acquired immunity (Greenwood 1997).

The human host plays an integral role in the successful transmission of malaria. For the parasite to transmit efficiently, there must be a large reservoir of young individuals or 'fresh susceptibles' in the human population at all times (Greenwood 1997). Therefore, younger children are the reservoir from which the parasite can maintain the level of endemicity. While it seems that children under the age of five years bear the brunt of malaria-related morbidity, their role in maintaining the stability of endemicity, and relative immunity of the wider community, is significant.

Vector ecology and the distribution of malaria in the Pacific Islands

The ecology and distribution of anopheline mosquitoes in the Pacific Islands may have played an integral role in the impact of this disease in prehistory (Fig. 13.1).

There are three primary vectors for malaria in the Pacific Islands: *Anopheles lungae*, *A. punctulatus* and *A. farauti*. The distribution of these vectors in the Pacific varies and is partly influenced by the different ecological environments of island groups. The most significant consequence of vector ecology variation for human prehistory is the complete absence of the vectors east of Vanuatu. It has been suggested that this is an artefact of the general zoological impoverishment in the Pacific Islands from west to east (Laird 1956). However, entomological studies have suggested that mosquito distribution was probably more a reflection of geological formation of the different island groups in the Pacific Ocean and independent of the west to east impoverishment of biodiversity that affected other species (Belkin 1962).

Regardless of the mechanism responsible, it is possible that vectors able to transmit malaria were present in the western Pacific Islands well before human colonisation. Certainly, *A. farauti* is not solely reliant on human blood for survival and may feed on birds or other animals (Spencer *et al.* 1974). Therefore, this species could have survived in the western Pacific Islands before human colonisation, although there is no documented evidence of *A. punctulatus* and *A. lungae* having zoophilic feeding behaviour (Spencer *et al.* 1974), it is likely that they also exploited animal blood before human arrival.

All vectors are present in Papua New Guinea (PNG) and the Bismarck archipelago; *A. farauti* is widespread throughout the Solomon Islands chain while *A. punctulatus* and *A. lungae* are restricted to the western islands (Belkin 1962, Avery 1974). The sole malaria vector in all island groups east of the mid Solomon Islands is *A. farauti* (Belkin 1962).

Prior to eradication attempts in PNG and nearby archipelagoes, *A. punctulatus* was largely restricted to inland regions away from river valleys. However, it was capable of rapid spread into coastal zones (Laird 1956). More recently, *A. punctulatus* has been recorded in the coastal plains of Guadalcanal in the Solomon Islands (Beebe *et al.* 2000). This species has the ability to increase its population density rapidly by seizing opportunities of environmental fluctuations such as river valley flooding and site disturbance accompanying vegetation clearance and pig wallowing (Spencer 1971).

A. farauti is widespread in its distribution but particularly dominates coastal zones (Laird 1956, Beebe *et al.* 2000). It is sensitive to site disturbance and prefers permanent to semi-permanent breeding sites. In site disturbance, the population will be temporarily reduced, allowing *A. punctulatus* to take advantage of the temporary decline in *A. farauti* numbers by colonising the recently disturbed breeding sites (Spencer 1971). In summary, *A. punctulatus* is more opportunistic than *A. farauti* (Spencer *et al.* 1974). As discussed below, this less-adaptable side of *A. farauti*'s nature probably had significant consequences for the history of malaria in the Pacific Islands.

Malaria endemicity in the Pacific islands

Generally, the endemicity of malaria is higher in the western island groups, with patchy levels east of the mid Solomon Islands. This is consistent with vector distribution. In PNG, malaria is holoendemic on the northern coasts, with lower levels on the southern coasts (Flint *et al.* 1986). Regions above 1300 m are entirely free of malaria because mosquitoes are unable to survive the colder temperatures of higher altitudes (Spencer 1971, Parkinson 1974).

Prior to eradication programmes in the 1960s, the Solomon Islands were generally mesoendemic while some of the outer islands and the central and eastern districts (i.e. northern Gaudalcanal, northern San Cristobel and Neggella) were hyperendemic. Some of the outer islands such as Tikopia, Anuta, Santa Ana and Santa Catalina experienced only hypoendemicity. Similarly, Rennell Island and most of the Reef Islands were mesoendemic, with the vector found only in very low frequencies (Avery 1974).

According to Maitland *et al.* (1996), malaria transmission in the Vanuatu chain of islands is constant throughout the year. However, other studies have suggested a seasonal fluctuation in endemicity (Flint *et al.* 1986). Malaria endemicity is more intense in the northern islands, while the disease is completely absent on Futuna (Flint *et al.* 1986, Williams *et al.* 1996, 1997).

It is evident that the endemicity of malaria in the western Pacific Islands is highly variable. This may be a consequence of the distribution and ecology of the vector. Variation in vector distribution corresponds to holoendemic malaria in the western island groups and a more variable endemicity in the eastern groups, with seasonal fluctuations of incidence. Furthermore, the two main mosquito vectors prefer different ecological niches for breeding, which means both coastal and inland zones of the western islands would be blanketed by vectors in the west. It would seem then that there is less opportunity for malaria transmission in the eastern islands because there is only a single vector, *A. farauti*. The pattern of breeding behaviour also means that environmental fluctuations and site disturbances would not be taken advantage of by *A. farauti* in the east. This pattern of malaria endemicity has probably changed through time, and it cannot be assumed that the disease affected people in prehistory in exactly the same manner as that observed in clinical studies reviewed here.

Given the profound consequences of a lack of malaria east of Vanuatu, it must be asked why *A. farauti* did not move east with the human colonisers of western Polynesia. As mentioned above, large populations of *A. farauti* have been recorded living 'wild' and feeding off animals. It is possible that wild populations of *A. farauti* were present in Vanuatu prior to human arrival. However, east of Vanuatu, the vector would have to be transported with the mixture of domesticated animals, plants, vermin and diseases accompanying the Lapita colonisers (Kirch 2000). It would seem that this did not occur for *A. farauti*. As discussed above, *A. farauti* has been described as 'the steady companion of the settled community and of the subsistence economy of native living' (Spencer 1971, p. 22). The archaeological record suggests that human movement into Vanuatu and eastwards was extremely rapid. Therefore, it is possible that the 'domestication' of *A. farauti* was not well established before the initial colonisers of Polynesia set forth from Vanuatu.

Similarly, it cannot be assumed that simply the presence of the mosquito and human populations would automatically lead to successful and rapid transmission of malaria. The disturbance of the environment associated with subsistence economy, a behaviour that would not have favoured the breeding habits of *A. farauti*, may have discouraged the rapid establishment of anthropophilic populations of this species of mosquito. It would also be expected that a period of adjustment may have been required for local vector populations to adapt from solely zoophilic to anthropophilic feeding behaviour.

Apart from these ecological issues, in order for the vector to be transported eastwards, larvae would need to be deposited within water containers or even the bilge water of the voyaging canoes. Entomological studies have shown that female *A. farauti* never lay their eggs in artificial containers (Laird 1956, Belkin 1962). So, it would seem that, because of this finicky breeding behaviour, the vector for malaria quite literally missed the boat to Polynesia.

The presence of certain globin gene variants in the Pacific Islands indicates a protracted association between humans and malaria in this region. However, β-thalassaemia has only been recorded from the northern coast of PNG and in Vanuatu (Bowden *et al.* 1985, Hill *et al.* 1989, Ganczakowski *et al.* 1995). In its heterozygous form, β-thalassaemia is thought to confer a relative advantage for survival in infection with *P. falciparum* (Kariks and Woodfield 1972, Yenchitsomanus *et al.* 1985, 1986, Fleming 1996). Mutations of the α-globin gene have also been recorded in the Pacific Islands (Fleming 1996). These globin gene variants are mostly asymptomatic and no clear mechanism for selection by malaria has been found (Weatherall 1997). The evidence for a relationship between α-thalassaemia and malaria is largely epidemiological in that a positive correlation has been found between higher frequencies of the mutation and intense malaria endemicity (Flint *et al.* 1986). The relationship between malaria and globin gene variants is an active field of research in the Pacific Islands.

Avenues for investigating the role of malaria on population health in prehistory

The epidemiology of malaria in the Pacific Islands is similar to other malarial areas of the world, where young women and children between the ages of one and five years are most at risk (Sayers 1943, Avery 1974, McMahon 1974, Parkinson 1974, van Dijk and Parkinson 1974, Zigas and Morea 1974, Genton *et al.* 1995, Maitland *et al.* 1996).

Given the distribution and epidemiology of malaria in the Pacific Islands, it would be expected that investigating skeletal indicators of growth disturbance and stress would reveal different patterns depending on the malaria status of the island group. Malaria does not directly affect the skeleton; however, the malaise associated with this disease is extreme. Therefore, a number of non-specific parameters for growth disruption and/or infection could be used for comparing infant and child health between island groups of differing malarial status. Evidence of ill-health in the infant and child samples of skeletal assemblages is particularly useful to investigate, as these individuals are considered the most sensitive indicator of adverse effects of the environment in prehistory (Lewis 2000). In the Pacific Island context, the study of infant and child health parameters provides the researcher with a barometer for levels of stress suffered by individuals when they are most vulnerable to malaria-associated morbidity; in the womb and between one and five years of age.

Because malaria does not directly affect the skeleton, the most pragmatic approach to this question is to examine the non-specific stress indicators of bones and teeth. While not ideal, this approach is based on the assumption that the morbidity experienced with malaria would also predispose individuals to more severe stress experiences from other sources, such as nutritional deficiency or other infectious diseases. This concept of a synergistic relationship between nutrition and infection as a cause of illness is well established in the clinical and palaeopathological literature (Scrimshaw *et al.* 1968, Infante and Gillespie 1974, Cook and Buikstra 1979, Scrimshaw 1981, Beisel 1984, Goodman 1993).

Therefore, it is hypothesised that higher rates of non-specific indicators of stress will be observed in skeletal samples from malarial islands compared with those from non-malarial islands. In order to test this hypothesis, a palaeopathological investigation of two skeletal samples from Pacific Islands of different malaria status was conducted (Buckley 2001). Buckley (2001) considered the prevalence of several parameters of ill-health in relation to malaria presence or absence using a skeletal sample from Taumako island in the Duff Group, southeast Solomon Islands, and two samples from Tongatapu (To-At-1 and To-At-2) in the Kingdom of Tonga. More recent preliminary data from re-appraisals of skeletal collections from Nebira in PNG and another sample from Tongatapu (To-At-36) have provided the opportunity to test the robusticity of the previous findings.

Materials and methods

Skeletal samples

Skeletal samples from three geographic regions of the Pacific Islands have been used to test whether indicators of non-specific stress may be higher in malarial areas than in non-malarial Polynesia (Fig. 13.1). Susan Bulmer excavated the skeletal sample of 44 individuals at Nebira in 1968. The site is located above a riverbank near the modern capital of Port Moresby. Tentative dates from this site suggest an occupation period from 1,000 to 500 years BP (Bulmer 1979). Unfortunately, infants and young children are poorly represented in this sample. Malaria is hyperendemic in this region of PNG (Flint *et al.* 1986) and was probably present in prehistory.

A large skeletal sample of 226 individuals was excavated from a burial mound on Taumako island, Duff group, southeast Solomon islands, in the early 1970s. Electron spin resonance analysis of a series of bone samples suggested the burial mound was in use for about 170 years, from 470 to 300 years BP (Whitehead *et al.* 1986). Malaria was hyperendemic in the Duff Islands prior to eradication programmes in the 1960s (Avery 1974). The local inhabitants live on an artificially constructed island in the lagoon 150 m off the shore of the main island. This was the case at initial contact in 1606 by the Spanish explorer Quiros (Kelly 1966). It has been suggested that the construction of artificial islands in the Solomon Islands region was an attempt to avoid the malarial vectors and may have provided some protection in prehistory (Parsonson 1966). This hypothesis has not been explicitly tested and has met with some scepticism (Chowning 1968). Considering that the main island of Taumako experienced hyperendemic malaria before eradication programmes (Avery 1974), it is possible that the people living on the artificial island were exposed to malaria in prehistory. The parameters used in this study for measuring levels of ill-health may aid in testing whether the people of Taumako were adversely affected by malaria in prehistory.

The three Tongan samples were excavated from the ’Atele region of Tongatapu island, Kingdom of Tonga. Two samples (To-At-1 and To-At-2) were excavated in the early 1960s and date from *c*. 1,000–300 years BP (Davidson 1969). The third sample (To-At-36) was from the Ha'ateiho region of Tongatapu, very close to the sites where the first two samples were found (Spenneman 1986). There are no published radiocarbon dates from this site; however, the burial mound construction and method of internment suggest that it dates from a similar time period as the other two samples. The type of mound construction suggests these were all communal burial sites of commoners rather than aristocracy (Davidson 1969). Collectively, the Tongan assemblages comprise a sample of 128 individuals. The summary data of the samples excavated by Davidson are presented as one group and the data of To-At-36 are presented separately. These data are also merged as a combined group to increase the sample sizes for statistical comparison with Nebira and Taumako.

Anaemia and cribra orbitalia

While the bony changes associated with anaemia are well described, the underlying aetiology of anaemia is considered to be multifactorial (Resnick 1995, Ryan 1997). Therefore, for the purposes of this study, the pathological bony changes associated with anaemia are considered as indicators of non-specific stress.

The pathological bony changes of anaemia are caused by an increase of haemopoietic marrow activity owing to low haemoglobin levels; this results in thinning of the cortex and a reduction in trabeculae (Resnick 1995). Such changes are most readily apparent on the orbital roof, where the outer plate of bone is resorbed and the internal spongiosa is exposed. This change in the orbital roof is termed cribra orbitalia in the palaeopathological literature (Hengen 1971).

Changes in the skeletal tissue develop during infancy and early childhood as a result of high marrow activity and rapid bone turnover during growth (Resnick 1995). This combination of normal physiological processes during bone growth predisposes the immature skeleton to more extensive and rapid development of bone lesions (Lewis 2000). Lesions have been recorded in adults, but these are thought to be representative of lesions developed in childhood that have not remodelled completely (Stuart-Macadam 1985).

The pathogenesis and morphology of cranial vault changes in thalassaemia are essentially the same as in acquired anaemia, but thalassaemia affects the postcranial skeleton more frequently than acquired anaemia (Weatherall and Clegg 1981, Ortner 2003). Disturbance of epiphyseal growth plates in the long bones and the inhibition of normal development of maxillary and sphenoid sinuses are also observed in some patients with thalassaemia (Caffey 1957, Middlemiss and Raper 1966, Ortner 2003). These growth disturbances caused by thalassaemia can lead to distinctive bone and joint deformities (Laor *et al.* 1982, Hershkovitz *et al.* 1991).

The causes of anaemia in the western Pacific Islands are complex and multifactorial, including nutritional deficiency of iron, hookworm infection and genetic anaemia (Bailey 1966, Bowden *et al.* 1985, Ganczakowski *et al.* 1995). However, one pervading theme in clinical studies of anaemia in children is the synergistic interaction of malaria with other infectious diseases and nutritional deficiency (Jelliffe 1968, 1970, Malcolm 1969, 1970, Wark and Malcolm 1969, Gray 1982, Thomason *et al.* 1986, Hendrickse and Brabin 1996).

The haemolytic anaemia of malaria is multifactorial and complex. With parasitaemia, destruction of red blood cells is an integral part of the parasite's life cycle. A precocious destruction of non-parasitised red cells also occurs (Wyler 1982, White 1996). The consequent haemolytic anaemia is compounded by a dysfunction of the bone marrow, in which erythropoiesis is defective for some time during and after infection (White 1996). Hyperplasia of the bone marrow has been shown to be an important factor in the pathogenesis of malarial anaemia (Hendrickse 1987). These pathophysiological changes are associated with both *P. falciparum* and *P. vivax*. Infection with *P. vivax* has also been strongly implicated as a cause of malnutrition (measured as low weight for age) in Vanuatu children less than five years of age who had suffered a recent clinical attack of this particular species (Williams *et al.* 1997). These factors suggest that the presence of malaria should be considered as a significant contributor to anaemia and its skeletal manifestations in the prehistory of the western Pacific Islands.

Evidence of orbital changes associated with anaemia was recorded in all of the samples, based on the protocol of Stuart-Macadam (1985). This is presented as presence/absence data, including all levels of severity and stages of remodelling.

Dental defects

Investigation of evidence in the dentition of growth disruption is another avenue for assessing patterns in health between populations from different environments. The researcher is provided with a permanent record of the health of the individual during the time of tooth crown development.

Dental enamel hypoplasia is a deficiency in enamel thickness caused by a disruption or cessation of amelogenesis during tooth crown formation (Goodman *et al.* 1980).

The term enamel hypoplasia is used for an enamel surface defect including grooves, pits or missing enamel. However, linear enamel hypoplasia (LEH) is the most commonly observed and recorded enamel defect by anthropologists (Goodman and Song 1999). It is characterised by a linear horizontal groove, caused by decreased enamel thickness (Goodman *et al.* 1980).

The cells that form enamel are sensitive to physiological stress such as infection, dietary deficiencies and psychological or physical trauma (Hillson 1996). Therefore, evidence of a disruption of enamel formation provides a non-specific indicator of a variety of stresses experienced by the individual during the period of enamel formation (Goodman and Song 1999). Observations of enamel defects can provide a retrospective record of physiological stress from five months in utero in the deciduous dentition and during the first seven years of life in the permanent dentition (Goodman *et al.* 1980). Therefore, defects of the deciduous teeth can be used as a means to infer maternal and fetal health, while permanent crown development encompasses the early childhood period. These developmental stages of tooth development correlate with the periods of life when an individual is most vulnerable to malaria-associated illness.

Since the early 1980s, numerous macroscopic observations of LEH have been carried out on prehistoric human dental remains (e.g. Goodman *et al.* 1980, Duray 1996, Stodder 1997). Many have been concerned with recording a shift in health patterns that may correlate with a shift in subsistence or technology (Cohen and Armelagos 1984). The prevailing theme has been that a change in economy from hunter–gatherer to agricultural economy would lead to an increase in enamel hypoplasia (Roberts and Manchester 1995). Some researchers have suggested caution when recording enamel defects because of the complex development of tooth crown enamel (e.g. Skinner and Goodman 1992, Hillson 2000). Despite these limitations, a comparative approach of using dental evidence of growth disruption during tooth development may have some application for considering the effects of malaria distribution in the Pacific Islands.

A dental defect observed in some studies of deciduous teeth is a carious lesion on the buccal surface of the crown, thought to develop over defective enamel (Cook and Buikstra 1979). These dental defects are known as 'circular caries' yet it is the cause of the underlying enamel defect that is of most interest to palaeopathologists (Cook and Buikstra 1979, Hanson 1990). Cook and Buikstra (1979) interpreted the occurrence of these lesions in prehistoric American sites as reflecting severe stress associated with malnutrition and infectious disease during the first few weeks of life. These conclusions were based on epidemiological studies of LEH in living children from disadvantaged backgrounds (Sweeney *et al.* 1969, Sweeney *et al.* 1971, Infante and Gillespie 1974). Circular caries and defective enamel of the deciduous teeth have been noted in clinical studies of Pacific Island populations from diverse ecological zones (Kirkpatrick 1935, Davies 1958, Baume and Meyer 1966, Barmes 1967, Baume and Vulliemoz 1970). These clinical studies attributed the defective enamel underlying carious lesions of the deciduous teeth more to maternal than perinatal ill-health.

In this study, LEH was recorded from macroscopic observation of all incisors and canines following the protocol of Goodman *et al.* (1980). The occurrence of circular caries was also recorded in all deciduous incisors and canines from each of the samples following the methods of Cook and Buikstra (1979). Only presence or absence of data is reported here but a more detailed analysis of the number of defects per tooth and age at occurrence was conducted previously (Buckley 2001).

The statistical tests employed to compare the proportional data from the small samples were non-parametric chi-square tests. A standard critical level of $p = 0.05$ was employed.

Results

The prevalence of cribra orbitalia is presented in Table 13.1. These data show that the sample from Nebira had a significantly higher level of cribra orbitalia compared with the separate and combined Tongan samples and

Table 13.1. *Prevalence of cribra orbitalia in western Pacific Island and Polynesian skeletal samples*

	Affected/observed (%)[a]				
	Nebira	Taumako	To-At-1 & 2	To-At-36	Tonga combined
Subadult	3/5 (60.0)	26/67 (38.8)	7/32 (21.8)	1/5 (20.0)	8/37 (21.6)
Adult	11/17 (64.7)*	38/111 (34.2)	5/30 (16.6)	1/6 (16.6)	6/36 (16.6)
Total	14/22 (63.6)*	64/178 (35.9)	12/62 (19.3)	2/11 (18.1)	14/73 (19.1)

To-At, specific sites in Tonga.

[a]Number of individuals with cribra orbitalia/number of individuals with appropriate cranial material for assessing the presence of cribra orbitalia.

*Statistically significant difference ($p \leqslant 0.05$) between Nebira and Tonga combined.

there was a clear pattern of this indicator decreasing in prevalence from the west towards Polynesia. The high prevalence of cribra orbitalia in the Nebira sample is similar to those reported in previous assessments (Houghton n.d., Pietrusewsky 1976). Cribra orbitalia is also more prevalent at Taumako compared with the Tongan samples, although this difference was not significant.

Similarly the prevalence of dental defects in the permanent teeth is significantly higher in the Nebira sample compared with both the Taumako and the combined Tongan samples (Table 13.2). It is unfortunate that the Nebira sample had no observable deciduous teeth to assess maternal and fetal health. However, the significantly higher number of defects in the deciduous teeth from Taumako compared with Tonga suggest that malaria may have contributed to more maternal and fetal ill-health in the western Pacific Islands than in Polynesia.

Discussion

It is recognised that some of these data are based on small sample sizes that may not be representative of the prehistoric situation in all islands of Melanesia and Polynesia. However, based on these rather crude measures, it would seem that the Melanesian malaria status is reflected in dental and skeletal indicators of non-specific stress. While the data presented above support the hypothesis that malaria affected health in prehistory, it is important to consider other ecological and cultural factors that may have contributed to the observed patterns.

For example, the diversity of the Pacific Island biota available for human subsistence is known to decrease dramatically from western to eastern islands, and this probably influenced the adequacy of the diet in prehistory (Green 1997). All of the cultivated plants and domestic animals of Polynesia were introduced by the initial settlers of these islands (Kirch 2000). Because of the relatively rich variety of subsistence foods available to populations in the more western islands, it would be expected that evidence of anaemia and growth disturbance would be lower than in Polynesia where access to a variety of food sources was limited. The data presented in this study do not support such an expectation, where an inverse relationship between biodiversity and evidence of non-specific stress was identified. Therefore, the higher frequencies of cribra orbitalia and LEH in PNG and Taumako cannot be explained by the difference in biodiversity.

However, access to available resources can be influenced by cultural practices, which may significantly affect the health of populations. In some

Table 13.2. *Prevalence of developmental defects in the deciduous and permanent teeth of western Pacific Island and Polynesian skeletal samples*

	Affected/observed (%)[a]				
	Nebira	Taumako	To-At-1 & 2	To-At-36	Tonga combined
Deciduous permanent	NA NA	34/138 (24.6)*	5/117 (4.2)	0/15 (0.0)	5/132 (3.7)
Subadult	23/67 (34.3)	37/126 (29.3)	10/57 (17.5)	1/5 (20.0)	11/62 (17.7)
Adult	56/138 (40.5)**	110/514 (21.4)*	11/155 (7.0)	9/44 (20.4)	20/199 (10.0)
Total permanent	79/205 (38.5)**	147/640 (22.9)*	21/212 (9.9)	10/49 (20.4)	31/261 (11.8)
Total	79/205 (38.5)**	181/778 (23.2)*	26/329 (7.9)	10/64 (15.6)	36/393 (9.1)

To-At, specific sites in Tonga.
[a]Number of incisors and/or canines with developmental defects/number incisors canines. The deciduous defects included linear enamel hypoplasia and circular caries.
*Statistically significant difference between Taumako and Tonga combined ($p \leqslant 0.05$).
**Statistically significant differences between Nebira and Taumako, and Nebira and Tonga combined ($p \leqslant 0.05$).

populations of the western Pacific Islands, pregnant women and young children are denied protein-rich foods, which inhibits growth and affects overall health (van der Hoeven 1958, Wark and Malcolm 1969, Malcolm 1969, 1970, Thomason *et al.* 1986, Lepowsky 1987). Surveys of infant and child feeding patterns in Polynesia do not report food taboos such as those in the Melanesian islands. However, Polynesian peoples would have felt different pressures. High rates of population growth were possible in the relatively disease-free environment, leading to the development of more hierarchical social systems (Kirch 1984). It is known that, in pre-European Tonga, the development of social hierarchy led to inequality, especially in terms of access to protein-rich resources (Pollock 1992). Therefore, access to resources may have been restricted in both Polynesian and Melanesian populations by cultural factors.

An overall lower incidence of other infectious diseases in Polynesia may account for the lower frequencies of cribra orbitalia and LEH in the Tongan samples. Early clinical surveys of Tonga noted negligible levels of hookworm, which is a ruthless cause of anaemia in the western islands (Lambert 1941). Therefore, it is possible that higher levels of intestinal parasites, coupled with endemic malaria, may have acted synergistically to exacerbate indicators of non-specific stress in the samples from the prehistoric western Pacific Islands.

Initial European observations suggest that yaws (*Treponema pertenue*) was rife throughout the Pacific Islands (Buxton 1928, Marples 1950, Marples and Bacon 1953, Mills 1955, van der Sluis 1969, Garner and Hornabrook 1970, Alemaena 1986, Harris *et al.* 1989, Fegan *et al.* 1990). This non-venereal cousin of syphilis is capable of causing significant mutilation and debilitation (Buckley and Tayles 2003a) and would undoubtedly have contributed to the frequencies of non-specific indicators of stress in the skeletal samples observed. Palaeopathological studies of skeletal remains have reported evidence of yaws in samples from many of the Pacific Islands (Snow 1974, Trembly 1996, Pietrusewsky *et al.* 1997, Stodder 1997, Buckley 2000, Buckley and Tayles 2003b). While most previous studies have been concerned with intraregional patterns of treponemal disease in prehistory, Buckley (2001) found a higher prevalence of skeletal lesions indicative of yaws in the Taumako sample compared with Tonga. It was argued that an impaired immune response caused by malaria would have exacerbated the clinical course of yaws at Taumako. It was also suggested that the high population density on the artificial island at Taumako and relatively arid environment of Tongatapu may have contributed to this pattern.

The comparative data presented here would seem to support the hypothesis that the presence of malaria in the western Pacific Islands is reflected in higher frequencies of non-specific indicators of stress in the skeletal samples. The higher prevalence of cribra orbitalia and LEH in Taumako compared with Tonga also suggests that these people were exposed to malaria despite the protective advantage of the artificial island. The lower prevalence of these indicators at Taumako compared with Nebira may be a reflection of the artificial island providing some protection from vectors compared with the riverine site of Nebira. However, as discussed above, the presence of a single vector east of the mid Solomon Islands may also have affected the transmission of the parasite at Taumako. Future research on skeletal samples from other island groups in the malarious and non-malarious Pacific Islands may aid in identifying inter-island variation of indicators of health and disease using the ecological approach adopted in this study. It would also be valuable to assess the evidence of health and disease in Lapita-associated skeletal samples to evaluate changes over time from initial colonisation to more established settlements in the western Pacific islands.

Conclusions

This chapter has explored the role of malaria in the settlement of the Pacific Islands. A review of the clinical and entomological literature has shown that this disease would not only have affected population growth, as Kirch (2000) and Groube (1993) suggested, but was probably a significant contributor to ill-health in the living. It has also been shown that the investigation of prehistoric dental and skeletal material can help to illuminate past adaptation to different island environments by examining the people themselves. There are few well-preserved skeletal samples from controlled archaeological excavations in the southwestern Pacific Islands, and many previous studies of these samples have concentrated on investigating the biological affinities of the people. However, the data presented here demonstrate the value of examining the skeletal evidence for clues regarding prehistoric environmental adaptation. The differences in malaria distribution in the Pacific Islands and its effect of population health are one avenue for addressing this question. In order to begin to address differences in prehistoric environmental adaptation and its influence on the success of human settlement in the Pacific Islands, it is necessary that more comparisons of disease parameters in existing skeletal samples be carried out. This study has begun to address this question, and it is hoped that a more biocultural investigation of new skeletal samples may be possible.

References

Alemaena O. 1986. Yaws situation in the Solomon Islands. *Southeast Asian Journal of Tropical Medicine and Public Health* **17**: 14–28.

Avery J. 1974. A review of the malaria eradication programme in the British Solomon Islands 1970–1972. *Papua New Guinea Medical Journal* **17**: 50–60.

Bailey K. 1966. Some aspects of anaemia, haemoglobin levels and iron metabolism in the New Guinea highlands. *Medical Journal of Australia* **5**: 356–393.

Barmes D. 1967. Dental and nutritional surveys of primitive peoples in the Pacific Islands. *Australian Dental Journal* **12**: 442–454.

Baume L. and Meyer J. 1966. Dental dysplasia related to malnutrition, with special reference to melanodontia and odontoclasia. *Journal of Dental Research* **45**: 726–741.

Baume L. and Vulliemoz J. 1970. Dietary fluoride uptake in the enamel of caries-susceptible 'yellow' permanent teeth and of caries-resistant permanent and primary teeth of Polynesians. *Archives of Oral Biology* **15**: 431–443.

Beebe N., Bakote'e B., Ellis J. and Cooper R. 2000. Differential ecology of *Anopheles punctulatus* and three members of the *Anopheles farauti* complex of mosquitoes on Guadalcanal, Solomon Islands, identified by PCR–RFLP analysis. *Medical and Veterinary Entomology* **14**: 308–312.

Beisel W. 1984. Synergism and antagonism of parasitic diseases and malnutrition. *Reviews of Infectious Diseases* **4**: 746–750.

Belkin J. 1962. *The Mosquitoes of the South Pacific (Diptera; Culcidae)*. Berkeley, CA: University of California Press.

Bloland P., Neafie R. C. and Marty A. M. 1998. Malaria: a re-emerging disease. In Nelson A. and Horsburgh C., eds., *Pathology of Emerging Infections*. 2. Washington, DC: American Society for Microbiology, pp. 283–316.

Bowden D., Hill A., Higgs D. and Weatherall D. 1985. Relative roles of genetic factors, dietary deficiency, and infection in anaemia in Vanuatu, south-west Pacific. *Lancet* **8463**: 1025–1028.

Buckley H. 2000. Subadult health and disease in prehistoric Tonga, Polynesia. *American Journal of Physical Anthropology* **113**: 481–506.

2001. Health and disease in the prehistoric Pacific islands. Ph.D. thesis, University of Otago, Dunedin.

Buckley H. and Tayles N. 2003a. The functional cost of tertiary yaws (*Treponema pertenue*) in a prehistoric Pacific island skeletal sample. *Journal of Archaeological Science* **30**: 1301–1314.

2003b. Skeletal pathology in a prehistoric Pacific island skeletal sample: issues in lesion recording, quantification and interpretation. *American Journal of Physical Anthropology* **122**: 303–324.

Bulmer S. 1979. Prehistoric ecology and economy in the Port Moresby region. *New Zealand Journal of Archaeology* **1**: 5–27.

Buxton A. 1928. *Researches in Polynesia and Melanesia: An Account of Investigations in Samoa, Tonga, the Ellice Group and the New Hebrides, in 1924, 1925*. London: London School of Hygiene and Tropical Medicine.

Caffey J. 1957. Cooley's anemia: a review of the roentgenographic findings in the skeleton. *American Journal of Roentgenology, Radium Therapy and Nuclear Medicine* **78**: 381–391.

Chowning A. 1968. The real Melanesia: an appraisal of Parsonson's theories. *Mankind* **6**: 641–652.

Clark J. and Kelly K. 1993. Human genetics, paleoenvironments, and malaria: relationships and implications for the settlement of Oceania. *American Anthropologist* **95**: 612–630.

Cohen, M. and Armelagos G. (eds.). 1984. *Paleopathology at the Origins of Agriculture*. London: Academic Press.

Cook D. and Buikstra J. 1979. Health and differential survival in prehistoric populations: prenatal dental defects. *American Journal of Physical Anthropology* **51**: 649–664.

Davidson J. 1969. Archaeological excavations in two burial mounds at 'Atele, Tongatapu. *Records of the Auckland Institute and Museum* **6**: 251–286.

Davies G. 1958. A comparative epidemiological study of diet and dental caries in three isolated communities. *Alabama Dental Review* **6**: 19–44.

Duray S. 1996. Dental indicators of stress and reduced age at death in prehistoric Native Americans. *American Journal of Physical Anthropology* **99**: 275–286.

Fegan D., Glennon M., MacBride-Stewart G. and Moore T. 1990. Yaws in the Solomon Islands. *Journal of Tropical Medicine and Hygiene* **93**: 52–57.

Fleming A. 1996. Haematological diseases in the tropics. In Cook G. C., ed., *Manson's Tropical Diseases*. London: Saunders, pp. 101–173.

Flint J., Hill A., Bowden D. *et al.* 1986. High frequencies of α-thalassaemia are the result of natural selection by malaria. *Nature* **321**: 744–750.

Ganczakowski M., Bowden D., Maitland K. *et al.* 1995. Thalassaemia in Vanuatu, south-west Pacific: frequency and haematological phenotypes of young children. *British Journal of Haematology* **89**: 485–495.

Garner M. and Hornabrook R. 1970. Treponematosis in New Guinea. *Papua New Guinea Medical Journal* **13**: 53–55.

Genton B., Al-Yaman F., Beck H. *et al.* 1995. The epidemiology of malaria in the Wosera area, East Sepik Province, Papua New Guinea, in preparation for vaccine trials. II Mortality and morbidity. *Annals of Tropical Medicine and Parasitology* **89**: 377–390.

Gilles H. and Warrell D. 1993. *Bruce-Chwatt's Essential Malariology*. London: Edward Arnold.

Goodman A. 1993. On the interpretation of health from skeletal remains. *Current Anthropology* **34**: 281–288.

Goodman A. and Song R. 1999. Sources of variation in estimated ages at formation of linear enamel hypoplasias. In Hoppa R. and Fitzgerald C., eds., *Human Growth in the Past: Studies from Bones and Teeth*. Cambridge, UK: Cambridge University Press, pp. 210–240.

Goodman A., Armelagos G. and Rose J. 1980. Enamel hypoplasias as indicators of stress in three prehistoric populations from Illinois. *Human Biology* **52**: 515–528.

Gray B. M. 1982. Enga birth, maturation and survival: physiological characteristics of the life cycle in the New Guinea Highlands. In McCormack C. P., ed., *Ethnography of Fertility*. London: Academic Press, pp. 75–113.

Green R. 1997. Linguistic, biological and cultural origins of the initial inhabitants of Remote Oceania. *New Zealand Journal of Archaeology* **17**: 5–27.

Greenwood B. 1997. The epidemiology of malaria. *Annals of Tropical Medicine and Parasitology* **91**: 763–769.

Groube L. 1993. Contradictions and malaria in Melanesian and Australian prehistory. In Smith M., Spriggs M. and Fankhauser B., eds., *Occasional Papers in Prehistory* No. 24. *Sahul in Review: Pleistocene Archaeology in Australia, New Guinea and Island Melanesia,* Canberra: Department of Prehistory, Australian National University, pp. 164–186.

Hanson D. B. 1990. Paleopathological observations on human skeletal remains from Rota, Mariana Islands: epidemiological implications. *Micronesica Supplement* **2**: 349–362.

Harris M., Nako D., Hopkins T. *et al.* 1989. Yaws infection in Tanna, Vanuatu. *Southeast Asian Journal of Tropical Medicine and Public Health* **22**: 113–119.

Hendrickse R. 1987. Malaria and child health. *Annals of Tropical Medicine and Parasitology* **81**: 499–509.

Hendrickse R. and Brabin B. 1996. Paediatrics in the tropics. In Cook G. C., ed., *Manson's Tropical Diseases*. London: Saunders, pp. 363–377.

Hengen O. 1971. Cribra orbitalia: pathogenesis and probable etiology. *Homo* **22**: 57–75.

Hershkovitz A., Ring B., Speirs M. *et al.* 1991. Possible congenital hemolytic anemia in prehistoric coastal inhabitants of Israel. *American Journal of Physical Anthropology* **85**: 7–13.

Hill A., O'Shaughnessy D. and Clegg J. 1989. Haemoglobin and globin gene variants in the Pacific. In Hill A. and Serjeantson S., eds., *The Colonization of the Pacific: A Genetic Trail.* Oxford: Oxford University Press, pp. 246–285.

Hillson S. 1996. *Dental Anthropology*. Cambridge, UK: Cambridge University Press.

2000. Dental pathology. In Katzenburg M. and Saunders S., eds., *Biological Anthropology of the Human Skeleton*. New York: Wiley-Liss, pp. 249–286.

Houghton P. n.d. *The people of Nebira, Papua New Guinea*: Manuscript on file. University of Otago.

Infante P. and Gillespie G. 1974. An epidemiological study of linear enamel hypoplasia of deciduous anterior teeth in Guatemalan children. *Archives of Oral Biology* **19**: 1055–1061.

Jelliffe D. 1968. *Infant Nutrition in the Subtropics and Tropics*. Geneva: World Health Organization.

(ed.). 1970. *Diseases of Children in the Subtropics and Tropics*. London: Edward Arnold.

Kariks J. and Woodfield D. 1972. Anaemia in Papua New Guinea: a review. *Papua New Guinea Medical Journal* **15**: 15–24.

Kelly C. 1966. *La Austrialia Del Espiritu Santo: the Journal of Fray Martin de Munilla and Other Documents Relating to the Voyage of Pedro Fernandez de*

Quiros to the South Sea (1605–1606). Cambridge, UK: Cambridge University Press.

Kirch P. 1984. *The Evolution of Polynesian Chiefdoms*. Cambridge, UK: Cambridge University Press.

2000. *On the Road of the Winds: An Archaeological History of the Pacific islands*. Berkeley, CA: University of California Press.

Kirkpatrick R. 1935. Dental caries and odontoclasia in New Guinea. *Dental Journal of Australia* **7**: 707–714.

Laird M. 1956. Studies of mosquito fresh water ecology in the South Pacific. *Royal Society of New Zealand Bulletin* **6**.

Lambert S. 1941. *A Doctor in Paradise*. Melbourne: Dent.

Laor E., Garfunkel A. and Koyoumdjisky-Kaye E. 1982. Skeletal and dental retardation in β-thalassemia major. *Human Biology* **54**: 85–92.

Lee R. 1988. Parasites and pregnancy: the problems of malaria and toxoplasmosis. *Clinics in Perinatology* **15**: 351–363.

Lepowsky M. 1987. Food taboos and child survival: a case study from the Coral Sea. In Scheper-Hughes N., ed., *Child Survival*. Dordrecht: Reidel, pp. 71–92.

Lewis M. 2000. Non-adult palaeopathology: current status and future potential. In Cox M. and Mays S., eds., *Human Osteology in Archaeology and Forensic Science*. London: Greenwich Medical Media, pp. 39–57.

Maitland K., Williams T., Bennett S. *et al.* 1996. The interaction of *Plasmodium falciparum* and *P. vivax* in children on Espirito Santo Island, Vanuatu. *Transactions of the Royal Society of Tropical Medicine and Hygiene* **90**: 614–620.

Malcolm L. 1969. Growth and development of the Kaiapat of the Markham Valley, New Guinea. *American Journal of Physical Anthropology* **31**: 39–52.

1970. *Growth and Development in New Guinea: A Study of the Bundi People of the Madang District*. Madang: Institute of Human Biology.

Marples M. 1950. The incidence of certain skin diseases in Western Samoa: a preliminary survey. *Transactions of the Royal Society of Tropical Medicine and Hygiene* **44**: 319–332.

Marples M. and Bacon D. 1953. Observations on yaws and certain skin diseases in Manono, Western Samoa. *Transactions of the Royal Society of Tropical Medicine and Hygiene* **47**: 141–147.

McMahon J. 1974. Malaria endemicity amongst the semi-nomadic people of the Karma area of Papua New Guinea. *Papua New Guinea Medical Journal* **17**: 99–107.

Middlemiss J. and Raper B. 1966. Skeletal changes in the haemoglobinopathies. *Journal of Bone and Joint Surgery* **48B**: 693–702.

Mills A. 1955. The incidence of yaws in the New Hebrides. *Transactions of the Royal Society of Tropical Medicine and Hygiene* **49**: 58–61.

Ortner D.J. 2003 *Identification of Pathological Conditions in Human Skeletal Remains*. Amsterdam: Academic Press.

Parkinson A. 1974. Malaria in Papua New Guinea 1973. *Papua New Guinea Medical Journal* **17**: 8–16.

Parsonson G. 1966. Artificial islands in Melanesia: the role of malaria in the settlement of the Southwest Pacific. *New Zealand Geographer* **22**: 1–21.

Pasvol G., Weatherall D. J., Wilson R. J. M., Smith D. H. and Gilles H. M. 1976. Fetal hemoglobin and malaria. *Lancet* **7972**: 1269–1272.

Pietrusewsky M. 1976. *Prehistoric Human skeletal Remains from Papua New Guinea and the Marquesas.* Honolulu: Social Sciences and Linguistics Institute, University of Hawaii.

Pietrusewsky M., Douglas M. and Ikehara-Quebral R. 1997. An assessment of health and disease in the prehistoric inhabitants of the Marianas islands. *American Journal of Physical Anthropology* **104**: 315–342.

Pollock N. 1992. *These Roots Remain: Food Habits in Islands of the Central and Eastern Pacific Since Western Contact.* Honolulu: Institute for Polynesian Studies.

Resnick D. 1995. Hemoglobinopathies and other anemias. In Resnick D. and Niwayama G., eds., *Diagnosis of Bone and Joint Disorders.* Philadelphia, PA: Saunders, pp. 2105–2295.

Roberts C. and Manchester K. 1995. *The Archaeology of Disease.* New York: Cornell University Press.

Ryan A. 1997. Iron-deficiency anemia in infant development: implications for growth, cognitive development, resistance to infection and iron supplementation. *Yearbook of Physical Anthropology* **40**: 25–62.

Sayers E. 1943. *Malaria in the South Pacific.* Wellington: E. V. Paul.

Scrimshaw N. 1981. Significance of the interactions of nutrition and infection in children. In Suskind R., ed., *Textbook of Pediatric Nutrition.* New York: Raven Press, pp. 229–240.

Scrimshaw N., Taylor C. and Gordon J. 1968. *Interactions of Nutrition and Infection.* Geneva: World Health Organization.

Skinner M. and Goodman A. 1992. Anthropological uses of developmental defects of enamel. In Saunders S. and Katzenberg M., eds., *Skeletal Biology of Past Peoples: Research Methods.* New York: Wiley-Liss, pp. 153–174.

Snow C. E. 1974. *Early Hawaiians: An Initial Study of Skeletal Remains from Mokapu, Oahu.* Lexington, KY: University Press of Kentucky.

Spencer M. 1971. Bionomics of vector anophelines in Papuan Islands. *Papua New Guinea Medical Journal* **14**: 14–23.

Spencer T., Spencer M. and Venters D. 1974. Malaria vectors in Papua New Guinea. *Papua New Guinea Medical Journal* **17**: 22–30.

Spenneman D. 1986. *Archaeological Fieldwork in Tonga 1985–1986.* Canberra: Australian National University Press.

Stodder A. 1997. Subadult stress, morbidity, and longevity in Latte period populations on Guam, Mariana Islands. *American Journal of Physical Anthropology* **104**: 363–380.

Stuart-Macadam P. 1985. Porotic hyperostosis: representative of a childhood condition. *American Journal of Physical Anthropology* **66**: 391–398.

Sweeney E., Cabrera J., Urrutia J. and Mata L. 1969. Factors associated with linear hypoplasia of human deciduous incisors. *Journal of Dental Research* **48**: 1275–1279.

Sweeney E., Saffir A. and de Leon R. 1971. Linear hypoplasia of deciduous incisor teeth in malnourished children. *American Journal of Clinical Nutrition* **24**: 29–31.

Taylor T. and Strickland G. 2000. Malaria. In Strickland G., ed., *Hunter's Tropical Medicine and Emerging Infectious Diseases*. Philadelphia, PA: Saunders, pp. 614–642.

Thomason J., Jenkins C. and Heywood P. 1986. Child feeding patterns amongst the Au of the West Sepik, Papua New Guinea. *Journal of Tropical Pediatrics* **32**: 90–92.

Trembly D. 1996. Treponematosis in pre-Spanish western Micronesia. *International Journal of Osteoarchaeology* **6**: 397–402.

van der Hoeven J. A. 1958. Taboos for pregnant women, lactating mothers and infants on the north coast of Netherlands New Guinea. *Tropical Geographical Medicine* **10**: 71–76.

van der Sluis I. 1969. *The Treponematosis of Tahiti: Its Origin and Evolution a Study of Sources*. Amsterdam: Peco.

van Dijk W. and Parkinson A. 1974. Epidemiology of malaria in New Guinea. *Papua New Guinea Medical Journal* **17**: 17–21.

Wark L. and Malcolm L. 1969. Growth and development of the Lumi child in the Sepik district of New Guinea. *Medical Journal of Australia* **2**: 129–136.

Weatherall D. 1997. Thalassaemia and malaria, revisited. *Annals of Tropical Medicine and Parasitology* **91**: 885–890.

Weatherall D. and Clegg J. 1981. *The Thalassaemia Syndromes*. Oxford: Blackwell.

White N. 1996. Malaria. In Cook G. C., ed., *Manson's Tropical Diseases*. London: Saunders, pp. 1087–1164.

Whitehead N., Devine S. and Leach B. F. 1986. Electron spin resonance dating of human teeth from the Namu burial ground, Taumako, Solomon islands. *New Zealand Journal of Geology and Geophysics* **29**: 359–361.

Williams T., Maitland K., Bennett S. *et al.* 1996. High incidence of malaria in α-thalassaemic children. *Nature* **383**: 522–525.

Williams T., Maitland K., Phelps L. *et al.* 1997. *Plasmodium vivax*: a cause of malnutrition in young children. *Quarterly Journal of Medicine* **90**: 751–757.

Wyler, D. 1982. Malaria: host–pathogen pathology. *Reviews of Infectious Diseases* **4**: 785–797.

Yenchitsomanus P., Summers K., Bhatia K., Cattani J. and Board P. 1985. Extremely high frequencies of α-globin gene deletion in Madang and on Kar Kar island, Papua New Guinea. *American Journal of Human Genetics* **37**: 778–784.

Yenchitsomanus P., Summers K., Chockkalingam C. and Board P. 1986. Characterization of G6PD deficiency and thalassaemia in Papua New Guinea. *Papua New Guinea Medical Journal* **29**: 53–58.

Zigas V. and Morea V. 1974. Pre-operational malariometric survey Kairuku subdistrict. *Papua New Guinea Medical Journal* **17**: 93–98.

Part III *Conclusions*

14 *Synthesising Southeast Asian population history and palaeohealth*

MARC OXENHAM
Australian National University, Canberra, Australia

NANCY TAYLES
University of Otago, Dunedin, New Zealand

This is the first book specifically to address issues relating to prehistoric Southeast Asian skeletal biology. The principal aim of this volume was the examination of both the population history of Southeast Asia, in terms of origins and micro-evolutionary development, and the health and quality of life of these people. This chapter will discuss (a) the competing views or, alternatively, the consensus of opinion regarding the population history of the region; (b) descriptions of health and trends in health over time in Southeast Asia; and (c) the manner in which the current findings from both population history and palaeohealth can be synthesised in addressing questions surrounding economic and demographic change in this portion of the globe.

Neither the population history nor the evidence for health in Southeast Asia can be understood without reference to archaeological and linguistic evidence for the emergence of agriculture and the spread of Austroasiatic and Austronesian languages into the region. The current archaeological and linguistic consensus supports a hypothesis of agriculturally driven demic expansion or migration into Southeast Asia from a more northerly source (see Bellwood 2004). Higham (2001) noted that clearly a neolithic revolution occurred in the Yangzi valley in what is now China, beginning as early as 10,000 years BP. A southward expansion of agricultural communities, speaking Austroasiatic languages (a group that includes modern Vietnamese and Khmer), is now well documented into southwest China and on into Southeast Asia following major river systems such as the Chao Praya (into Thailand), Red (into northern Vietnam) and Mekong (into Laos, Cambodia and southern Vietnam) (Higham 2001). For instance, Higham noted that the important site Baiyangcun in Yunnan (southwest China), dated to approximately 4,500 years BP, was an important staging

Bioarchaeology of Southeast Asia. Marc Oxenham and Nancy Tayles.
Published by Cambridge University Press.

point for subsequent expansion into Southeast Asia, with contemporary and subsequent sites in northern Vietnam (Phung Nguyen culture period sites for instance) and Thailand (Ban Chiang and Ban Lum Khao being two important sites examined in this volume). A related expansion, involving Austronesian-speaking agriculturalists, has also been recorded for island Southeast Asia and is documented by Bellwood (1997) with a well-dated timeline that sees the Philippines colonised by agriculturalists by 4,000–3,500 years BP and subsequently Indonesia between 3,500 and 3,000 years BP.

Southeast Asian bioarchaeology has the potential to address the agriculture/language dispersal hypothesis both in terms of population history and palaeohealth accounts. While not necessarily a specific aim of the contributors to this volume, the results of many of these studies lend themselves to an interpretation of the archaeologically and linguistically formulated model for an agriculturally driven colonisation of Southeast Asia in the neolithic (see Bellwood 2004).

Population history

How does the skeletal biology of past and current Southeast Asians contribute to accounts of prehistory as proposed by archaeologists and linguists? Contributions to this volume demonstrate that there is not a consensus view, and two models, with various permutations regarding details, currently serve to explain the Holocene population history of Southeast Asia. These hypotheses closely parallel, in structure at least, the models of 'Regional Continuity' and 'Out of Africa' postulated to explain the origins and distribution of anatomically modern humans, who would appear to have emerged as a distinct taxonomic category between 200,000 and 100,000 years ago. Given that the earliest anatomically modern human appeared within the current borders of southern China at least by approximately 67,000 years BP (Liujiang, see Ch. 5), regional Southeast Asian and global population histories would seem to be intimately related.

The chief advocate in this book of the 'Out of Northeast Asia' hypothesis, usually referred to as the migration, colonisation or Two-layer model (synonymous with the Agricultural Demic Expansion model), is Hirofumi Matsumura (Ch. 2). According to Matsumura, late Pleistocene and early Holocene skeletal material can be attributed to members of the Australo-Melanesian lineage. Moreover, Australo-Melanesians are still represented as relict populations in modern Southeast Asia in the form of the various so-called Negrito populations (see Chs. 6 and 7 for fuller accounts of

Negritos). Matsumura explains modern Southeast Asian morphology in terms of either admixture between these original Australo-Melanesian peoples and immigrants originating to the north of the region (probably from southern China/Taiwan) or local *in situ* evolutionary development from the original Australo-Melanesian inhabitants. His non-metric and metric dental analyses revealed a close affinity between modern Southeast and Northeast Asians, which he interprets as evidence supporting the colonisation model, although he admits this does not necessarily refute the alternative. Interestingly, Matsumura sees the strongest evidence for a Northeast Asian genetic influence in the postneolithic samples, which would suggest that demic expansion need not necessarily be coupled with agricultural expansion. Further, Matsumura notes that Northeast Asian movements into Southeast Asia were likely occurring as early as the late Pleistocene (the Tam Hang skeletal series in Laos for instance).

Every contributor to Part I of this volume, the population history section, supports the view that there are close similarities between Northeast and Southeast Asians. Fabrice Demeter (Ch. 5) has argued that the similarities have greater antiquity, with Northeast Asians moving into Southeast Asia around 30,000 years ago and mixing with earlier anatomically modern inhabitants of Southeast Asia. While suggesting that the similarities between Southeast and Northeast Asians can be explained using either model, Michael Pietrusewsky (Ch. 3) sees evidence for long-term continuity within Southeast and Northeast Asian populations. Moreover, he specifically stresses the evidence for long-term continuity among inland populations, citing early Holocene Laos to mid-to-late Holocene Ban Chiang as an example. The greatest level of differentiation within Southeast Asian samples is between coastal and inland groups, a separation that may have great antiquity and/or be influenced by the greater mobility of coastal groups.

Tsunehiko Hanihara (Ch. 4), more in agreement with Pietrusewsky and counter to Matsumura, argues that, as other samples cluster and fall between the similar, albeit clearly separated, Northeast and Southeast Asian samples, complete or near complete replacement of Southeast Asians by more northerly situated populations did not occur. Interestingly, Hanihara sees considerable intraregional variation in both Northeast and Southeast Asia, which could have been caused by several non-mutually exclusive factors to a greater or lesser degree. The first cause explored involves the effects of long-term multiple migrations into each region. For the north, this includes early migrations from Southeast Asia, detailed in the sundadont/sinodont hypothesis of Turner (e.g. 1989, 1990), as well as later migrations from central Asia. For Southeast Asia, there is the possibility of substantive gene flow originating from early times from the west (see Lahr 1996, p. 318). The second cause relates to the potential for deep population histories and/or large populations sizes for each region. The deep population history of each region is stressed by Demeter. Finally, Hanihara notes that high mutation and drift rates will serve to maintain intraregional variation, and this would be particularly relevant in Northeast Asia given the strong selective pressures associated with the harsh environment.

David Bulbeck and Adam Lauer (Ch. 6) address the larger questions at issue by way of a detailed examination of one of the more important aboriginal groups, the Orang Asli, in what is now Malaysia. They have developed a model for the Orang Asli that acknowledges clear biological continuity between earlier populations and this group, notably the Temiar, while at the same time explaining differences between the various contemporary Orang Asli and between some members of these modern Orang Asli and Hoabinhian and late Holocene presumed forebears (Semang for instance) in terms of (a) *in situ* evolutionary change caused by migrants fragmenting the original gene pool, with resulting subpopulation specific drift and adaptation; and (b) changes to subsistence regimens (reduced protein diets particularly) leading to smaller stature and various observed paedomorphic features. The importance of this model lies in the observation that, while major population movements into Southeast Asia are a necessary requirement for splintering the original gene pool, major gene flow between indigenes and migrants is not a requirement for subsequent micro-evolutionary change. The model thus allows for an imported agricultural revolution and also significant evolutionary continuity between early and later Southeast Asians. Others also support such deep continuity, with Storm (1995) arguing that the Wajak material from Indonesia can be seen as prefiguring the appearance of later modern Southeast Asians. Interestingly, Christy Turner and James Eder (Ch. 7) argue that the Batak, a modern Negrito population from the Philippine Island Palawan, can likewise best be characterised as not so much a relict of an original Southeast Asian form but rather descendant, with extensive micro-evolutionary modification, of such an early population. Turner and Eder see the Batak as likely descendant of a proto-sundadont population that has a considerable antiquity in the region.

As noted, there is a mixture of mutual compatibility and considerable disparity among the findings of the population history proponents. Nonetheless, there is also a level of consensus in terms of the bioarchaeological evidence for the agriculture/language dispersal model. None of the contributors examining population history from a broad temporal and geographic perspective found evidence necessarily supporting the view for significant migration into Southeast Asia with the adoption and/or spread of agriculture. Matsumura was the chief proponent of the position that the region witnessed significant amounts of migration from Northeast Asia, but this was seen to occur only after the neolithic. Some migration into the region was also proposed in the late Pleistocene, which suggests the possibility of continuous migration into Southeast Asia from more northeastern sources for much of the Holocene, with certain heightened pulses occurring at times other than when agriculture was developing and/or spreading. Both Pietrusewsky and Hanihara argue for continuity among Northeast and Southeast Asian communities with little if anything to suggest major movements of people into Southeast Asia. However, Pietrusewsky did argue for the possibility of high mobility among coastal groups throughout antiquity. While not the aim of their chapter either, Bulbeck and Lauer were the only authors in the population history section to support the agriculture/language hypothesis. Nonetheless, support of this model was not contingent upon their argument as they simply used the possibility of significant levels of migration into the region during prehistory as one mechanism for fragmentation of the preexisting Negrito gene pool. In fact, they still argue for significant evolutionary continuity between earlier and later Southeast Asians. In summary, overall the current interpretations of Southeast Asian population history, as advanced in this volume, do not support a hypothesis whereby agriculture was established in the region following significant levels of migration from any source, including neighbouring regions directly north and east of Southeast Asia proper.

Palaeohealth

Before anatomically modern humans

While discussions of the population history of our study area is limited to anatomically modern humans living during the late Pleistocene and Holocene, we expanded the temporal range when looking at palaeohealth in order to include the earliest evidence in Southeast Asia for a compromised quality of life. Etty Indriati's (Ch. 12) discussion of the Ngandong 7 (*Homo erectus*) cranial lesions is important given that the palaeohealth of this species is very poorly understood globally and virtually unknown in Southeast Asia. The probable traumatic aetiology of the Ngandong 7 lesions is especially interesting given the nature and frequency of cranial lesions described for other Asian *H. erectus* specimens from northern China, Zhoukoudian (Weidenreich 1943) and Gongwangling (Caspari 1997). The aetiology of the Gongwangling lesions included trauma and infection. Weidenreich described what appear to be at least two cases of healed depressed cranial fractures in skulls X and XII of the Zhoukoudian collection and considered cave roof falls and interpersonal violence (Weidenreich 1943, p. 188) as potential ultimate causes of these lesions. Boaz and Ciochon (2004) went as far as to suggest that interpersonal violence was so rife that the thick cranial walls of these hominins was an evolutionary adaptation to high levels of cranial trauma. Indriati (Ch. 12) concludes that the most probable explanation for the morphology of the Ngandong 7 lesions was an infection subsequent to some form of traumatic injury. While the cause of the injury could not be determined, several plausible scenarios were discussed including interpersonal violence, trephination and, perhaps more parsimoniously, accidental trauma by misadventure. At the very least, these cases indicate that the earliest inhabitants of the region suffered serious traumatic injuries that some of them, at least, were able to survive. The relatively thick cranial walls of these people no doubt facilitated survival from otherwise mortal injuries, which subsequently healed and eventually contributed to an apparent elevated level of such trauma in the fossil record. The only other vault lesions described for Asian *H. erectus* is a case of hyperostosis calvariae interna in Sangiran 2 (Antón 1997). The aetiology of the disease is not clear and it may have been asymptomatic in Sangiran 7 and not have had any obvious health implications.

While cranial lesions seem to be rather common in Asian *H. erectus*, the first description of a pathological condition in this species was provided by Eugene Dubois (1894) in reference to a femur. The so-called pithecanthropine femur exhibits ossification of the tendinous adductor insertions. Soriano (1970) argued that the lesion was caused by toxic levels of fluorine presumably caused by consuming foods present in the volcanic environment of Java at the time. While few would now accept that the femur is attributable to *H. erectus*, it is still a very early, if not the earliest, instance of compromised anatomically modern human health in what is now Indonesia. Finally, it is worth noting von Koenigswald's (1968, his Fig. 2) illustration of trauma to the *Meganthropus* 1953 mandible ostensibly caused perimortem by a crocodile. Von Koenigswald went on to suggest that such a fate was not uncommon for our less-fortunate forebears in the region.

In summary, the palaeohealth of Asian *H. erectus* would appear somewhat compromised in terms of cranial trauma relative to later

Homo sapiens inhabitants of the region. This perhaps reflects important behavioural differences between the two species of human, but whether this is reflective of preferences in accommodation (caves versus open sites), social relations (interpersonal violence for instance) or some other factor is unclear. The occurrence of hyperostosis calvariae interna in at least one early hominin in Southeast Asia is important in extending the temporal span of this condition, while the apparent crocodile-induced traumatic injury to the meganthropine mandible invites speculation regarding the dangers of living in mid Pleistocene Java.

Holocene humans

Apart from those few examples from species other than *H. sapiens*, and a limited late Pleistocene anatomically modern human sample, studies of palaeohealth in Southeast Asia have been restricted to larger Holocene skeletal assemblages and such samples form the health and disease focus of Part II of this volume. The relative wealth of Holocene skeletal materials has enabled a number of different approaches to viewing palaeohealth in Southeast Asia. Michele Toomay Douglas (Ch. 8) focuses on temporal changes within a single assemblage with a particular emphasis on changes in the dental health of males and females. Douglas concludes that the dental profile of the Non Nok Tha sample, which spans the pre-metal and bronze periods, is consistent with a mixed foraging and cultivating economy and appears not unlike the profiles seen in other pre-metal assemblages from the region. Intriguingly, Douglas notes that male dental health was consistently worse than that of females over time, although female dental health worsened with time. This, combined with increasing female stature, led Douglas to suggest female behavioural changes associated with their greater relationship with and benefit from agriculture. Other potential evidence for the increasing economic importance of agriculture at Non Nok Tha comes from a reported increase in cervical and upper thoracic vertebral osteoarthritis with the transition into the bronze age. Clearly, the biological signature of the adoption and/or intensification of agriculture at Non Nok Tha was subtle.

Kate Domett and Nancy Tayles (Ch. 9) chose to examine two assemblages where environmental variables could be controlled for, but which represented different socio-economic conditions in terms of the bronze (Ban Lum Khao) and iron (Noen U-Loke) ages. One finding was that fertility was higher in the bronze age assemblage than in the iron age series. Low fertility also characterised a number of skeletal samples from

Thailand, including Ban Chiang, Non Nok Tha and Ban Na Di (Pietrusewsky and Douglas 2002). The relatively elevated level of fertility at Ban Lum Khao is contrasted with increased levels of stress in terms of attained male height and enamel defects, with the suggestion of differential access to resources in the bronze age. This appears to contradict the archaeological evidence (O'Reilly 2001) and presumably this is a reflection of some level of social hierarchy not hitherto visible in the archaeological record of the site. Better health but lower fertility in the iron age series was seen in the context of improved technology and environmental conditions. Alternative explanations may include an increased level of population fission and associated establishment of new communities. This model is certainly consistent with the evidence for increasing settlement density during the iron age (O'Reilly 2001) and may perhaps explain the lower level of infant mortality as infants were simply being buried elsewhere.

Marc Oxenham, Nguyen Lan Cuong and Nguyen Kim Thuy (Ch. 11) chose a yet broader-brush approach with an examination of oral health across all major Holocene skeletal assemblages recovered from the region to date. They have found a remarkable degree of homogeneity in the markers of oral health and behaviours they examined. The maintenance of this status quo, admittedly associated with some degree of noise, is argued to be caused by two factors. The first is the low cariogenicity of rice itself. The second factor suggests that the social and economic structure of bronze age communities was not consistent with hierarchical structures built upon agricultural surpluses. Rather, the oral health of these communities is more consistent with very broad subsistence economies and socially fluid, albeit structured, heterarchy. There was no big bang in terms of the adoption of agriculture or its intensification in the region.

In Ch. 10, Christopher King and Lynette Norr's study of diet using stable isotopes was important not only in being the first of its kind using a mainland Southeast Asian sample, Ban Chiang, but also for contributing to the interpretation of health changes both over time and within the assemblage. They found that C_3 plants were the dominant plant group targeted over time and that important dietary differences are apparent between the bronze and iron age samples, consistent with a move to wet-rice agriculture and consumption of domesticated animals. Further they note numerous differences in diet, as measured over time and/or between each sex, thus highlighting a kaleidoscope of important behavioural changes at Ban Chiang (Pietrusewsky and Douglas 2002).

Hallie Buckley's work (Ch. 13) is particularly important in explicitly bridging the topical divide in this volume with her analysis of the relationship between palaeohealth and population history in terms of one of the most debilitating and dangerous of tropical infectious diseases, malaria. Colonisation of the Pacific by Austronesian-speaking peoples originating from Southeast Asia is the last and arguably the most spectacular movement of people in the prehistory of the region and forms an important counterpoint to the other contributions on Southeast Asian palaeohealth in this volume. Buckley tests the hypothesis that malaria had a significantly negative impact on population growth, settlement and migration in Near Oceania in comparison with Polynesia. Using cribra orbitalia and enamel defects as proxies for the effects of malaria, significantly elevated levels of skeletal and dental signatures of stress were shown in the Near Oceanic samples in comparison with the Polynesian samples. Polynesia, being free from malaria, health-debilitating food taboos and infectious diseases experienced accelerated population growth and the development of hierarchical societies in stark contrast to their Near Oceanic cousins.

The combined evidence for palaeohealth in this book is not in conflict with the adoption and/or intensification of agriculture over time. The evidence from Southeast Asia is more subtle than that seen in other collective studies (e.g. Cohen and Armelagos 1984, Steckel and Rose 2002.). There is no evidence for a wholesale decline in health moving from the neolithic through the bronze age and on to the iron age. At some sites, the evidence for changes in subsistence are as subtle as differential male and female dental health and an increase in vertebral osteoarthritis at Non Nok Tha. Increases in fertility are generally regarded as signs of the adoption or intensification of agriculture. Assuming these samples are representative of their respective cemetery populations, fertility was relatively low in a number of ostensibly agricultural sites, including the iron age sample from Non Nok Tha, while being higher in others, such as the neighbouring bronze age assemblage of Ban Na Di. Intriguingly, these findings are consistent with modelling from nutritional ecology, which predicts higher fertility in populations with varied diets and lower fertility in populations potentially consuming more calories but having a less-diverse diet (Hockett and Haws 2003). Other assemblages with clear signs of subsistence changes over time include Ban Chiang, which was examined in terms of a stable isotopic study in Ch. 10. King and Norr provide convincing evidence in support of Pietrusewsky and Douglas' (2002) interpretations of subsistence changes over time, despite findings for low fertility at Ban Chiang. Somewhat paradoxically, the move to agriculture may have been accompanied by decreased fertility in many Southeast Asian populations, in the early stages at least. Further, the bioarchaeological approach also uncovered health changes at Ban Lum Khao, thus highlighting potential deficiencies in the archaeological record with the suggestion by Domett and Tayles of otherwise hidden social hierarchy in this sample.

Studies of the palaeohealth of these Southeast Asian assemblages can also be used to address the agriculture/language dispersal model, although the signs are subtle. There is clearly skeletal evidence for the emergence and spread of infectious disease in the metal period (Oxenham *et al.* 2005, Tayles and Buckley 2004). However, general health seems to have remained fairly homogeneously stable for much of the period examined in this volume. Nonetheless, while the emergence and spread of such disease is consistent with the timing of the adoption and/or intensification of agriculture, it does not necessarily indicate the introduction of a package of genes, languages and agriculture. Migration can certainly facilitate the introduction and spread of disease and was an important factor in the emergence of infectious disease in prehistoric Japan (Suzuki 1991) and perhaps northern Vietnam in the metal period (Oxenham *et al.* 2005). However, other factors particularly relevant in the context of the adoption and/or intensification of agriculture requiring scrutiny include the reduction in dietary diversity and increased exposure to pathogens through land-modification behaviours. Further, it is now becoming clear that many pathogens will evolve to greater levels of virulence over time rather than evolving toward states of equilibrium with the host (e.g. Ewald 1994). Whether or not pathogens were behaving in such a manner in prehistoric Southeast Asia is difficult to test at present. What is clear, and we prefigure the findings in the following section, is that the evidence from palaeohealth is consistent with that from population history: there is no necessity at present to argue for significant levels of gene flow into Southeast Asia at any period in the Holocene.

A synthesis

Is a synthesis of the population history and palaeohealth of ancient Southeast Asians possible? Do the debates concerning the origins and movements of people in Southeast Asia during the late Pleistocene and much of the Holocene as well as the emerging picture of health, disease and well-being inform each other's respective debates? The two major population history models for the region lend themselves to different sets of predictions concerning patterns of health and disease. The health effects of significant levels of migration, documented for prehistoric Japan (Suzuki 1991) and Vietnam (Oxenham *et al.* 2005) for instance, would suggest that the Southeast Asian colonisation model would predict elevated levels of stress and poorer health in general among the samples that were either new migrants to the region or indigenous populations being impacted by the effects of large-scale migration. An important and extensively studied site (Higham and Bannanurag 1990) and skeletal assemblage (Tayles 1999) particularly relevant to this issue is 4,000 to 3,500 years BP Khok Phanom Di in the Gulf of Thailand, which is situated within the range of 4,000 to 3,000 years BP that Peter Bellwood (personal communication) regards as the period during which agriculture was adopted in mainland Southeast Asia. Regardless of whether Khok Phanom Di is regarded as a foraging community with a marine/estuarine focus or the vanguard of migrating agriculturalists, the skeletal sample suggests high fertility, a more than adequate diet, and no evidence for infectious disease except indirectly with respect to biological adaptations to malaria. The skeletal biology of this sample is more representative of a well-adapted foraging community than either a newly established agricultural community or a population undergoing major changes in either economic orientation or exposure to immigrant-borne pathogens.

Two other assemblages from Thailand have considerable temporal depth and span the period during which agriculture was adopted. While Ban Chiang has been studied in great detail by Douglas elsewhere (1996, Pietrusewsky and Douglas 2002), Non Nok Tha is examined in this volume. Douglas' dental pathology profile is entirely consistent with a mixed foraging–cultivating economy over the entire extent of the assemblage. Nonetheless, changes in sex-based markers of health over time suggested a shift to a greater reliance on agriculture. However, there is nothing to suggest any phenotypic changes in the community over time consistent with an intrusive population's new genes.

With respect to Ban Chiang, there are clear biological indicators for a significant reorientation in subsistence economy over time. The Ban Chiang sample shows evidence of an increase in enamel defects consistent with increasing physiological stress (Douglas 1996, Pietrusewsky and Douglas 2002), an increase in mortality and a decrease in average age at death, which equates with an increase in fertility (see Sattenspiel and Harpending 1983). In terms of the evidence for infectious disease, less than 10% of the sample displayed lesions consistent with conditions such as mastoiditis and pulmonary disease, although most lesions were non-specific infections identified by way of localised periostitis (Pietrusewsky and Douglas 2002). The authors suggested that the frequency and patterning of infectious lesions at Ban Chiang is not consistent with population-level transmission. While these skeletal signatures, in total, are consistent with a move to or intensification in agriculture, the dental health profile is relatively silent on this issue. However, this may not be unexpected and Ch. 11 reviewed potential reasons for the lack of a dental signature with the adoption or intensification of agriculture. It is also important that these changes occurred at Ban Chiang without any obvious genetic changes to the population. Pietrusewsky and Douglas (2002) have argued for genetic continuity over the entire human occupation of the site and do not see any evidence for immigration.

The changes between bronze phase Ban Lum Khao and iron phase Noen U-Loke have been summarised above. While there are clear indicators for a shift in subsistence orientation over time, notwithstanding the ambiguous evidence from an examination of fertility patterns, there is little to suggest the introduction of migrant genes into the region with the adoption or intensification of agriculture. The same can be said for other skeletal series such as Ban Na Di and Nong Nor when viewed in a broader comparative context (see Domett 2001). However, there is some evidence in the form of infectious disease for population movements and/or contacts in the region, although the question of gene exchange cannot be addressed. Noen U-Loke displays evidence of an elevated level of infectious disease, and specific ones such as tuberculosis and leprosy, in comparison with Ban Lum Khao (Tayles and Buckley 2004). Further, infectious diseases do not appear in the bioarchaeological record from northern Vietnam until large demographic changes, principally immigration and population increase, occurred at the close of the first millennium BCE (Oxenham *et al.* 2005). Further, major demographic change in the manner of rapid population growth in Polynesia is demonstrated to be predicated upon the lack of infectious disease (Ch. 13). Clearly, the presence or absence of disease in this part of the world is sensitive to migration and population growth as well as contributing to these factors.

The evidence for a shift toward significant agricultural components in the economies of these communities from the pre-metal through bronze and iron periods is clear, even if it does not necessarily mean uniform declines or improvements in health and well-being. Nonetheless, this pattern needs to be considered within the context of the evidence for a lack of morphological differentiation among prehistoric samples in the region with the exception of the earliest assemblages from Khok Phanom Di in Thailand and Con Co Ngua in Vietnam. Even if Khok Phanom Di and Con Co Ngua are seen as the thin edge of the agricultural wedge, there is no evidence that these communities contributed in a significant manner to the gene pool of the following bronze and iron age populations. The most parsimonious interpretation of a synthesis of population history and palaeohealth is that agriculture was adopted and/or intensified beginning in the bronze phase with a mixed 'bag' of both negative and positive health effects. However, this subsistence reorientation occurred without any obvious changes to the physical appearance (and by proxy the genetic composition) of these communities. In short, the skeletal biology of the past inhabitants of the region is not consistent with significant immigration following agricultural diffusion, although admittedly, a number of the sites discussed in this book do not span the very beginnings of agriculture in Southeast Asia.

What's next?

Ten years ago, perhaps even five years ago, this book in its present form would not have been possible. While the study of the population history of the region has a deep pedigree, studies in palaeohealth and disease with a focus on Southeast Asia are only now beginning to emerge. Both approaches to past human skeletal biology are necessary and mutually informative in developing an understanding of how humans have lived and behaved in the past. This volume needs to be treated for what is: the vanguard of a growing interest and a research focus on the human experience as revealed through skeletal biology in Southeast Asia. This volume should be used as a source of hypotheses to test, questions to ask and models to develop and build on over the next several years. The contributors have literally only scraped the surface of what is one of the most diverse and fascinating regions of the globe, both currently and in the past. Even as we write, there are new projects with an emphasis on questions relating to human skeletal biology being developed and progressing in Cambodia and Burma (Myanmar). Discussions develop on how best to tackle the Holocene of Indonesia and the Philippines. Established research programmes continue, and new ones are being formed, in Thailand and Vietnam. We also hope that more researchers will pursue investigations of population history by way of DNA and stable isotopic studies in an effort to seek a consensus between skeletal morphology, palaeohealth, genetics and bone chemistry. It is our desire that this volume will catalyse and inspire further work in this very under-researched region of the globe.

References

Antón S.C. 1997. Endocranial hyperostosis in Sangiran 2, Gibraltar 1, and Shanidar 5. *American Journal of Physical Anthropology* **102**: 111–122.

Bellwood P. 1997. *Prehistory of the Indo-Malaysian Archipelago*, revised edn. Honolulu: University of Hawaii Press.

2004. The origins and dispersals of agricultural communities in Southeast Asia. In Glover I. and Bellwood P., eds., *Southeast Asia: From Prehistory to History*. London: Routledge Curzon, pp. 21–40.

Boaz N. T. and Ciochon R. L. 2004. *Dragon Bone Hill: An Ice-age Saga of* Homo erectus. Oxford: Oxford University Press.

Caspari R. 1997. Brief communication: evidence of pathology on the frontal bone from Gongwangling. *American Journal of Physical Anthropology* **102**: 565–568.

Cohen M. N. and Armelagos G. J. (eds.). 1984. *Paleopathology at the Origins of Agriculture*. Orlando, FL: Academic Press.

Domett K. M. 2001. *British Archaeological Reports International Series*, No. 946: *Health in Late Prehistoric Thailand*. Oxford: Archaeopress.

Douglas M. T. 1996. Paleopathology in human skeletal remains from the pre-metal, bronze, and iron ages, northeastern Thailand. Ph.D. thesis, University of Hawaii-Manoa. Ann Arbor: University Microfilms.

Dubois E. 1894. Pithecanthropus erectus. *Eine menschenaehliche Ubergangsform aus Java*. Batavia: Landsdrukkerij.

Ewald P. W. 1994. The evolution of virulence and emerging diseases. *Journal of Urban Health* **75**: 480–491.

Higham C. F. W. 2001. Prehistory, language and human biology. In Jin L., Seielstad M. and Xiao C., eds., *Genetic, Linguistic and Archaeological Perspectives on Human Diversity in Southeast Asia*. Singapore: World Scientific, pp. 3–16.

Higham C. F. W. and Bannanurag R. (eds.). 1990. *Report of the Research Committee* XLVII. *The Excavation of Khok Phanom Di, a Prehistoric Site in Central Thailand*. London: Society of Antiquities.

Hockett B. and Haws J. 2003. Nutritional ecology and diachronic trends in Paleolithic diet and health. *Evolutionary Anthropology* **12**: 211–216.

Lahr M. M. 1996. *The Evolution of Modern Human Diversity: A Study of Cranial Variation*. Cambridge, UK: Cambridge University Press.

O'Reilly D. J. W. 2001. From the bronze age to the iron age in Thailand: applying the heterarchical approach. *Asian Perspectives* **39**: 1–19.

Oxenham M. F., Nguyen K. T. and Nguyen L. C. 2005. Skeletal evidence for the emergence of infectious disease in bronze and iron age northern Viet Nam. *American Journal of Physical Anthropology*, **126**: 359–376.

Pietrusewsky M. and Douglas M. T. 2002. *University Museum Monograph 111: Ban Chiang, a Prehistoric Village Site in Northeast Thailand*. Philadelphia, PA : University of Pennsylvania Press.

Sattenspiel L. and Harpending H. 1983. Stable populations and skeletal age. *American Antiquity* **48**: 489–498.

Soriano M. 1970. The fluoric origin of the bone lesion in the *Pithecanthropus erectus* femur. *American Journal of Physical Anthropology* **32**: 49–58.

Steckel R. H. and Rose J. C. 2002. *The Backbone of History. Health and Nutrition in the Western Hemisphere*. Cambridge, UK: Cambridge University Press.

Storm P. 1995. *The Evolutionary Significance of the Wajak Skulls*. Leiden: *Scripta Geologica* **110**: 1–247.
Suzuki T. 1991. Paleopathological study on infectious disease in Japan. In Ortner D. J. and Aufderheide A. C., eds., *Human Paleopathology: Current Synthesis and Future Options*. Washington, DC: Smithsonian Institution Press, pp. 128–139.
Tayles N. 1999. *Report of the Research Committee* LXI. *The Excavation of Khok Phanom Di, a Prehistoric Site in Central Thailand*, Vol. V: *The People*. London: Society of Antiquaries.
Tayles N. and Buckley H. R. 2004. Leprosy and tuberculosis in iron age Southeast Asia. *American Journal of Physical Anthropology*, **125**: 239–256.
Turner C. G. II. 1989. Teeth and prehistory in Asia. *Scientific American* **260**: 88–96.
1990. Major features of sindodonty and sundadonty. *American Journal of Physical Anthropology* **82**: 295–317.
von Koenigswald G. H. R. 1968. Observations upon two *Pithecanthropus* mandibles from Sangiran, central Java. *Proceedings of the Academy of Science of Amsterdam* **B71**: 99–107.
Weidenreich F. 1943. The skull of *Sinanthropus pekinensis*: a comparative study on a primitive hominid skull. *Palaeontologica Sinica New Series D* **10** (whole series **127**): 1–298.

Index